CAD-Based Optical Design with Quadoa

In *CAD-Based Optical Design with Quadoa*, we explore how optical design entails evaluating application requirements and creating optical systems specifically engineered to meet the exact needs of optical instruments. This extensive system may include optics, light sources, sensors, electronic boards, display devices, and other components. Primarily, optical design focuses on determining the shapes, positions, and materials of lenses and mirrors to ensure the optical system adheres to indicated criteria and specifications. To meet specific requirements, lens designers must use optical design software as an aid for their challenging task.

Most of the mentioned optical design programs are based on the idea of a spreadsheet that contains the information on each surface, radius, thickness, materials, and position among other parameters. This paradigm has its positive aspects and negative aspects. But optical design is a field under constant evolution and new ways to design optical systems are emerging. In this book, we are going to construct, optimize, and analyze optical systems under the fresh paradigm of Computer-aided design (CAD) with the help of Quadoa Optical CAD, a novel optical design software. Within the chapters of the book, we are going to see why this paradigm can be very useful in designing optical systems in comparison with the one proposed by spreadsheets.

The purpose of this book is to teach optical design with a fresh and reliable tool, Quadoa Optical CA. This book mixes the classical books in optical design and a manual of Quadoa Optical CAD. The purpose of this blend is to give readers enough knowledge to start a practical optical design, with solid theory and robust tools in Quadoa Optical CAD.

Rafael G. González-Acuña studied his PhD at the Tecnologico de Monterrey (2020). He has 38 optics papers in index journals. He has been awarded several prizes to mention a few Romulo Garza Science Awards, and the Recognition of Civic Merit, which is the highest distinction that the State Nuevo León grants. SPIE has awarded Rafael the 2019 Optical Design and Engineering Scholarship and is the co-author of three books: Analytical Lens Design, Stigmatic Optics, and Optical Path Theory. He is known for analytically solving a 2400-year-old unsolved problem, by obtaining the general equation of the stigmatic lens. He currently works as a senior lens designer for Huawei Technologies.

Series in Optics and Optoelectronics

Detection of Optical Signals
Antoni Rogalski & Zbigniew Bielecki

Handbook of Optoelectronics, Second Edition
Applied Optical Electronics (Volume Three)
John P. Dakin, Robert G. W. Brown

Handbook of Optoelectronic Device Modeling and Simulation
Lasers, Modulators, Photodetectors, Solar Cells, and Numerical Methods, Vol. 2
Joachim Piprek

Handbook of Optoelectronics, Second Edition
Concepts, Devices, and Techniques (Volume One)
John P. Dakin, Robert Brown

Handbook of GaN Semiconductor Materials and Devices
Wengang (Wayne) Bi, Haochung (Henry) Kuo, Peicheng Ku, Bo Shen

Handbook of Optoelectronic Device Modeling and Simulation (Two-Volume Set)
Joachim Piprek

Handbook of Optoelectronics, Second Edition (Three-Volume Set)
John P. Dakin, Robert G. W. Brown

Optical MEMS, Nanophotonics, and Their Applications
Guangya Zhou, Chengkuo Lee

Thin-Film Optical Filters, Fifth Edition
H. Angus Macleod

Laser Spectroscopy and Laser Imaging
An Introduction
Helmut H. Telle, Ángel González Ureña

Fourier Optics in Image Processing
Neil Collings

Holography
Principles and Applications
Raymond K. Kostuk

An Introduction to Quantum Optics, Second Edition
Photon and Biphoton Physics
Yanhua Shih

Polarized Light and the Mueller Matrix Approach, Second Edition
José J. Gil, Razvigor Ossikovski

Introduction to Holography, Second Edition
Vincent Toal

Optical Coherence Tomography in Dentistry
Scientific Developments to Clinical Applications
Anderson S. L. Gomes, Denise M. Zezell, Cláudia C. B. O. Mota, John M. Girkin

Holography: Principles and Applications, Second Edition
Raymond K. Kostuk

CAD-Based Optical Design with Quadoa
Rafael G. González-Acuña

For more information about this series, please visit: *https://www.crcpress.com/Series-in-Optics-and-Optoelectronics/book-series/TFOPTICSOPT*

CAD-Based Optical Design with Quadoa

Rafael G. González-Acuña

CRC Press
Taylor & Francis Group
Boca Raton London New York

CRC Press is an imprint of the
Taylor & Francis Group, an **informa** business

Designed cover image: Rafael G. González-Acuña

First edition published 2025
by CRC Press
2385 NW Executive Center Drive, Suite 320, Boca Raton FL 33431

and by CRC Press
4 Park Square, Milton Park, Abingdon, Oxon, OX14 4RN

CRC Press is an imprint of Taylor & Francis Group, LLC

ISBN: 978-1-032-82809-1 (hbk)
ISBN: 978-1-032-82810-7 (pbk)
ISBN: 978-1-003-50635-5 (ebk)

DOI: 10.1201/9781003506355

Typeset in CMR10 font
by KnowledgeWorks Global Ltd.

To Jesus Christ and the Virgin Mary.

Contents

Acknowledgments

Almighty God, this book aims to honor your glory. Thank you, Lord, for giving me the intelligence, the desire, the faith, and the means to complete this book. Lord, please help me to be a faithful servant of your will. Help me to deserve the promises of your son Jesus Christ. The man without God is not owning himself. You are the path, the truth, and the life.

I want to thank my family, but especially my mom, Carmen Leticia Acuña Medellín, for all the love and my dad, Rogelio González Cantú, for encouraging me to read at a very early age.

To several Professors, that in my path, they have advised me wisely R. Barry Johnson, Simon Thibault, Reinhard Kettle, Julio C. Gutierrez-Vega, Bernardino Barrientos-García, Abundio Dávila-Álvarez, and Benjamin Perez-Garcia,

I would also like to thank my friends and family Rolando González Acuña, Sonia Villareal, Maria Isabel González Villareal, Jose Miguel González Villareal, Juan Pablo González Villareal, Lilia Acuña Medellín, Leopoldo Fernandez, Daniel Lomas Satarain, Eduardo de Villa, Adrian Lozano Cortez, Zane Sibley, Mira Kankkonen, Gerardo Lozano Cortez, Lorena Patricia, Yoshio Catillejos, Israel Meléndez, Roberto Martinez, Fernando Tamayo, Roberto Vera, Luis Garza Soto, Jorge Sánchez, Lionel Gonzalez, Miguel Ianilli, Mauricio Reyes Arroyo, Guillermo Tamargo, Oona Niklas-Salminen, Eero Salmelin, Ghulam Murtaza, and Héctor A. Chaparro-Romo.

To `Quadoa Optical CAD` team, Jens Siepmann, Goran Bastian Baer, Julian Baer, for encouraging me and giving the confidence and support to write this book.

Finally, I want to thank Taylor and Francis's editorial team for their support.

Rafael G. González-Acuña
27 June 2024
Tampere, Finland.

Preface

Optical design entails evaluating application requirements and creating optical systems specifically engineered to meet the exact needs of optical instruments. This extensive system may include optics, light sources, sensors, electronic boards, display devices, and other components. Primarily, optical design focuses on determining the shapes, positions, and materials of lenses and mirrors to ensure the optical system adheres to specified criteria and specifications.

The process of optical design involves highly nonlinear optimization problems due to the numerous parameters that must be considered, each having complex, nonlinear interrelationships. These parameters are governed by physical laws that dictate the behavior of light rays. The relationships between these parameters are expressed through intricate mathematical equations, which typically lack straightforward analytical solutions and instead require numerical approximation methods. Navigating the design space in optical design is challenging due to its high dimensionality, with numerous variables interacting nonlinearly and complex boundary conditions impacting most variables. This makes optical design a demanding task that requires significant patience and computational effort. To meet specific requirements, lens designers must use optical design software as an aid for their challenging task.

Most of the mentioned optical design programs are based on the idea of a spreadsheet that contains the information on each surface, radius, thickness, materials, and position among other parameters. This paradigm has its positive aspects and negative aspects. But optical design is a field under constant evolution and new ways to design optical systems are emerging. In this book we are going to construct, optimize, and analyze optical systems under the fresh paradigm of Computer-aided design (CAD) with the help of `Quadoa Optical CAD`, a novel optical design software. Along the chapters of the book, we are going to see why this paradigm can be very useful in designing optical systems in comparison with the one proposed by spreadsheets.

The purpose of this book is to teach optical design with a fresh and reliable tool, `Quadoa Optical CAD`. This book pretends to mix the classical books in optical design (just focus on the theory) and the manual of `Quadoa Optical CAD`. The purpose of this mix is to give readers the enough knowledge to start a practical optical design, with solid theory and robust tools (`Quadoa Optical CAD`).

Blub

Optical design entails evaluating application requirements and creating optical systems specifically engineered to meet the exact needs of optical instruments. This extensive system may include optics, light sources, sensors, electronic boards, display devices, and other components. Primarily, optical design focuses on determining the shapes, positions, and materials of lenses and mirrors to ensure the optical system adheres to specified criteria and specifications. To meet specific requirements, lens designers must use optical design software as an aid for their challenging task.

Most of the mentioned optical design programs are based on the idea of a spreadsheet that contains the information on each surface, radius, thickness, materials, and position among other parameters. This paradigm has its positive aspects and negative aspects. But optical design is a field under constant evolution and new ways to design optical systems are emerging. In this book, we are going to construct, optimize, and analyze optical systems under the fresh paradigm of Computer-aided design (CAD) with the help of Quadoa Optical CAD, a novel optical design software. Along the chapters of the book, we are going to see why this paradigm can be very useful in designing optical systems in comparison with the one proposed by spreadsheets.

The purpose of this book is to teach optical design with a fresh and reliable tool, Quadoa Optical CA. This book pretends to mix the classical books in optical design (just focus on the theory) and the manual of Quadoa Optical CAD. The purpose of this mix is to give readers enough knowledge to start a practical optical design, with solid theory and robust tools in Quadoa Optical CAD.

Bio

Rafael G. González-Acuña studied his PhD at the Tecnologico de Monterrey (2020). He has 38 optics papers in index journals. He has been awarded several prizes to mention a few Romulo Garza Science Awards, and the Recognition of Civic Merit, which is the highest distinction that the State Nuevo León grants. SPIE has awarded Rafael the 2019 Optical Design and Engineering Scholarship and is the co-author of three books: Analytical Lens Design, Stigmatic Optics, and Optical Path Theory. He is known for analytically solving a 2400-year-old unsolved problem, by obtaining the general equation of the stigmatic lens. He currently works as a senior lens designer for Huawei Technologies.

Part I

Initial Considerations for an Optical System

1

Getting Started with `Quadoa Optical CAD`

> All the mathematical sciences are founded on relations between physical laws and laws of numbers so that exact science aims to reduce the problems of nature to the determination of quantities by operations with numbers.
>
> James C. Maxwell

1.1 What is Optical Design?

Optical design involves analyzing application prerequisites and crafting optical systems tailored to fulfill the precise requirements of optical instruments [1, 2]. This comprehensive system may encompass optics, light sources, sensors, electronic boards, display devices, and/or additional components [3]. But mostly optical design is based on the shape, positions, and materials of lenses/mirrors such that an optical system meets specific criteria and specifications [4, 5].

The process of optical design consists of extremely nonlinear optimization problems since several parameters must be considered and the relation between them is highly nonlinear [6, 7]. Therefore, we have some parameters with some relations these relations are physical laws to which light rays behave according to [8]. The relations are mathematical expressions that are not simple and usually, there is no analytical solution but rather a numerical approximate solution. Given that we navigate into a design space characterized by numerous dimensions, where multiple variables interrelate with image quality in a nonlinear fashion and where cumbersome boundary conditions typically affect most variables, optical design is not an easy task as it requires a lot of patience and computations.

Therefore, lens designers need to optimize numerically some merit functions to meet some specifications. Computationally speaking this means a lot of computations which is mostly impossible for a human to do it. So, there

DOI: 10.1201/9781003506355-1

have been developed packages/software that help with the construction, optimization, and analysis of an optical design [9].

There are several optical design programs. Several of these programs were originally developed for exclusive use within the industry. To mention a few we enlist their name with the last name of their author [10],

- Slams (C G Wynne)
- Ordeals (Tropel)
- Flair (Radkowski)
- COP (Grey)
- Lead (Kodak)
- Spade (Sperry)
- Optik V (Texas Institute)
- Alsie (Osaka, Suzuki)
- SIGMA (Kidger)
- Bathos (Blandford)
- ACCOS (Spencer)
- CERCO (French)
- Cool Genii (Genesee)
- CODE n (Harris)
- Oslo (Sinclair)
- ZEMAX (Moore)
- SYNOPSYS (Dilworth)

Most of the mentioned optical design programs are based on the idea of a spreadsheet that contains the information on each surface, radius, thickness, materials, and position among other parameters [11, 12, 13, 14]. This paradigm has its positive aspects and negative aspects. To mention a few positives, it is well known by the community of lens designers and it can be considered the standard [15, 16, 17]. But optical design is a field under constant evolution and new ways to design optical systems are emerging.

For example, instead of considering the system as a list of surfaces, consider the optical elements like lenses and mirrors. We can consider them as solid elements and not describe them just by the properties of the light we want to trace through the system (the case of a spreadsheet). These solid

FIGURE 1.1
`Quadoa Optical CAD` logo.

elements can be described with the help of Computer-aided design (CAD). This book is going to construct, optimize, and analyze optical systems under the fresh paradigm of CAD with the help of `Quadoa Optical CAD`. Along the chapters of the book, we are going to see why this paradigm can be very useful in designing optical systems [18].

1.2 The Underlying Philosophy of Quadoa

`Quadoa Optical CAD` stands out as an optical design, simulation, and analysis software boasting a distinctive architecture (see figure 1.1). What sets it apart from other software in this field is its departure from describing the optical system merely as a list of surfaces traversed by light. Instead, `Quadoa Optical CAD` employs a hierarchical structure of high-level elements, such as lenses or beamsplitters, organized into assemblies.

Moreover, `Quadoa Optical CAD` emphasizes strict segregation between optical system description and ray tracing. While the model of the optical system delineates all physical components like lenses, mirrors, prisms, and beam splitters, it refrains from encapsulating the properties of the light intended for tracing through the system. The configuration and setup of light properties occur separately during the establishment of sequences for tracing.

This unique approach of segregating optical system geometry and light sequence confers numerous advantages upon `Quadoa Optical CAD`, including:

1. **Multiple sequences:** Unlike other optical design software, `Quadoa Optical CAD` offers the convenience of incorporating multiple sequences within a single model. This feature proves highly advantageous for tasks such as ghost analysis or for designing and simulating optical systems with multiple paths of interest, such as interferometers.
2. **Non-sequential raytracing:** Non-sequential raytracing, while powerful, can be computationally intensive compared to sequential raytracing. This discrepancy arises because the time complexity of the non-sequential algorithm grows exponentially with the number of surfaces. Consequently, when dealing with a large number of surface interactions, non-sequential raytracing can become significantly slower.
3. **Easy setup of double path systems:** However, many tasks, such as ghost analysis, can be efficiently tackled using sequential raytracing. By examining all potential optical paths up to a specified number of reflections at optical surfaces, sequential raytracing can provide valuable insights. In `Quadoa Optical CAD`, this capability is seamlessly integrated into a single model through auto-generated ghost sequences, thereby conserving considerable computation time. Streamlined setup for double-path systems: Designing double- or multi-path systems is straightforward in `Quadoa Optical CAD`. Users can define the optical system using the `Optical System Editor` and then specify the precise path for ray traversal through the system. This methodology allows for the tracing of rays through a surface multiple times without the necessity of duplicating the surface properties within the surface properties list.
4. **Integration of mechanics:** An optical system relies on mounting mechanics to secure lenses and mirrors in position. The optimal performance of a high-end system hinges on well-designed mechanical mounts. Hence, collaboration between optical and mechanical designers is crucial. To mitigate the risk of errors during the exchange between optical and mechanical design processes, `Quadoa Optical CAD` facilitates the direct inclusion of mechanical components into the optical design model. Additionally, the optical system can be exported to CAD data for import into mechanical CAD software. These data exchange capabilities enable swift identification and resolution of issues such as vignetting or incorrect dimensions in mechanical parts. As a result, this streamlines troubleshooting, saving considerable time and effort.

`Quadoa Optical CAD` employs a hierarchical structure of high-level elements, such as lenses or mirrors, organized into assemblies. Quadoa emphasizes strict separation between optical system description and ray tracing. The configuration and setup of light properties occur separately during the establishment of sequences for tracing.

1.3 Download and Installation

The first step in our journey is to download and install Quadoa. In this book, we are going only to cover the installation of `Quadoa Optical CAD` of `Single License Installation (USB-dongle) for Windows`.

To install `Quadoa Optical CAD` on your personal computer, simply follow these steps:

Download the `Quadoa Optical CAD` installation package from the official website `https://quadoa.com/download` (see figure 1.2). Then, agree to `General Terms and Conditions` and download the `Quadoa Optical CAD` for `Windows`.

Locate the downloaded file on your computer and double-click to run it. This action will start the installation wizard.

Follow the prompts provided by the installation wizard to proceed with the installation process. You may need to agree to the terms and conditions,

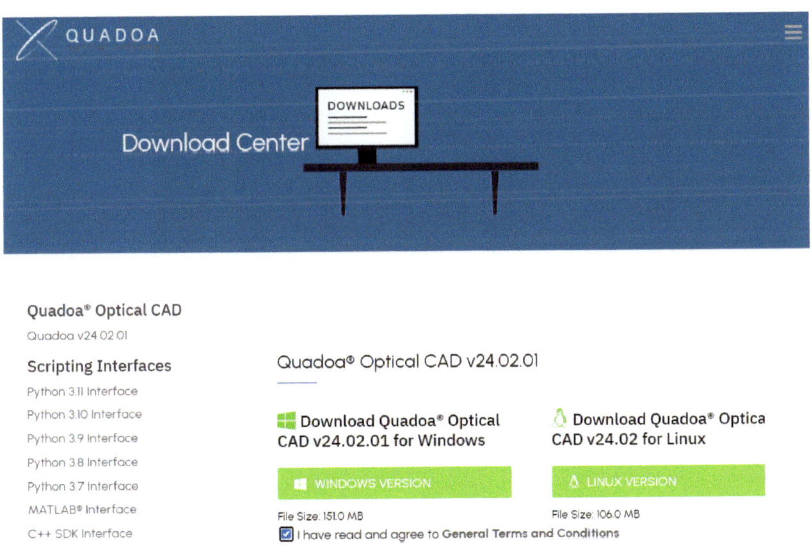

FIGURE 1.2
`Quadoa Optical CAD` Download Center.

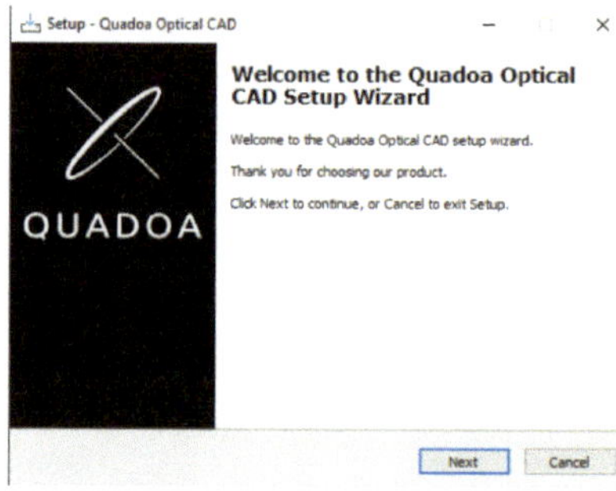

FIGURE 1.3
`Quadoa Optical CAD` Setup Wizard.

choose the installation directory, and select any additional options or settings as required. See figure 1.3.

Once you have completed the installation steps, click `Close` to exit the installation wizard.

Connect the `Quadoa Optical CAD` USB dongle to your personal computer.

`Quadoa Optical CAD` should now be installed on your personal computer and ready to use. You can launch the software from the desktop shortcut or the Start menu.

That's it! You have successfully installed `Quadoa Optical CAD` on your personal computer.

1.4 Beginning with `Quadoa Optical CAD`

One primary goal of developing `Quadoa Optical CAD` is to ensure the software and user interface are as intuitive as possible. However, while many features are designed to be self-explanatory, not every aspect may be immediately clear. Therefore, it is useful to see the manual, video tutorials, and some examples and access the official forum to communicate questions to other `Quadoa Optical CAD` users.

1.4.1 Manual

The first thing we should do with `Quadoa Optical CAD` is to familiarize ourselves with the `Quadoa Optical CAD` manual. To get the manual, just go to the toolbar and press help and you can obtain the manual in PDF format, see figure 1.4. The manual provides a comprehensive overview of the parameters for each object or surface property item available within the software.

FIGURE 1.4
The manual of `Quadoa Optical CAD` can be obtained by going to the toolbar, pressing help, and pressing the PDF button.

1.4.2 Video Tutorials

You can press in `Video Tutorials` and the following web page will open `https://www.quadoa.com/mediacenter`. It contains several video tutorials that contain some basic aspects, construction, raytracing, polarization, mechanics, analysis optimization, tolerancing, scripting, and design examples in `Quadoa Optical CAD`. `Video Tutorials` offer an excellent method for learning and enhancing skills, particularly when acquainting oneself with new software. I highly recommend the reader see the mentioned videos before continuing with this book.

1.4.3 Examples

Upon installation, `Quadoa Optical CAD` provides a folder containing example models. To access these models, simply go to the `File` tab and click on the `Open Examples` button. Within this folder, you'll discover a wide array of examples demonstrating various features. These include basic showcases of available surface properties and coordinate system options, as well as advanced topics like multi-sequential ray tracing. Moreover, you'll find real-world examples of optical systems such as camera lenses, zoom lenses, telescopes, microscope objectives, interferometers, spectrometers, free-form optics, and many others.

1.4.4 Online Forum

You can visit the online user forum at `https://www.forum.quadoa.com`. This platform allows users to interact with one another, sharing optical designs, asking questions about `Quadoa Optical CAD` usage, and seeking advice on optical design, simulation, and related topics. Users can also contribute by providing answers to questions posted by others. Registration is required to participate in the forum. The forum is accessible to all users and is free of cost.

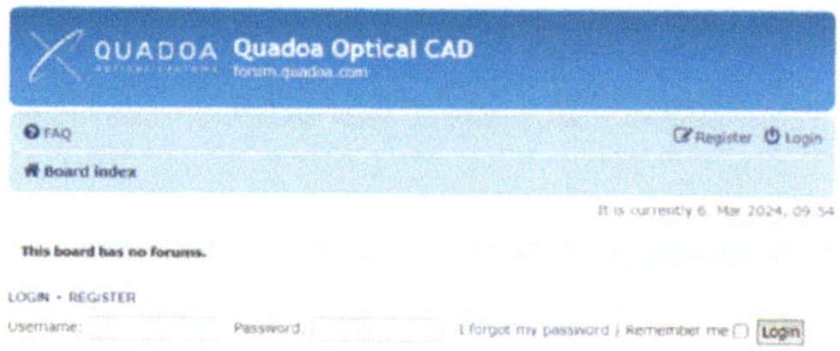

FIGURE 1.5
The online user forum, at `https://www.forum.quadoa.com`, provides a space where users to interact with one another, share optical designs, ask questions about `Quadoa Optical CAD` usage, and seek advice on optical design, simulation, and related topics.

Before we start with `Quadoa Optical CAD`, I need to mention that the intention of this book is to be as intuitive as possible so all the concepts related to `Quadoa Optical CAD` will have this `font`. `Quadoa Optical CAD` is very graphical software if the concept relates to `Quadoa Optical CAD` has a figure in the `Graphical User Interface`. It will be placed after the concept's name. For example: `Optimization`.

1.5 User Interface

The `Graphical User Interface` (GUI) comprises various main areas, as depicted in figure 3.1.

From the figure, we can see that the GUI is divided into several toolbars. `Quick Access Toolbar` is displayed at the *top-left* of the GUI. Besides the `Quick Access Toolbar` is the `Ribbon Bar Tabs`. Below the `Ribbon Bar Tabs` is the `Ribbon Group`, which contains The `Ribbon Bar` offers a tab for each category to navigate within `Quadoa Optical CAD`, housing most of the functions. Additionally, many functions can be accessed through the right-click menu of items within the views. `The Ribbon Bar` consists of the file, setup, construction, coating, sequence, analysis, optimization, tolerancing, wave optics, mechanics, and help.

At the center of the GUI is the `3D-View`, which shows the elements of the design and the ray-tracing. The system setup editor is called in the GUI as system setup. The `System Setup Editor` is comprised of multiple categories about the system in general like the environment (temperature and pressure), sequences (ray-tracing), optimization, tolerancing, plots, and reports) of items, all grouped under top-level container items for organization. Please take time to analyze figure 1.6 to recognize all the mentioned `Toolbars`.

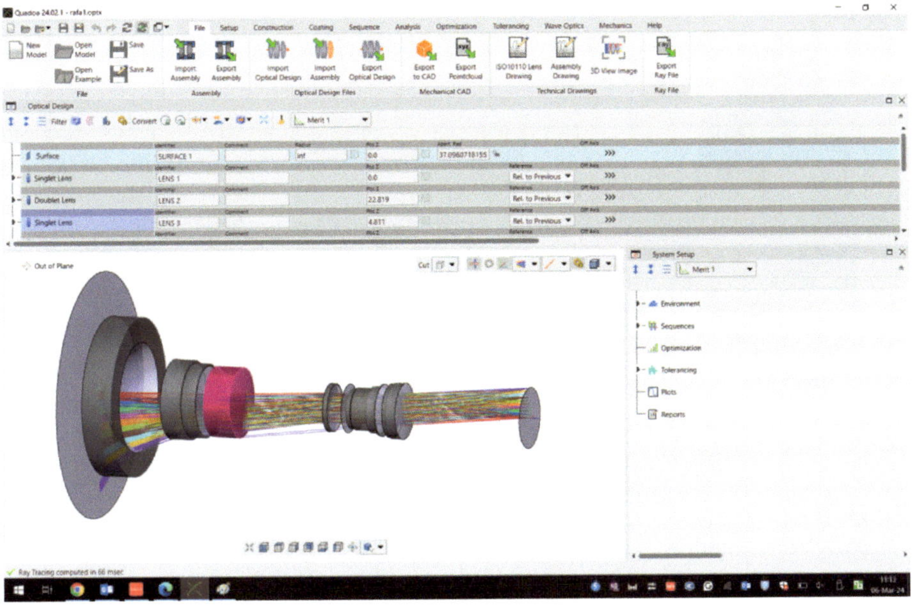

FIGURE 1.6
The Quadoa Optical CAD Graphical User Interface comprises various main areas: Quick Access Toolbar, the Ribbon Bar Tabs, 3D-View, The System Setup Editor, and Optical design editor.

Finally, the Optical Design Editor shows information optical elements of the optical system. Within the Optical Design Editor, users define the optical design, detailing parameters such as the position, thickness, and materials of lenses, as well as surface shapes and other relevant attributes.

1.6 Quick Access Toolbar

The Quick Access Toolbar offers swift access to frequently used functions and remains visible at all times. It includes functions like a new file, opening a file, and opening a recently closed model file. Save the model file, save the model file with a specific name and path.

- **New:** (Ctrl + N): Generate a new empty model file.
- **Open:** (Ctrl + O): Open a file.
- **Open:** (Ctrl + O): Open a recent file.

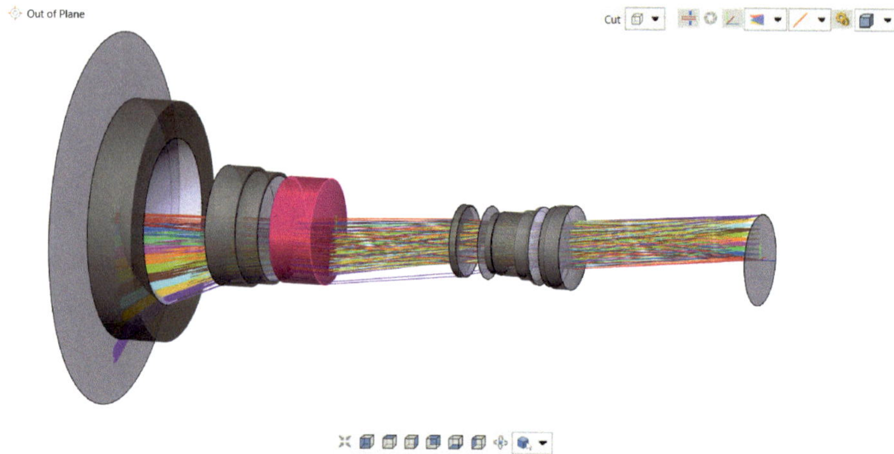

FIGURE 1.7
The 3D-View `Quadoa Optical CAD` shows the raytracing and the elements of the optical system. The user can manipulate the view of the system by rotating, translating, and zooming.

- **Save:** (Ctrl + S): Save a recent file.
- **Save As:** (Ctrl + Shift + S): Save model file with specific name and path.

Also, useful commands such as the previous action is undone and previous undo is undone, refresh current mode, and auto refresh current model.

1.7 3D-View

The `3D-View` serves as the focal point of `Quadoa Optical CAD` user interface. Here, the optical system, as defined in both the `Optical Design Editor` and the `System Setup Editor`, is displayed. Users can manipulate the view by rotating, translating, and zooming. Various drawing styles and cutaways are also available for customization. Additionally, users can visualize the path of rays and the polarization state of each ray segment. Real rays are depicted as solid lines, while virtual rays are represented with dashed lines.

The right-click tool menu within the `3D-View` enables direct editing of lenses within the optical system.

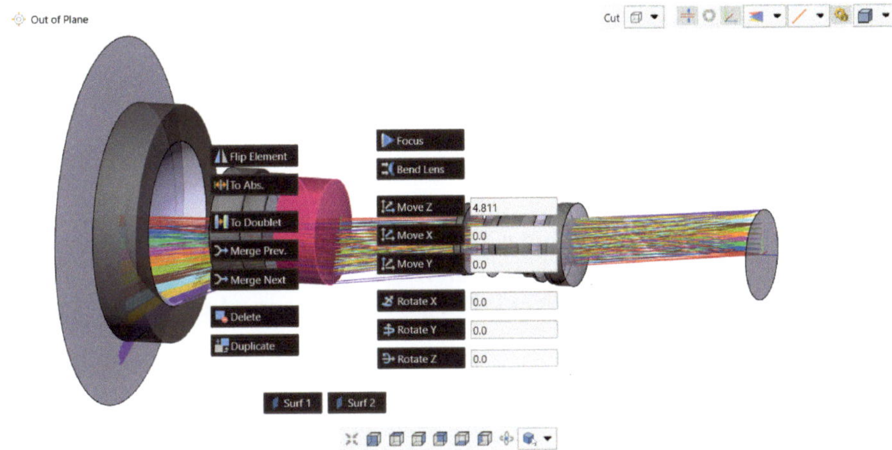

FIGURE 1.8
`Quadoa Optical CAD` logo.

1.8 Ribbon Bar Tabs

The `Ribbon Bar` features individual tabs corresponding to different categories in `Quadoa Optical CAD`, providing easy navigation and hosting most of the functions. Moreover, many of these functions can also be accessed conveniently through the right-click menu while interacting with items in the views.

1.8.1 File Tab

The `File Tab` encompasses various functions related to files, including loading and saving models, importing and exporting third-party file formats, and exporting technical drawings. It is categorized into the following sections

1.8.1.1 File

The `File` section encompasses all functions related to loading or saving `Quadoa Optical CAD` model files. It functions it has very similar functionality to `Quick Access Toolbar`.

- **New:** (Ctrl + N): Generate a new empty model file.
- **Open:** (Ctrl + O): Open a file.
- **Open:** (Ctrl + O): Open a recent file.

- **Save:** (Ctrl + S): Save a recent file.
- **Save As:** (Ctrl + Shift + S): Save model file with specific name and path.

1.8.1.2 Assembly

The `Assembly` category includes functions specifically for importing or exporting assemblies or lenses between different `Quadoa Optical CAD` models. `Assemblies` in `Quadoa Optical CAD` are a group of elements that are considered as a single unit.

- **Import Assembly:** Imports an `Assembly` to `Quadoa Optical CAD`.
- **Export Assembly:** Exports an `Assembly` from `Quadoa Optical CAD`.

1.8.1.3 Mechanical CAD

The `Mechanical CAD` category provides the capability to export the model as a `CAD file`, along with surface shapes in the form of 3D point clouds. These point clouds can be utilized, for instance, for diamond turning control of complex free-form surfaces. The formats that can be exported can be `.stl`, `.step` and `.iges`. Also, the designs can be exported as a point cloud.

- **Import to CAD:** Import `CAD file` to `Quadoa Optical CAD`.
- **Export to CAD:** Export `CAD file` from `Quadoa Optical CAD`.

1.8.1.4 Optical Design Files

The `Optical Design Files` category empowers users to import and export third-party `Optical Design Files`, facilitating compatibility with other optical design software. The formats that can be imported/exported can be `.csv`, `.seq` and `.zmx`.

- **Import Optical Design:** Import `Optical Design Files` to `Quadoa Optical CAD`.
- **Export Optical Design:** Export `Optical Design Files` from `Quadoa Optical CAD`.

1.8.1.5 Technical Drawings

Quadoa also can create Technical Drawings according to ISO 10110 norm.

- **ISO 10110 Lens Drawing:** launches the Lens Drawing Wizard.

1.8.2 Set up Tab

The Setup Tab encompasses the primary project settings and oversees the editors, catalogs, and scripting wizards.

1.8.2.1 Settings

The Settings category houses all global or project-related settings within the user interface. For example the Ray Coloring Scheme, Plot Line Thickness, Plot Font Size, Default Plot Color, etc. Also, it considers the decimal points in the editor. I invite the reader to take a look at this section and play with it a little bit.

1.8.2.2 Editors

The Editors category provides access to open the main editors. The editors are:

- **Optical Design:** Launches the Optical Design Editor.
- **System Setup:** Launches the System Setup Editor.
- **Multi Config:** Opens the Multi Config Lookup Table.
- **Slider Interface:** Launches the Slider Interface.

1.8.2.3 Catalogs

The Catalogs category contains all catalogs that are included with Quadoa Optical CAD or have been added by the user. The catalogs are the Material Catalog and the Lens Catalog which is going to be studied in more detail in chapter 3.

1.8.3 Construction Tab

In the Construction Tab, Optical Elements within the optical system can be added and modified. Any Optical Element can be incorporated into the Optical Design Editor by clicking the corresponding icon in the ribbon menu. When an element in the Optical Design Editor is selected, the new item will, if feasible, be added as a child of the selected element.

The `Basic Elements` category comprises buttons to add various `Optical Elements` to the optical system.

- **Surfaces** form the fundamental components of any optical system.
- **Singlet:** Launches the singlet lens wizard.
- **Doublet** Launches the singlet lens wizard.
- **Triplet** Opens the triplet lens wizard.
- **Custom** The `Custom Element` is an element entirely composed of a single material with a number of surfaces desired by the user.
- **Assembly:** Assemblies serve to group various `Basic Elements`, with the position of the entire group defined within the Assembly's Local Coordinate System. Assemblies can be nested to accommodate other sub-assemblies, enabling users to model mechanical hierarchies in alignment with real-world structures.

The properties of the `Basic Elements` can be assigned by clicking right-click on them.

Also in the `Construction Tab`, `Quadoa Optical CAD` can display the form and shape of different surfaces. This topic will be covered in detail in its respective chapter, chapter 6.

1.8.4 Analysis Tab

The `Analysis Tab` provides access to various analysis plots and reports, essential to deeply analyze and evaluate an optical system. We are going to spend whole chapters with this tab, therefore we are not going to see it in detail for the moment.

1.8.5 Optimization Tab

The `Optimization Tab` contains all features related to optimization and is one of the keys in `Quadoa Optical CAD`. Thus, chapter 5 and 6 will be only focused on the strategy behind the optimization performed in `Quadoa Optical CAD`.

1.8.6 Tolerancing Tab

The `Tolerancing Tab` is utilized to define tolerances for elements or assemblies within the optical system. Additionally, it enables further analysis of tolerances. For the moment we are not going to pay too much attention to this particular tab, we are going to see it in detail in its respective chapter, chapter 11.

1.8.7 Other Tabs

Other interesting tabs are going to be considered in later the following chapters like the sequence tab and coating tab. Thus we just limit ourselves just to mention them.

The `Wave Optics Tab` and `Mechanics Tab` will not be considered in the book.

Finally, the help tab has already been briefly discussed in this chapter.

1.9 System Setup Editor

The `System Setup Editor` encompasses several categories: `Environment`, `Sequences`, `Optimization`, `Tolerancing`, `Plots`, and `Reports`. Each category contains a variety of items organized within their respective top-level container items.

- **Environment:** settings encompass the two environmental conditions: `Temperature` and `Pressure`. `Temperature`: Specifies the temperature of the environment in degrees Celsius. `Pressure`: Specifies the pressure of the environment in bar. The temperature and pressure settings of the environment apply to any lens in the optical system that is not marked as thermal and has its temperature. When a `Sequence` is defined over a range of surfaces, the material between any two consecutive surfaces, which do not belong to the same element or lens, is assumed to be air at the temperature and pressure defined by the environment. This also applies to surfaces located at the same position. To model two elements that are kitted without an air gap, the `Doublet` or any other n-let lens must be used.

- **Sequences:** object encompasses all `Sequential Ray Traces` defined within the model. The `Sequence` serves as the fundamental object for defining a sequential ray trace through the system. It outlines the sequential path for the rays traced through the system. Additionally, properties of the rays, such as fields, wavelengths, aperture type, and other ray attributes, are defined via the `Source Object` of each `Sequence`.

- **The Optimization Container** comprises all `Merit Functions` defined by the user. A merit function quantifies the agreement between data and the fitting model for a specific set of parameters. Typically, the merit function is minimized when the agreement between data and model is optimal.

- **Tolerancing Container:** houses all tolerances defined for the optical system, including any compensators. `Tolerance` refers to the permissible variation in a dimension, representing the difference between the upper (maximum) and lower (minimum) limits. Given the impossibility of achieving exact sizes in manufacturing, tolerances are specified on production drawings to regulate the dimensions of parts.

- **The Plots Container:** displays any plots that are currently open for a model. Additional plots can be added through the right-click menu or via the `Analysis Tab`.

- **The Reports Container:** presents any reports that are currently open for an `Optical Design File`.

1.10 Optical Design Editor

In the `Optical Design Editor`, the optical design parameters are specified, encompassing the position, thickness, and materials of lenses, as well as the surface shapes and related attributes. The sequential ray tracing setup, including fields and wavelengths, is established through the `System Setup Editor`. All components are organized in a hierarchical structure akin to a tree. Any object added as a child element inherits the properties of its parent element. For instance, a surface can be part of a lens, which in turn can be part of an assembly. Furthermore, surfaces may possess child elements representing specific property items such as shape, aperture, phase function, etc. Each column in the interface corresponds to a distinct parameter of the object, facilitating the manipulation of parameters like lens position or surface radius. See figure 1.6.

To expand the hierarchical structure of the element and reveal its child elements, simply click the left mouse button on the arrow icon. Conversely, to collapse the tree structure, click the left mouse button on the same arrow icon.

The top menu of the `Optical Design Editor` offers filter options for the tree view and functions applicable to the entire model. The most useful available filter options include:

- **Convert Position:** Enables global conversion between position definitions, such as converting all positions to absolute or relative values.

- **Conversion or Import Form:** Facilitates the conversion between surface form definitions, such as transitioning between various types of aspheric definitions or polynomials. Additionally, it allows fitting a surface form to imported data.

- **Scale System:** Adjusts the entire system uniformly by a specified factor.

The rest of the functions in the `Optical Design Editor` can be consulted directly from the `Manual`.

1.11 Summary

In this chapter, we have covered what is the main purpose of optical design, which is to get the best possible design meeting certain constraints. The problem is that to achieve that goal we usually end in a highly nonlinear system without an analytical close-form solution. Therefore they are several commercial optical software programs have appeared, whose main goal is to assist the design in the computations of construction, evaluation, and optimization of a design.

Most of the optical design software is based on a spreadsheet that contains information on each surface, radius, thickness, materials, and position among other parameters. This paradigm has intrinsic limitations to the design, like making some computations very cumbersome like ghost images, or multiple sequences. Therefore, a new paradigm for commercial optical software has appeared in the horizontal. this paradigm is based on Computer-aided design. The software for excellence for this paradigm is `Quadoa Optical CAD`.

`Quadoa Optical CAD` distinguishes itself as a comprehensive optical design, simulation, and analysis software, featuring a unique architecture. What sets it apart from other software in this field is its departure from describing the optical system merely as a list of surfaces traversed by light. Instead, `Quadoa Optical CAD` employs a hierarchical structure of high-level elements, such as lenses or beamsplitters, organized into assemblies.

In this book, we are going to study optical design with the guidance of `Quadoa Optical CAD`. So, in the last sections of this chapter, we show how to install it and present the main components of the `Graphical User Interface`. Most of the functionalities of the GUI will be covered in the book in parallel as the optical design concepts are explained. For this chapter, we just introduce them so the user can have a map in their minds of how the information, resources, and functionalities are organized in `Quadoa Optical CAD`.

2

Geometrical Optics and Wave Optics

> It is not knowledge, but the act of learning, not the possession of but the act of getting there, which grants the greatest enjoyment. Life stands before me like an eternal spring with new and brilliant clothes.
>
> Carl Friedrich Gauss

2.1 Geometrical Optics

As outlined in the introductory chapter, geometrical optics stands scientific discipline used to design optical systems. It is heavily reliant on principles from Euclidean geometry [19, 20, 21]. This chapter aims to provide a concise overview of the physics governing geometrical optics, focusing on the fundamental concepts central to this discourse.

At the core of geometrical optics lies the concept of a ray. An optical ray represents a line perpendicular to the light's wavefront, tracing the path of light propagation [22]. Rays serve as a useful abstraction, particularly when the optical elements interacting with light far exceed the wavelength of light, effectively disregarding diffraction in geometrical optics [23]. In `Quadoa Optical CAD` rays are obtained by a sequence. The sequence is the basic object that is used to define a sequential ray trace through the system. It defines the sequential path for the rays traced through the system.

Throughout this discourse, lenses, mirrors, and optical systems are presumed to vastly surpass the wavelength of the interacting light. Consequently, diffraction is overlooked, aligning with the principles of geometrical optics adopted herein [8].

Geometrical optics posits that optical elements can alter light's trajectory and ray directions through two principal mechanisms: reflection and refraction. Reflection entails the redirection of incident light rays upon striking polished surfaces, typically exemplified by mirrors. Refraction, on the other

DOI: 10.1201/9781003506355-2

hand, signifies the bending of light as it transitions into a medium of differing speed. The refractive index of a medium, a dimensionless quantity, quantifies the speed at which light traverses said medium, a parameter central to geometrical optics,

$$\boxed{n \equiv \frac{c}{v},} \tag{2.1}$$

in this context, c denotes the speed of light in vacuum, while v represents the phase velocity of light within the medium. A medium is deemed homogeneous if it maintains a constant refractive index throughout its entirety. Within homogeneous media, light rays propagate along straight paths [7].

Refraction surfaces encompass entities possessing a refractive index distinct from their surrounding environment. Such entities may include lenses or prisms. Lenses are designed to either converge or diverge light rays, altering their trajectory. Prisms, meanwhile, aim to manipulate light through refraction without inducing convergence or divergence in the light rays [24].

In essence, geometrical optics focuses on mirrors (reflective surfaces), lenses, and prisms (refractive bodies). However, this simplicity does not equate to ease of understanding. Much of the mathematical framework within geometrical optics involves nonlinear equations, often intricate and lengthy to manage. Consequently, optical practitioners frequently resort to approximations to streamline their analyses. These approximations stem from fundamental definitions involving trigonometric functions such as $\sin\theta$ and $\cos\theta$.

$$\sin\theta \equiv \sum_{m=0}^{\infty} \frac{(-1)^m}{(2m+1)!}\theta^{2m+1} = \theta - \frac{\theta^3}{3!} + \frac{\theta^5}{5!} - \frac{\theta^7}{7!} + \cdots \tag{2.2}$$

and,

$$\cos\theta \equiv \sum_{m=0}^{\infty} \frac{(-1)^m}{(2m)!}\theta^{2m} = 1 - \frac{\theta^2}{2!} + \frac{\theta^4}{4!} - \frac{\theta^6}{6!} + \cdots \tag{2.3}$$

The paraxial approximation, the most commonly used in optics, is applicable when the angles involved are sufficiently small [4], such tha

$$\sin\theta \approx \theta, \tag{2.4}$$

and,

$$\cos\theta \approx 1. \tag{2.5}$$

Typically, approximations are classified based on the number of elements included in the calculation. For instance, the third-order approximation of $\sin\theta$ is denoted when $m = 3$, indicating the inclusion of terms up to the third order.

$$\sin\theta \approx \theta - \frac{\theta^3}{3!}. \tag{2.6}$$

Despite the paraxial approximation denies a lot of the richness and information of an optical system it can be used to classify a system by its nature [1, 5].

2.1.1 The Principle of Least Action

To grasp Fermat's principle, let's delve into mechanics, a discipline within physics dedicated to studying objects in motion or at rest. Through mechanics, once we ascertain the initial position, velocity, and the forces acting upon these objects, we gain a comprehensive understanding of their behavior [25, 26]. This formulation owes much to Sir Isaac Newton, the English polymath, who demonstrated the laws governing moving bodies, employing vector calculus for rigorous analysis [27, 28]. While vector calculus offers a thorough analysis, enabling predictions about the evolution of physical systems, it can be challenging to derive equations of motion in certain cases. Moreover, these equations may form a nonlinear system, complicating matters further. To circumvent such challenges, some physicists proposed an alternative approach, leveraging the law of conservation of energy. This fundamental principle dictates that energy is neither created nor destroyed but rather transformed. In 1662, Pierre de Fermat, the French mathematician and lawyer, introduced the initial concepts of the principle of minimum action [29]. Fermat posited that light rays traveling from one point to another follow a path that minimizes the time taken. This principle, known as the principle of least action, suggests that nature tends to expend the least amount of energy possible.

Applied to optics, this principle is termed Fermat's principle, which asserts that light rays traverse paths that minimize the time taken to travel between two points [6].

> The optical path length between two distinct fixed points, traversed by light, corresponds to the global minimum. It is calculated by multiplying the physical length of the path by the refractive index of the medium, defining the optical length.

Then, let us consider the notion of the *global minimum*. This term refers to the smallest value within an entire set. Imagine a scenario where we have a set comprising all possible optical paths between two points. Each path possesses its optical length. According to Fermat's principle, the physically valid path among this set is the one with the minimum optical path length.

Throughout this book, we will consistently refer to Fermat's principle, as it serves as a cornerstone for designing stigmatic systems.

Fermat's principle can be conceptualized as an integral across time, representing the time required for light to travel from one point, x_1, to another point, x_2. Let t_1 and t_2 denote the times when light is located at points x_1 and x_2, respectively,

$$\int_{t_1}^{t_2} dt = \frac{1}{c}\int_{t_1}^{t_2} \frac{c}{v}\frac{ds}{dt}dt = \frac{1}{c}\int_{x_1}^{x_2} nds, \tag{2.7}$$

where c represents the speed of light in a vacuum, ds stands for an infinitesimal displacement along the ray, $v = ds/dt$ denotes the speed of light in a medium, and $n = c/v$ signifies the refractive index of the medium.

Given these definitions, the optical path length (OPL) of light traveling from a point x_1 to a point x_2 can be expressed as:

$$\boxed{\text{OPL} = \int_{x_1}^{x_2} nds.} \tag{2.8}$$

Notice that the optical path is a geometrical quantity, since,

$$\text{OPL} = c\int_{t_1}^{t_2} dt. \tag{2.9}$$

Consider a set comprising all possible travel times between points x_1 and x_2, as well as a set encompassing all optical path lengths (OPL) between the same points. The global minimum travel time between points x_1 and x_2 for light corresponds to the global minimum optical path length between these points.

2.1.2 Reflection

Fermat's principle predicts how the path of light will behave when it is reflected. Observe the figure 2.1; it shows an indecent ray of light on a surface. The surface reflects the ray. We can see if the ray starts from point x_1, y_1 arrives in a straight line to the point where it is reflected on the surface and then reaches point x_2, y_2. Taking this into account we can write the time that light takes from point x_1, y_1 to point x_2, y_2 as the following expression,

$$T(x) = \frac{1}{c}\left[\sqrt{(x-x_1)^2 + y_1^2} + \sqrt{(x_2-x)^2 + y_2^2}\right], \tag{2.10}$$

where x is the place on the plane in the horizontal direction where the light strikes. Now, we know that $T(x)$ is correct in the way for places with constant refractive index the light moves in straight lines, all this from the ray equation.

But we do not know yet how the light reflects. The point $(x, 0)$ is crucial here. The value of x will tell us the slope of the straight lines. From the Fermat principle, we know that the light moves in the least time. Thus, we can derived respect x and equal to zero,

$$\frac{\partial T(x)}{\partial x} = 0, \tag{2.11}$$

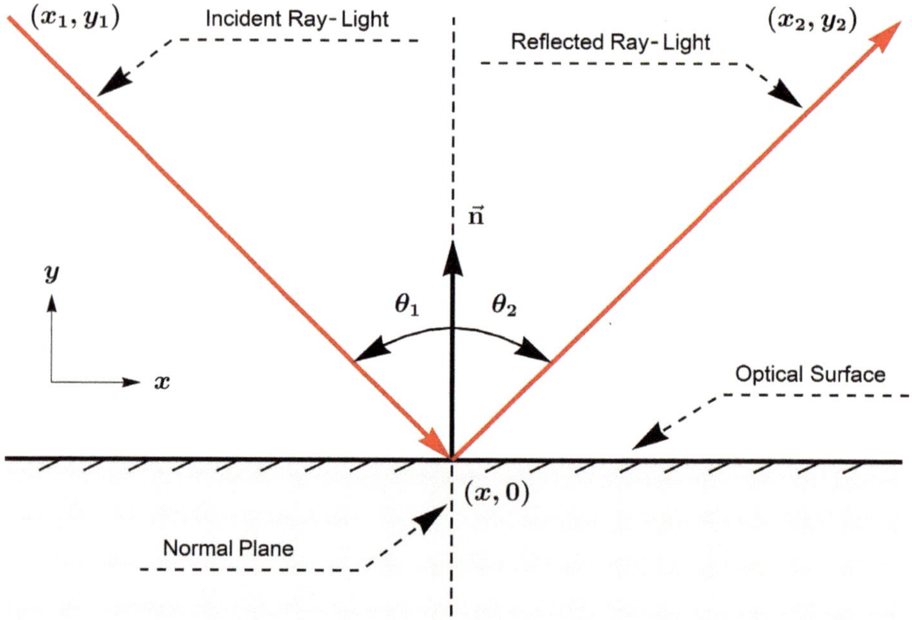

FIGURE 2.1
Law of reflection, the incident angle θ_1 is equal to the reflected angle θ_2.

computing the derivative,

$$\frac{x - x_1}{\sqrt{(x - x_1)^2 + y_1^2}} = \frac{x_2 - x}{\sqrt{(x_2 - x)^2 + y_2^2}}. \tag{2.12}$$

From figure 2.1, we can see that the last equation can be re-formulated as,

$$\boxed{\theta_1 = \theta_2}. \tag{2.13}$$

As in figure 2.1 can be seen as a simple diagram of a reflection taking place on a flat mirror, the incident angle θ_1 is equal to the reflected angle θ_2, in other means $\theta_1 = \theta_2$. We can show that is a real minimum by computing the second derivative,

$$c\frac{\partial^2 T}{\partial x^2} = (1 - \sin^2 \theta_2) \left[\frac{1}{\sqrt{(x - x_1)^2 + y_1^2}} + \frac{1}{\sqrt{(x_2 - x)^2 + y_2^2}} \right]. \tag{2.14}$$

Since $\sin \theta_2^2 \leq 1$ the above's expression is positive for all values of x, which means, we have an absolute minimum.

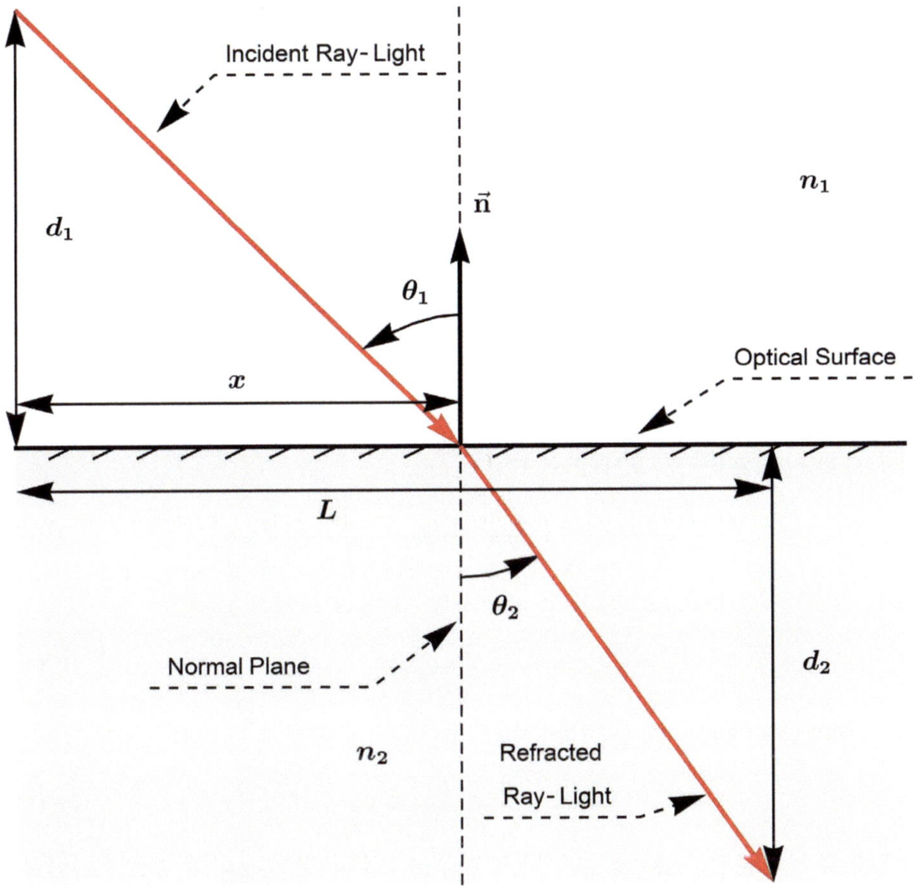

FIGURE 2.2
Refraction takes place on a flat interface between two mediums with different refractive indexes n_1 and n_2. The indecent vector is $\vec{\mathbf{s}}_1$, the refracted vector is $\vec{\mathbf{s}}_2$ and the normal vector of the surface is $\vec{\mathbf{n}}$.

2.1.3 Refraction

In figure 2.2, we present a diagram of a refraction taking place on a flat interface between two mediums with different refractive indexes.

Snell's law seems inoffensive, but it is tough to manipulate. It is so hard to manage that paraxial approximations are commonly used. But we decide in the introduction of this chapter that we are going to negate the paraxial approximations at all cost.

Refraction occurs when light travels through a region that has a changing index of refraction. The simplest case of refraction occurs when there is an interface between a uniform medium with an index of refraction n_1 and another medium with an index of refraction n_2. In such situations, **Snell's Law**, also called the **Law of Refraction**, describes the resulting deflection of the light ray.

In figure 2.2 we can see a simple diagram of a reflection taking place on a flat interface between two mediums with constant refraction indexes. From the figure we can put the trajectory of light as,

$$R(x) = n_1\sqrt{d_1^2 + x^2} + n_2\sqrt{d_2^2 + (L-x)^2}, \tag{2.15}$$

where d_1 is the height of the initial position of the light ray. x is the horizontal distance from the initial position of the light ray to the origin of the coordinate system. L is the horizontal distance from the initial position to the final position. d_2 is the height of the final position of the light ray.

To take the path that the light completes with the least time we deviate the equation (2.15) with respect to x.

$$\frac{dR(x)}{dx} = n_1\frac{x}{\sqrt{d_1^2 + x^2}} - n_2\frac{(L-x)}{\sqrt{d_2^2 + (L-x)^2}}. \tag{2.16}$$

From figure 2.2, we can see that precisely $\sin\theta_1$ is given by,

$$\sin\theta_1 = \frac{x}{\sqrt{d_1^2 + x^2}}, \tag{2.17}$$

also from the figure, we see that $\sin\theta_2$ is

$$\sin\theta_2 = \frac{L-x}{\sqrt{d_2^2 + (L-x)^2}}, \tag{2.18}$$

therefore, replacing equations (2.16) and (2.17) in equation,

$$\boxed{n_1\sin\theta_1 = n_2\sin\theta_2.} \tag{2.19}$$

When the light rays are in a homogeneous medium, their paths are straight lines. When the light changes from one homogeneous medium to another medium; the refraction (deviation) is expressed by above's equation. Which is Snell's law for refraction in angle notations.

2.1.4 Vectorial form of the Snell's Law

We need to express Snell's law in a more manageable way, in a vector form. Even it is in two dimensions because, in many chapters, we work with rotationally symmetric surfaces.

Let $\{\vec{\mathbf{e}}_1, \vec{\mathbf{e}}_2, \vec{\mathbf{e}}_3\}$ be an orthonormal basis for the Euclidean space $\mathbb{R}^3$. Let $\vec{\mathbf{s}}_1 \in \mathbb{R}^2$ be the incident unit vector, let $\vec{\mathbf{s}}_2 \in \mathbb{R}^2$ be the refracted unit vector and let $\vec{\mathbf{n}} \in \mathbb{R}^2$ be the normal unit vector of the surface. θ_1 incident angle respect to $\vec{\mathbf{n}}$ and θ_2 is the refracted angle respect to $-\vec{\mathbf{n}}$.

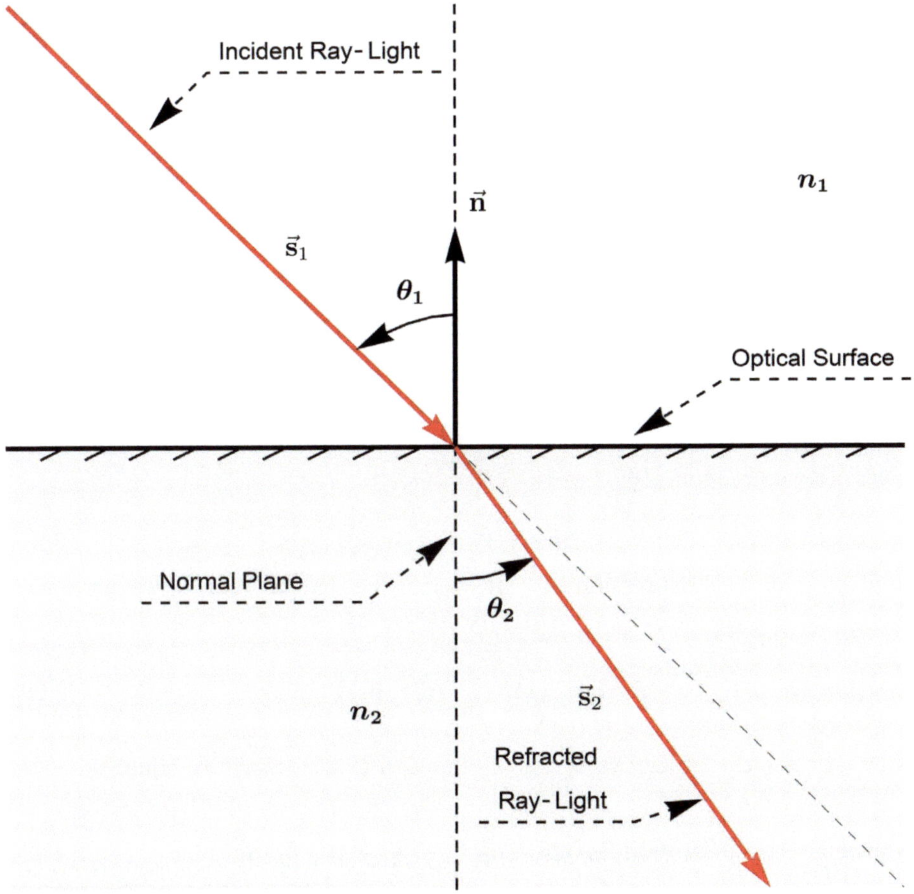

FIGURE 2.3
Diagram of the refraction of light of the vectorial Snell law.

Then, the refracted vector is given by

$$\vec{\mathbf{s}}_2 = \frac{n_1}{n_2}\vec{\mathbf{s}}_1 + \left(\frac{n_1}{n_2}|\vec{\mathbf{n}}||\vec{\mathbf{s}}_1|\cos\theta_1 - |\vec{\mathbf{n}}||\vec{\mathbf{s}}_2|\cos\theta_2\right)\vec{\mathbf{n}}. \tag{2.20}$$

Looking figure 2.3 we can see the vertical component of the refract red ray $\vec{\mathbf{s}}_2$ is given by

$$\vec{\mathbf{n}}\cdot\vec{\mathbf{s}}_2 = |\vec{\mathbf{n}}||\vec{\mathbf{s}}_2|\cos\theta_2, \tag{2.21}$$

Notice that since they are unit vectors $|\vec{\mathbf{n}}| = |\vec{\mathbf{s}}_2| \equiv 1$,

$$\vec{\mathbf{n}}\cdot\vec{\mathbf{s}}_2 = \cos\theta_2, \tag{2.22}$$

we square it,

$$(\vec{\mathbf{n}} \cdot \vec{\mathbf{s}}_2)^2 = \cos\theta_2^2 = 1 - \sin\theta_2^2 = 1 - \frac{n_1^2}{n_2^2}\sin\theta_1^2. \tag{2.23}$$

Using the cross-product,

$$\vec{\mathbf{n}} \times \vec{\mathbf{s}}_1 = |\vec{\mathbf{n}}||\vec{\mathbf{s}}_1|\sin\theta_1(\vec{\mathbf{e}}_1 \times \vec{\mathbf{e}}_2), \tag{2.24}$$

also we have, $|\vec{\mathbf{n}}| = |\vec{\mathbf{s}}_1| = 1$.

$$\vec{\mathbf{n}} \times \vec{\mathbf{s}}_1 = \sin\theta_1(\vec{\mathbf{e}}_1 \times \vec{\mathbf{e}}_2), \tag{2.25}$$

squaring,

$$(\vec{\mathbf{n}} \times \vec{\mathbf{s}}_1)^2 = \sin\theta_1^2(\vec{\mathbf{e}}_1 \times \vec{\mathbf{e}}_2)^2. \tag{2.26}$$

Since, $(\vec{\mathbf{e}}_1 \times \vec{\mathbf{e}}_2) = \vec{\mathbf{e}}_3$, $(\vec{\mathbf{e}}_3)^2 = 1$, thus,

$$(\vec{\mathbf{n}} \times \vec{\mathbf{s}}_1)^2 = -\sin\theta_1^2. \tag{2.27}$$

Replacing above's expression in $(-\vec{\mathbf{n}} \cdot \vec{\mathbf{s}}_2)^2 = 1 - (n_1^2/n_2^2)\sin\theta_1^2$, we get,

$$\vec{\mathbf{n}} \cdot \vec{\mathbf{s}}_2 = \sqrt{1 - \frac{n_1^2}{n_2^2}(\vec{\mathbf{n}} \times \vec{\mathbf{s}}_1)^2}. \tag{2.28}$$

Now let's see the other two terms of equation (2.20), they are equal to,

$$\frac{n_1}{n_2}[\vec{\mathbf{s}}_1 + |\vec{\mathbf{n}}||\vec{\mathbf{s}}_1|\cos\theta_1\vec{\mathbf{n}}] = \frac{n_1}{n_2}[\vec{\mathbf{s}}_1 - (\vec{\mathbf{n}} \cdot \vec{\mathbf{s}}_1)\vec{\mathbf{n}}]. \tag{2.29}$$

Therefore the refracted ray is given by

$$\boxed{\vec{\mathbf{s}}_2 = \frac{n_1}{n_2}[\vec{\mathbf{s}}_1 - (\vec{\mathbf{n}} \cdot \vec{\mathbf{s}}_1)\vec{\mathbf{n}}] - \vec{\mathbf{n}}\sqrt{1 - \frac{n_1^2}{n_2^2}(\vec{\mathbf{n}} \times \vec{\mathbf{s}}_1)^2} \quad \text{for } \vec{\mathbf{s}}_2, \vec{\mathbf{s}}_1, \vec{\mathbf{n}} \in \mathbb{R}^3.} \tag{2.30}$$

2.1.5 Stigmatism

There exists an inherent connection between Snell's law and Fermat's principle. Snell's law dictates the redirection of light, ensuring that it travels along the path of least optical path length (OPL). Similarly, Fermat's principle states that light follows the path of least OPL. In both principles, the shape of optical elements such as lenses and mirrors plays a crucial role, as it determines the orientation of the normal vector $\vec{\mathbf{n}}$.

The primary objectives of mirrors and lenses are to either converge or disperse light through reflection or refraction, depending on the scenario. Converging light aims to bring rays together to a single point, forming a focus,

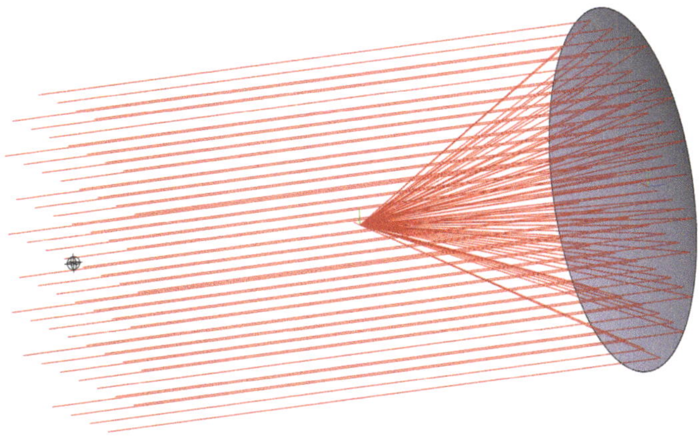

FIGURE 2.4
A parabolic mirror is stigmatic for an infinite object and a finite image.

while dispersing light spreads it as if emanating from a single point source. If the image point is a single point and not a region the system is called astigmatic [24].

Observing segments meeting the conditions set for each conic section, we find that they intersect at their respective foci. Viewing these segments as optical paths, it becomes evident that conic mirrors, mirrors shaped like conics, can concentrate light at their foci. This interpretation aligns with the notion that these paths represent the routes taken by light to minimize travel time.

The elliptic mirror is stigmatic for a finite object and a finite image. The ray-tracing of the elliptic mirrors is displayed in figure 2.5.

A parabolic mirror, for instance, reflects light from an object at a nearly infinite distance toward its focus. Refer to figure 2.4 for visualization.

If the object is real and the image is virtual the hyperbolic mirrors for a stigmatic pair. Please see figure 2.6. Notice that the ray that goes from the virtual image point are dashed lines. `Quadoa Optical CAD` draws the ray as dashed lines if the object or the image is virtual. A virtual ray is the extension of the ray if the light is moving in the positive direction. For the case of the hyperbolic mirror since the mirror reflects light, due to the law of reflection the extension of the rays (dashed lines) forms the image point. When an image or an object is virtual depends on the global coordinates of the system. In this section, we are going to study how `Quadoa Optical CAD` how defined the coordinate system and its implications in real and virtual images.

The elliptic mirror, the parabolic mirror, and the hyperbolic mirror are closed-form solutions and implications of the Fermat principle. When I said, closed-form solutions is that they can be expressed in finite mathematical terms. The mathematical derivation (implication of the Fermat principle) can

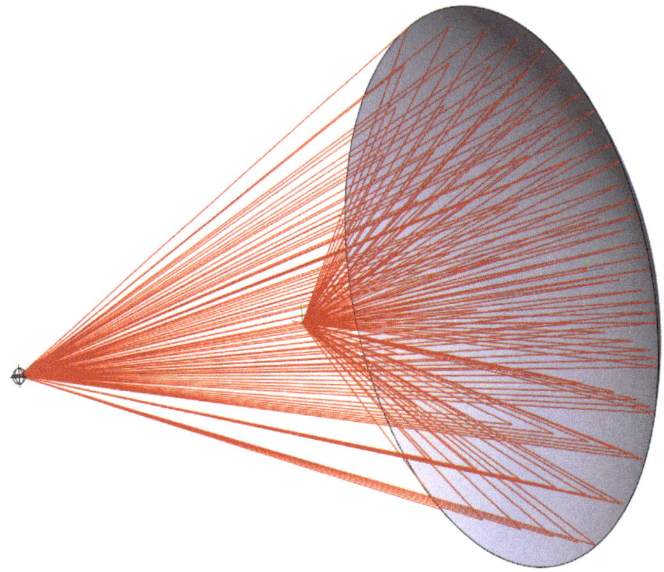

FIGURE 2.5
An elliptic mirror is stigmatic for a finite object and a finite image.

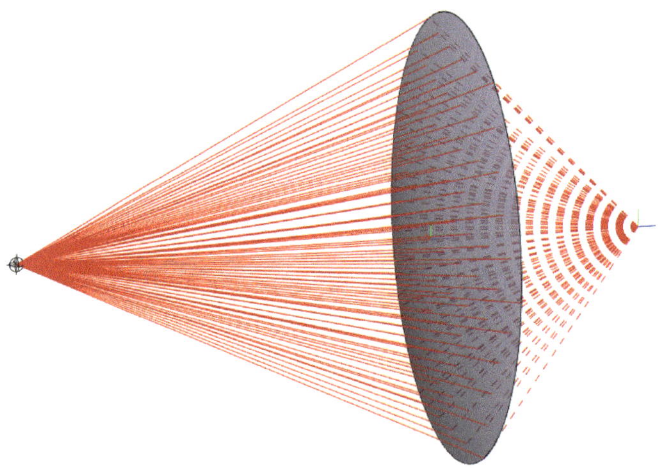

FIGURE 2.6
An elliptic mirror is stigmatic for a finite object and a finite image. But in this case, the image is virtual.

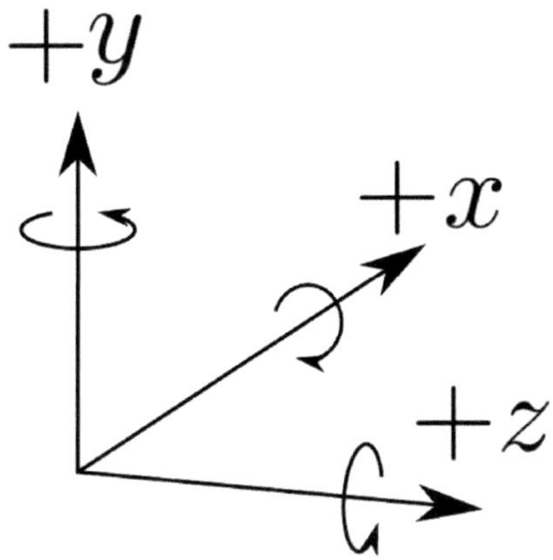

FIGURE 2.7
Coortdinate System in `Quadoa Optical CAD`. Arrows show in positive directions. In `Quadoa Optical CAD`, every coordinate system is right-handed. The z-axis, commonly representing the optical axis, extends to the right. The x-axis extends outward from the user towards the monitor, while the y-axis extends vertically upwards.

be found in the book `Stigmatic Optics` [24]. But this topic is out of the scope of this book. We show these results to define the object and image points.

2.2 Notation and Style Convention

Most optical systems have some associated symmetry. Identifying this symmetry is useful to identify the optical property of the optical system in question. When the system is radially symmetric the axis of symmetry is at the center of the system, and it is usually called the optical axis. The optical axis refers to an imaginary line traversing through the geometric center of an optical system, which may include devices such as camera lenses, microscopes, or telescopic sights. For of axis system the definition of an optical axis may differ but for the moment let us focus on radially symmetric systems.

In `Quadoa Optical CAD`, every coordinate system is right-handed. The z-axis, commonly representing the optical axis, extends to the right. The x-axis extends outward from the user towards the monitor, while the y-axis extends vertically upwards. When plotting data over a surface, the default coordinate system aligns with the perspective from the $+z$ direction when viewing the surface. The direction of rotation can be determined using the right-hand rule: when the thumb points along the positive axis direction, the fingers indicate the positive direction of rotation [18]. See figure 2.7.

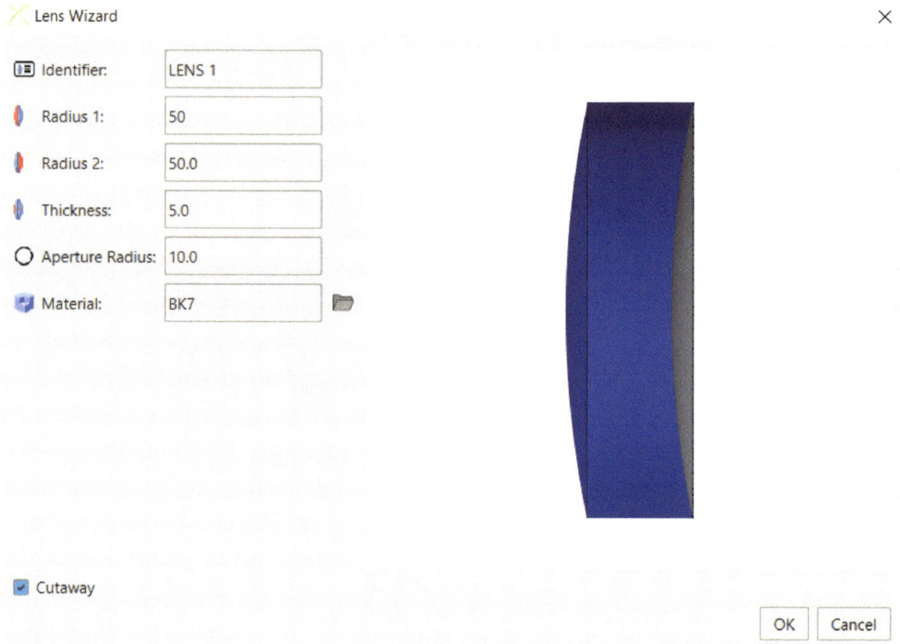

FIGURE 2.8
Display of a lens in the `Lens Wizard` in `Quadoa Optical CAD`. The first radius of the lens is positive and the second radius of the lens is positive too, as defined by the style convention.

There are many shapes of lenses, but historically the most popular lenses are the ones with spherical surfaces. A positive radius indicates that the surface is curved towards the right, in the $+z$ direction. Conversely, a negative radius of curvature signifies that the surface is curved towards the left, in the -z direction. In general, positive values for form objects denote curvature in the $+z$ direction. Most optical software follows this notation. Figure 2.8 describes the scenario where the first radius of the lens is positive and the second radius of the lens is positive too, as defined by the style convention.

To follow the example presented in figure 2.8, we go to the `Construction Tab` and `Press` the icon of the singlet to launch the singlet lens wizard. The case of a lens with a positive radius and a negative radius is displayed in figure 2.9

Now, lenses can be broadly classified into two main categories: positive and negative. A positive lens converges light rays passing through it, effectively focusing them. Such lenses possess a positive focal length. Conversely, a negative lens diverges the light rays, resulting in dispersion. Negative lenses are characterized by a negative focal length. Figure 2.10 (top) and (bottom) illustrate examples of positive and negative lenses.

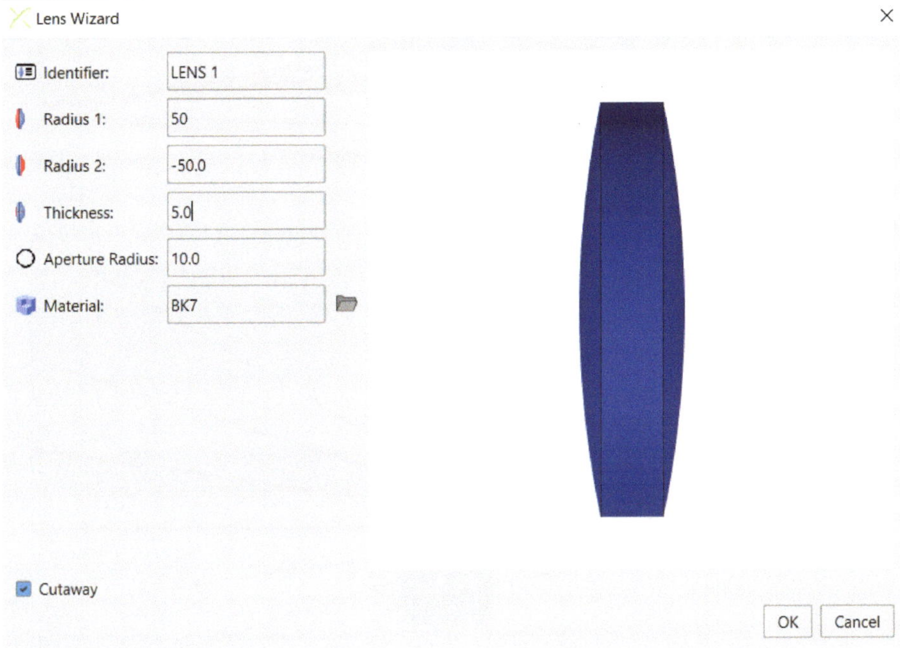

FIGURE 2.9
Display of a lens in the `Lens Wizard` in `Quadoa Optical CAD`. A lens with a positive radius and a negative radius

When optical rays parallel to the optical axis traverse a negative lens from left to right, the refracted rays are diverged and do not intersect across the optical axis. However, the backward extension (dashed lines) of these refracted rays will intersect across the optical axis at a specific point, as depicted in figure 2.10. This point represents one of the two focal points of the negative lens. Similarly, the other focal point can be found on the right-hand side of the negative lens. Notice that the images real or virtual in 2.10 are not a point but a region, this is because in general lenses are not stigmatic.

In `Quadoa Optical CAD`, the following convention is taken for rays. A ray transitions to a virtual state (or real, if it was previously virtual) when its path from one surface to the next indicates a change in direction opposite to its previous propagation direction (sign flip). The initial real or positive direction aligns with the positive z-coordinate of the object surface. Consequently, rays originating from the negative z-direction start as virtual rays, while those originating from the positive direction start as real rays. The state of a ray (real or virtual) does not affect ray tracing unless virtual rays are explicitly restricted by the `Sequence Parameter`, although the optical path length of virtual rays will exhibit a negative sign.

So let's recall the example of the stigmatic mirror for a real object and virtual image, the hyperbolic mirror. This mirror is displayed in figure 2.6.

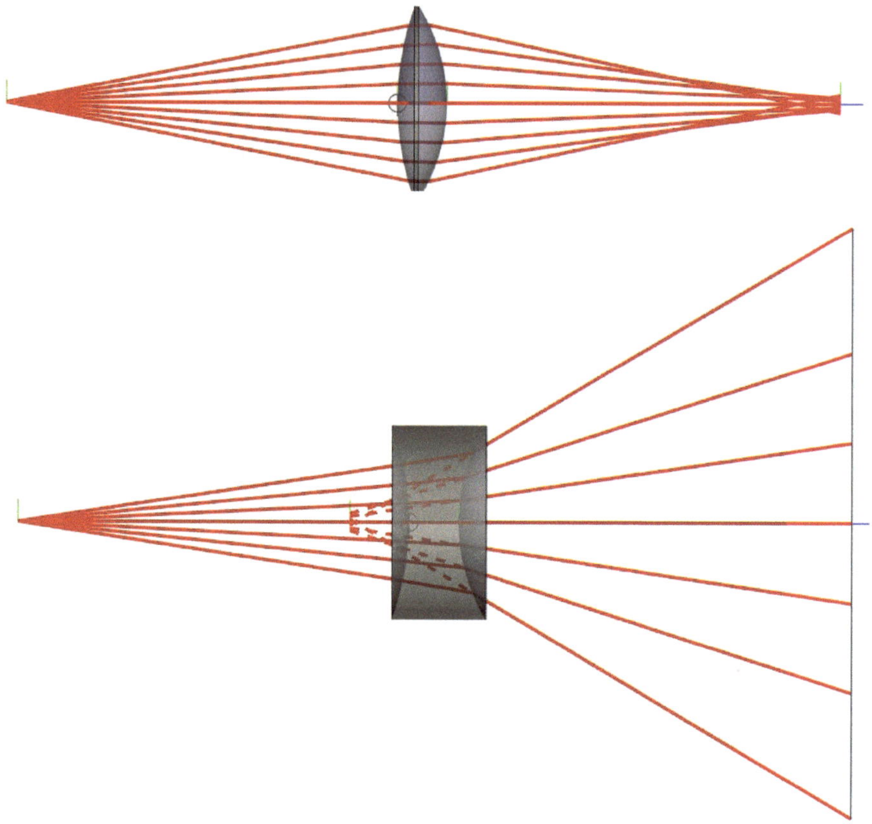

FIGURE 2.10
A positive lens converges light rays passing through it, effectively focusing them. Such lenses possess a positive focal length. Conversely, a negative lens diverges the light rays, resulting in dispersion. Negative lenses are characterized by a negative focal length. *Top:* a positive lens. *Bottom:* a negative lens.

This ray-tracing has been done in `Quadoa Optical CAD`. As you may guess the ray for the virtual image point is where the dashed converge.

2.3 Concepts in Optical Design

In this section, we explore the optical design concepts used by lens designers in practice. these concepts are crucial to characterizing an optical system. Some of the concepts are interrelated therefore it is useful to the reader to read this

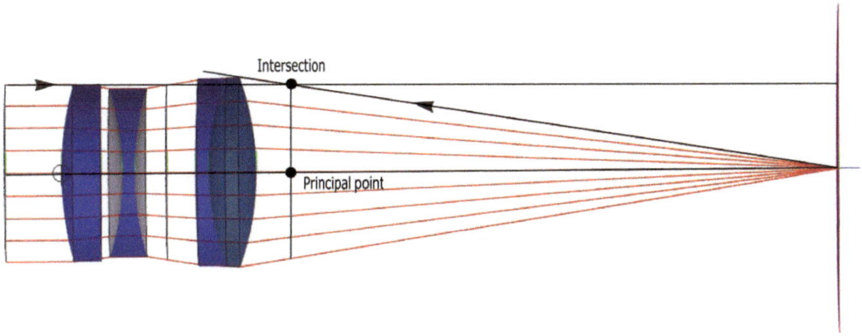

FIGURE 2.11
Principal Point is the point where the propagation of the incident ray touches the propagation of its corresponding outcome ray. It is protection in the optical axis of the point where the propagation of the ray in object space meets the propagation of its corresponding ray from image space.

section several times, since introducing this concept is not simple in a linear way.

2.3.1 Cardinal Points

The next concepts we are going to study are the cardinal points. In paraxial geometrical optics, the cardinal points comprise three pairs of points situated along the optical axis of a rotationally symmetric focal optical system. These include the focal points, principal points, and nodal points, each with two corresponding points [1].

The focal points are the points where the light focuses or converges, if the system is stigmatic the focus is an infinitesimally small point. If the image has aberrations, (not stigmatic) then the focus becomes a region rather than a point. We are going to consider these scenarios in chapter four [1].

Principal Point is the point where the propagation of the incident ray touches the propagation of its corresponding outcome ray. In other words, it is the point where the propagation of the ray in object space meets the propagation of its corresponding ray from image space. The principal point is easier to describe graphically, in figure 2.11 is displayed. Notice that it is the point where the propagation of the input/output rays meet. So each ray has a principal point, but usually, the principal point that is considered is the one of the ray at this in the margin of the field. We will study more on rays later in this chapter [1]. The principal point is also considered the center of the optical system. The principal point of a system composed of three lenses is shown in figure 2.11. Notice that the point lies in the optical axis. The principal plane lies in the principal place and is perpendicular to the optical axis.

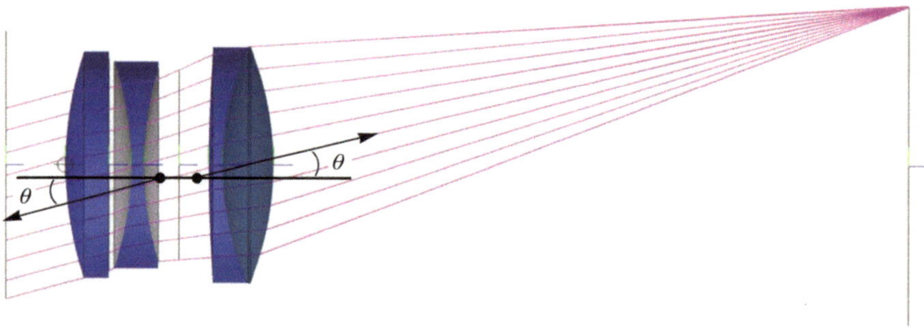

FIGURE 2.12
The front and rear nodal points of a lens exhibit a unique property: when a ray is directed towards one of these points, the lens refracts it in a manner that makes it appear as though it originated from the other nodal point, maintaining the same angle relative to the optical axis.

An optical system has two principal points, one for the standard configuration, object space, system, and image space. The other one is if the system is reversed. This last case is rarely used in practice. Figure 2.11 does not show the reverse case. But the procedure is the same.

The front and rear nodal points of a lens exhibit a unique property: when a ray is directed towards one of these points, the lens refracts it in a manner that makes it appear as though it originated from the other nodal point, maintaining the same angle relative to the optical axis. The nodal points play a similar role for angles as the principal planes do for transverse distance. When the medium on both sides of an optical system is identical (for example in vacuum), the front and rear nodal points align with the front and rear principal points, respectively.

Figure 2.12 shows the nodal points of an optical system composed of three lenses. This picture highlights a ray from a bunch of rays in the field. This ray has the particular that it seems is not being refracted by the optical system composed of three lenses. Since the entrance and exit angles are the same.

2.3.2 Focal Length and Back Focal Length

The focal length stands out as one of the most critical parameters of a lens, significantly influencing the performance of optical systems. The focal length of an optical system gauges the extent to which the system converges or diverges light; it represents the inverse of the system's optical power. A positive focal length denotes a system that converges light, whereas a negative focal length indicates a system that diverges light. A system with a shorter

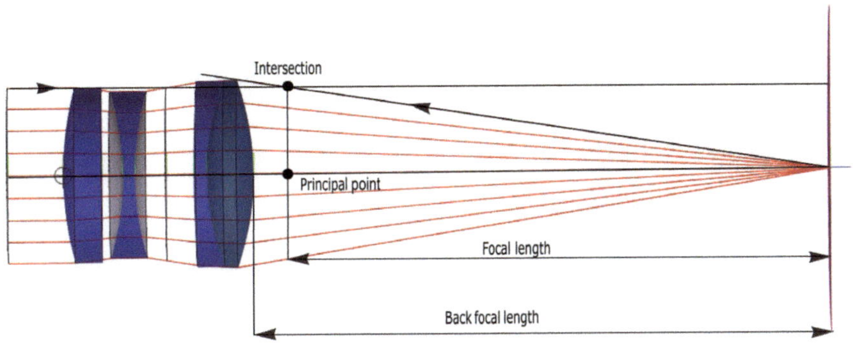

FIGURE 2.13
The effective focal length (also called just focal length) is defined as the distance between the rear principal point and the rear focal point of the lens. The back focal length is the distance between the rear vertex of the lens and the rear focal point. The principal point is protection in the optical axis where the propagation of the incident ray touches the propagation of its corresponding outcome ray.

focal length bends rays more sharply, either bringing them to a focus over a shorter distance or diverging them more rapidly [5].

The effective focal length (also called just focal length) is defined as the distance between the rear principal point and the rear focal point of the lens [5]. The back focal length is the distance between the rear vertex of the lens and the rear focal point [5]. Figure 2.13 shows the same optical system composed of three lenses, just like in figure 2.11, but now the distance between the principal point and the focal point is expressed as focal length. Also, the distance between the vertex and the focal point is defined as the back focal length.

2.3.3 Aperture Stop and Field Stop

An aperture is a hole through which light passes. The aperture stop, which limits the angular range from which an imaging system can receive light, is a crucial component [1]. This limitation is observed from the central point of the object plane, where the instrument is focused. The aperture stop can take various forms. It can be a physical element, such as a lens, with a finite size. Alternatively, it may be an intentionally introduced optical aperture, such as an iris. Notice that in an optical system with multiple apertures, the aperture stop is identified as the one that imposes the most stringent limitation on the angular range mentioned earlier. The other aperture are considered field stops, they just stop some rays of certain fields.

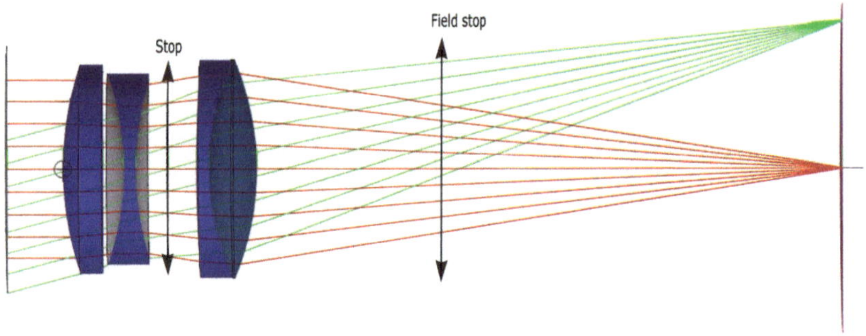

FIGURE 2.14
The stop of the optical system is composed of three lenses. The stop is placed between the second element and the third element, In this particular example the stop is an element on its own and not a surface of the other three lenses. At the right side of the system, there is a field stop. It just blocks rays from the off-axis field but not from the on-axis field.

Figure 2.14 highlights the stop of the optical system composed of three lenses. The stop is placed between the second element and the third element, In this particular example the stop is an element on its own and not a surface of the other three lenses. Also in figure 2.14 is displayed a field stop. This field (at the right side of the system) stops just affects the off-axis field since the on-axis field pass through it without any block of rays.

The number of field stops can change from one system to another but every system should have at least one stop. This stop is crucial to define several properties of the optical system. We will study this in the following sections of this chapter.

2.3.4 Entrance Pupil

The entrance pupil of an optical system is defined as the image of the aperture stop in the object space. This means that from object space, the aperture stop has the appearance of the entrance pupil [4].

To locate the entrance pupil, we need to find the ray that passes (after suffering refraction/reflections) through the center of the aperture of the incident field with the largest angle. This ray (before suffering refraction/reflections) is propagated until it reaches the optical axis. The entrance pupil diameter is the diameter of the entrance pupil. It is given by propagation of the incident on-axis field.

In figure 2.15 it is displayed once the optical system is composed of three singlet lenses. For this illustration, we have two fields, the on-axis field in the and the off-axis field in green. The field with the largest incident angle is the green field. We didn't show other off-axis fields to keep the image clear.

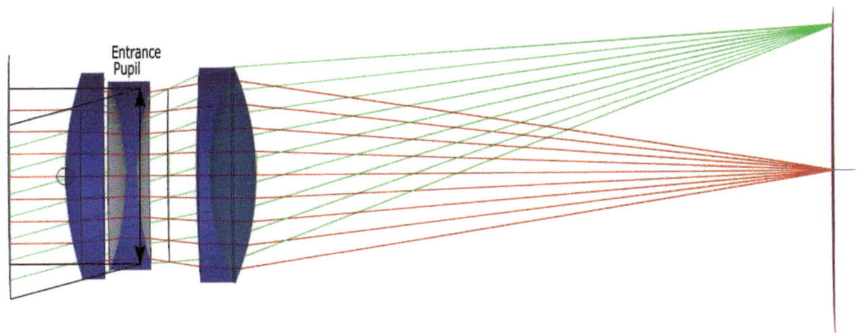

FIGURE 2.15
The entrance pupil of an optical system is defined as the image of the aperture stop in the object space. This means that from object space, the aperture stop has the appearance of the entrance pupil. From the green field, we select the ray that passes (after suffering refraction/reflections) through the center of the aperture. Before this ray suffers refractions/reflections, we extend it until it reaches the optical axis.

From the green field, we select the ray that passes (after suffering refraction/reflections) through the center of the aperture. Before this ray suffers refraction/reflections, we extend it until it reaches the optical axis. Its diameter is given by the two margin rays of the on-axis field (red field) reaching the plane of the entrance pupil position. This plane is perpendicular to the optical axis.

2.3.5 Exit Pupil

The exit pupil of an optical system is defined as the image of the aperture stop in the image space. In other words, the appearance of the aperture stops from the image space given by the exit pupil [1].

The process to locate the exit pupil has similarities with the process to locate the entrance pupil. The ray that passes through the center of the aperture of the outcome field with the largest angle is propagated until it reaches the optical axis. The entrance pupil diameter is the diameter of the entrance pupil. It is given by propagation of the outcome on-axis the field.

In figure 2.16, we can be observed as an example of an exit pupil. We first identify the ray of the field with the largest inside angle that passes at the center of the aperture stop. Then, we extended this ray unit until it reached the optical axis (from the image space). At that point, we can draw (not shown in the picture) a perpendicular plane to the optical axis which can be called the plane of the exit pupil. Then the exit pupil lies on the mentioned planed but until it covers the extension of marginal rays of the outcome on-axis field.

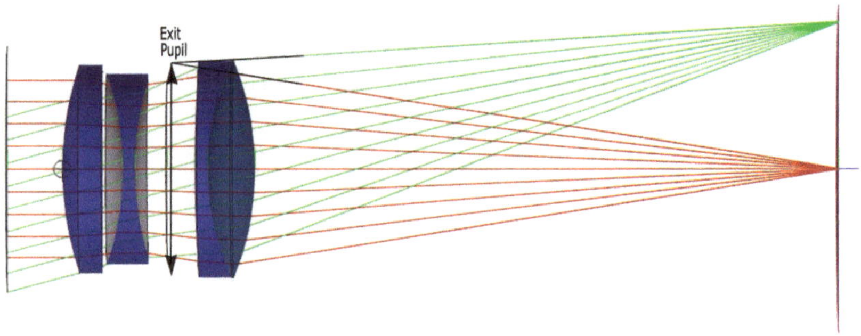

FIGURE 2.16
The exit pupil of an optical system is defined as the image of the aperture stop in the image space. The ray that passes through the center of the aperture of the outcome field with the largest angle is propagated until it reaches the optical axis. The entrance pupil diameter is the diameter of the entrance pupil. It is given by propagation of the outcome on-axis the field.

The rays through the center of the aperture are called chief rays and the rays at the edges are called marginal rays. We are going to explore more about several types of rays in the following subsections.

2.3.6 Raytracing

Ray tracing serves as the fundamental method for optical design and analysis. It relies on geometric optics, which disregards the wave nature of light and treats light propagation as straight rays. While geometric optics is an approximation of wave optics, it is sufficiently accurate in numerous scenarios. By tracing a few rays from an object through a lens, we can ascertain the position, dimensions, and orientation of the resulting image [9, 30, 31].

At this point, the reader can notice that we have been using several concepts related to rays, without defining the rays at lot. We can just do it all the way around: first study the rays and then the pupils. But both concepts are related to each other so we would use the stops and pupils to define the rays without mentioning them before. So now let's pay attention to the rays. To properly define rays we should mention the planes where the rays lay.

2.3.7 Tangential and Sagittal Plane

There are two main planes where we can usefully define ray that gives information about the properties of an optical system. The sagittal plane and the tangential plane. Using the coordinate system provided by `Quadoa Optical CAD` We know that the optical axis runs along $+z$ direction. In this context,

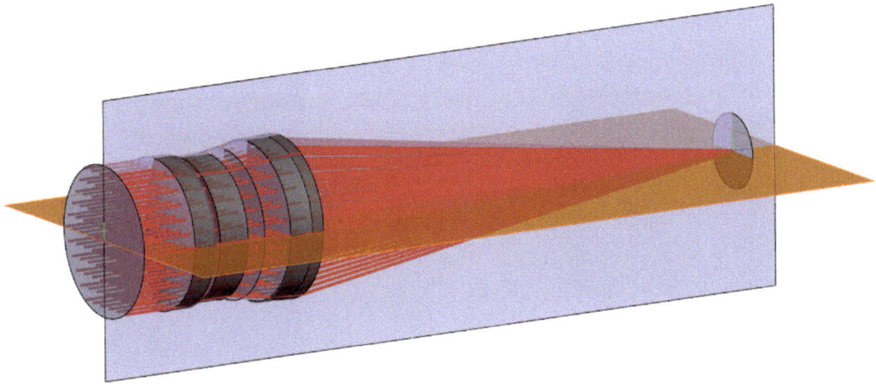

FIGURE 2.17
The tangential plane y/z is in gray and the sagittal plane is in orange x/z.

the tangential plane corresponds to the plane of this y/z. In contrast, the sagittal plane is perpendicular to the tangential plane. So the sagittal plane is the x/z plane [4].

Think once again about our optical system of three elements (remember all these concepts are not just valid for this system but all on-axis optical systems in general). The system is shown in figure 2.17 and the the tangential plane is in gray and the sagittal plane is in orange [4].

2.3.8 Meridional Rays

Another way to call the tangential plane is called the meridional plane. In a system with rotational symmetry, meridional rays are confined to the plane formed by the lens axis and an object point situated to one side of the axis. This plane is known as the meridional plane. When the object point aligns with the axis, all rays must be meridional [4]. For our notation, the tangential plane and/or the meridional plane is the y/z plane of the global coordinate system in `Quadoa Optical CAD`.

Therefore the meridional rays that all their trajectories are in the meridional plane. Figure 2.18 shows the meridional rays of the optical system composed of three elements. These rays are presented in the color orange.

2.3.9 Sagittal Rays

In a system with rotational symmetry, meridional rays are confined to the plane formed by the lens axis and an object point situated to one side of the

FIGURE 2.18
The meridional rays (orange) that all their trajectories are in the meridional plane, y/z plane. The sagittal rays (pink) are the rays that all their trajectories are in the sagittal plane, x/z plane

axis. This plane is known as the meridional plane. When the object point aligns with the axis, all rays must be meridional. For us, the sagittal plane is the x/z plane of the global coordinate system in `Quadoa Optical CAD`.

Therefore the sagittal rays are the ray that all their trajectories are in the sagittal plane. Figure 2.18 displays the sagittal rays of the optical system composed of three singlets. These rays are presented in the color pink.

2.3.9.1 Chief Ray

The chief ray of a lens is the path taken by a ray originating from a point off the optical axis that passes through the center of the aperture stop in the optical system. Figure 2.19 shows the chief ray (in orange) of an on-axis field (in pink) of the optical system with three elements [1].

2.3.10 Marginal Ray

The marginal ray of a field is the ray that is on the edge of the boulder of rays of the field. Figure 2.20 highlights in orange the marginal rays and the chief ray of a three-element optical system [1].

The marginal rays along with the chief ray are crucial to define the optical properties of a system. Like for example know the effective focal length of the system, and where are placed the entrance pupil and the exit pupil of the system. At this point, it should make the scene know why we use the marginal and chief ray in the previous subsections to define the mentioned concepts [1].

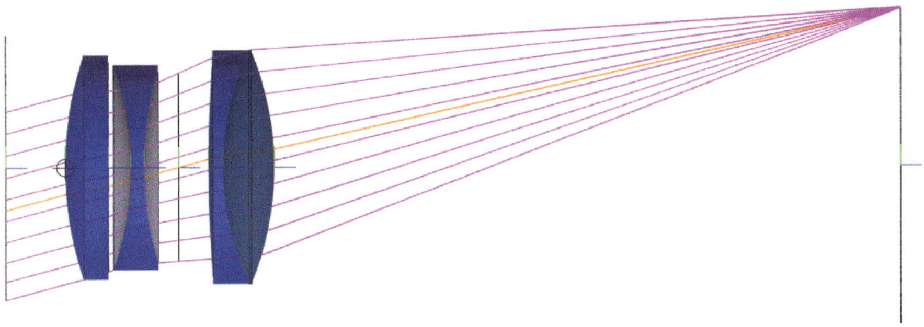

FIGURE 2.19
Chief ray (in orange) of an on-axis field (in pink) of the optical system with three elements.

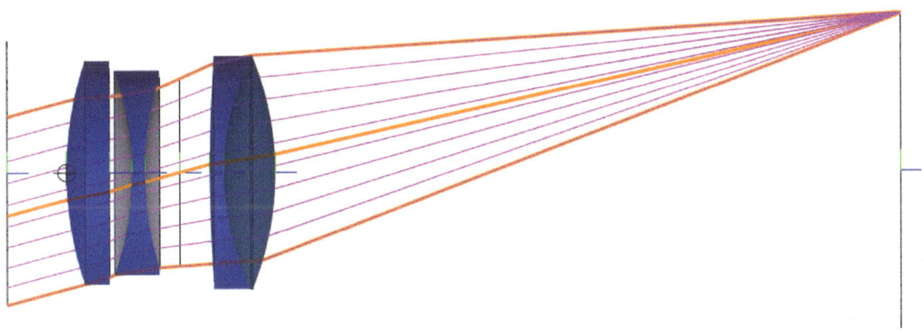

FIGURE 2.20
In orange the marginal rays (edge rays) and the chief ray (center ray).

2.3.11 Skew Ray

A ray in a symmetrical optical system that does not run parallel to or intersect any tangential (meridional) or sagittal plane is called a skew ray [1].

Figure 2.21 shows an optical system comprised of three singlet lenses and two fields. The fields are an on-axis field in blue and an off-axis field in red. The rays of the off-axis field can be considered skew rays since their trajectories do not lie in the tangential (meridional) or sagittal planes.

2.3.12 Paraxial Ray

A paraxial ray is a ray that travels at a small angle to the optical axis and remains close to it within the optical system. These rays are often effectively

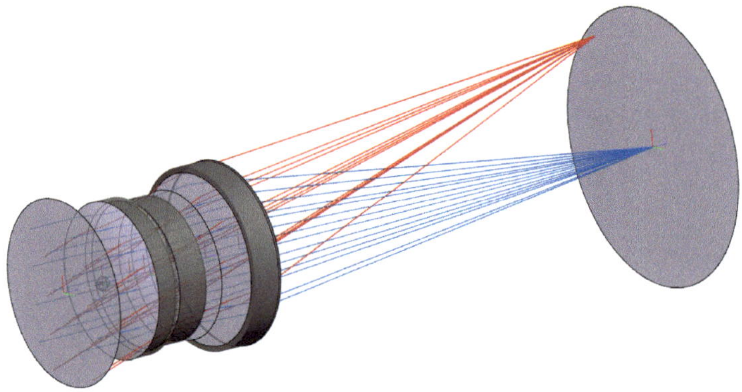

FIGURE 2.21
The rays of the off-axis (red rays) field are skew rays since their trajectories do not laid in the tangential (meridional) or sagittal planes.

modeled using the paraxial approximation. The previously mentioned rays may or may not fall into the category of paraxial rays. The rays of the on-axis field (blue) of figure 2.21 which are very close to the optical axis can be considered paraxial rays [4].

2.3.13 Rays, Object and Aperture Space

A final comment that needs to be addressed for all these rays, All the planes and rays discussed are associated with a specific object point and aperture. For instance, if the object's point of interest lies in the x/z plane, then the x/z plane becomes the tangential plane. Similarly, if the object point is in the y/z plane, then the y/z plane is considered the tangential plane. When choosing a coordinate system, it is advisable to position the object point either in the x/z plane or the y/z plane.

2.3.14 Field of View, Objects and Images Sizes

To define the entrance pupil and exit pupil we use the rays with larger incident angles. The field of view refers to the observable area that an optical system can capture in an image. It represents the portion of the object that fills the image plane. The half field of view is a term also used. A representation of the half field of view is shown in figure 2.22. θ is the angle between the optical axis and the ray with the largest incident angle [4].

When designing a lens system several fields are considered. Usually, it's a finite number of fields that cover the range of incident angle of the field of

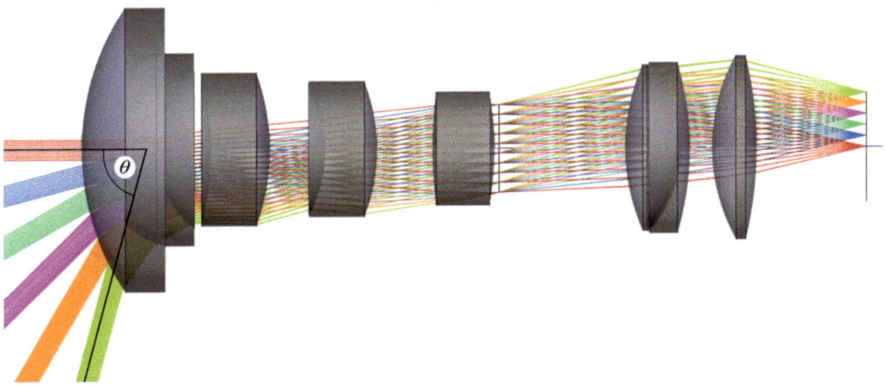

FIGURE 2.22
The half field of view θ is the angle between the optical axis and the ray with the largest incident angle.

view. In `Quadoa Optical CAD` the `Fields` and their rays are part of a larger object called sequence. The `Sequence` is the basic object that is used to define a sequential ray trace through the system. It defines the sequential path for the rays traced through the system.

Each sequence includes a source object as its initial child item. The source object serves to specify the characteristics of the traced rays, encompassing factors such as wavelengths, fields, polarization, apodization, and the distribution of the rays.

- **New Auto Sequence:** The button adds a new `Auto Sequence` to the model. The `Auto Sequence` will appear inside the `System Setup Editor` .

- **New User Sequence:** Clicking the button adds a new `User Defined Sequence` to the model. This `User Defined Sequence` will be displayed within the `System Setup Editor`.

- **Add Surface:** Clicking the button will add a new `Surface` to a `User Sequence`. To add the `Surface`, you must first select the `User Sequence` to which it should be added in the `System Setup Editor` . Failure to select a `User Sequence` will result in an error message. Another method to add a `Surface` to a `Sequential Ray Trace` is by selecting the `Surface List` of the `Sequence` within the `System Setup Editor` and using the right-click menu, or by pressing the insert key. It's important to note that a

`Surface` can only be added to a `User Sequence`. The `Auto Sequence` will automatically generate the surface list.

- **Add Filed:** To incorporate a new field, click the `Add Field Button`. If multiple `Sequences` exist, you must first select the `Sequence` or the `Source` to which the field should be added. Failure to select a `Sequence` or `Source` will prompt an error message. Alternatively, you can utilize the right-click menu on the field for this purpose.

- **Add Wavelength:** Clicking the button will add a new wavelength to the `Source` . If multiple `Sequences` exist, you must first select the `Sequence` or the `Source` to which the wavelength should be added. Failure to select a `Sequence` or `Source` will prompt an error message. Another option is to utilize the right-click menu on the wavelength. Additionally, a wide range of frequently used wavelengths is predefined for convenience.

The object distance is given by the distance between the object plane where the source of light is coming. The image plane or sensor seize in `Quadoa Optical CAD` the size of the image surface, the size of the last surface in a sequence.

The image height can be computed with the following expression,

$$I = f\tan(\theta) \tag{2.31}$$

where I is the height of the image, f is the effective focal length and θ is the half field of view. This last expression is obtained from paraxial approximation and the height of the image I can be affected by distortion which is an optical aberration that will be considered in chapter 4.

2.3.15 F-Number and Numerical Aperture

The f-number $f\#$ setting on a lens regulates the amount of light entering the lens, and the lens's ability to produce contrast at a specific resolution [4]. Essentially, the f-number $f\#$ is the ratio of the effective focal length f of the lens and the diameter of the entrance pupil D,

$$f\# = \frac{f}{D} \tag{2.32}$$

In many lenses, the f-number $f\#$ is adjusted by rotating the iris adjusting ring, which opens or closes the iris diaphragm inside the lens. Figure 2.23 shows the parameter used to compute the f-number in an optical system.

The numerical aperture NA is mathematically defined as

$$NA = \frac{D}{2f} = \frac{1}{2f\#} \tag{2.33}$$

NA is a representation of the ratio between the entrance pupil size and the focal length of a lens [4].

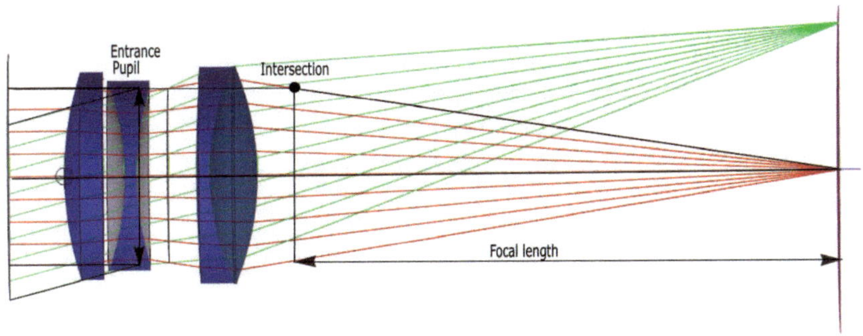

FIGURE 2.23
The f-number is the ratio of the effective focal length and entrance pupil diameter.

2.4 Concepts in Diffraction

A diffraction integral is a mathematical expression that describes the behavior of light waves as they propagate through or interact with obstacles or apertures. It is a fundamental concept in the study of optics and wave phenomena [23].

In essence, the diffraction integral describes how a wavefront evolves as it passes through an aperture or encounters an obstacle. It takes into account the wave nature of light, considering factors such as interference and diffraction patterns that arise due to the bending of light around edges or obstacles.

Mathematically, the diffraction integral is typically expressed as an integral over the aperture or obstacle, involving the complex amplitude of the wavefront at various points in space. Solving this integral allows one to determine the resulting wave pattern or intensity distribution on a screen or observation plane beyond the diffracting element [32].

Overall, diffraction integrals play a crucial role in understanding various optical phenomena, including the formation of images in optical systems, the behavior of light in diffraction gratings, and the analysis of optical instruments.

In this section, we explore the diffraction concepts used by lens designers in practice. these concepts are crucial to characterizing the performance of optical systems. This section gives just a quick summary of some concepts but it does not intend to prove them, several well-known texts are given to the user in the bibliography for a more deep analysis of the topic.

Geometrical optics, in general, ignores two fundamental phenomena of light interference and diffraction. Diffraction, stemming from the wave nature of light, is an observable phenomenon. Its effects become particularly pronounced when light interacts with an interface or an optical system whose dimensions

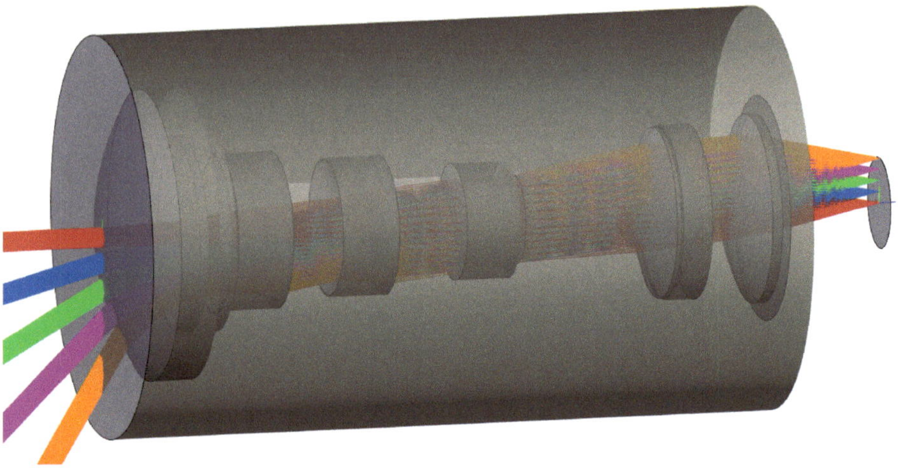

FIGURE 2.24
An optical system surrounded by a black box (the gray cylinder).

closely align with the incident wavelength. Diffraction produced by an optical element is highly important as a tool to measure the quality of the image that it produces. Utilizing diffraction theory enables the calculation of alterations in wavefronts and their subsequent propagation post-interaction with such elements.

Consider an imaging system comprising not just a single thin lens, but possibly multiple lenses, including both positive and negative ones, arranged at different intervals. We'll presume that the system ultimately generates a tangible image in space since our optical system is an imagining system (it produces an image). To define the characteristics of the lens system, we take the approach that all imaging components can be consolidated into a single so-called *black box* [32]. We assert that the essential attributes of the system can be fully elucidated by solely outlining the terminal properties of this system. Then, we only consider the entrance and exit pupils. So, let be z to represent the distance of the plane of the exit pupil from the image plane [32]. See figure 2.24.

There are two main diffraction theories the scalar and the vector diffraction theory [33]. The vector diffraction equations offer a comprehensive method for conducting a thorough analysis of a diffractive optical system [34, 35]. However, vector diffraction equations are usually not considered in optical design, but only in very specialized applications that are not in the scope of this book [36]. Here we are going to mention the scalar diffraction formulation, which is described in several textbooks. The scalar diffraction theory considers the $E(u, v)$ electric field to be a scalar quantity [8, 37]. The main purpose of this is just to simplify the computation of the diffraction patterns. The fundamental equation in scalar diffraction formulation is the Huygens integral. The Huygens integral is constructed on the principle that each point of a wavefront is the source of spherical waves and it is given by,

$$E(u,v) = \frac{z}{\lambda j}\int\int_{\sigma} A(x,y)\frac{\exp(jkr)}{r^2}dxdy \tag{2.34}$$

where (x, y) is the diffraction plane (the exit pupil), and (u, v) is the observation plane, also called the image plane. σ is the of the aperture. The radius r is,

$$r = \sqrt{z^2 + (u - x^2) + (v - y)^2} \tag{2.35}$$

At smaller angles, spherical wavefronts can be approximated as parabolic wavefronts, leading to what is known as the Fresnel approximation [8, 37], then,

$$r \approx z\left[1 + \frac{1}{2}\left(\frac{u-x}{2}\right)^2 + \frac{1}{2}\left(\frac{u-x^2}{2}\right)\right], \tag{2.36}$$

with this approximation, the Huygens integral becomes the so-called, Fresnel diffraction integral which is given by,

$$E(u,v) = \int_{-\infty}^{\infty}\int_{-\infty}^{\infty} A(x,y)\exp\left[\frac{jkr(x^2+y^2)}{2}\right]\exp\left[-j\frac{2\pi}{\lambda z}(xu+yv)\right]dxdy \tag{2.37}$$

Solving the above integral is simpler than The Huygens integral, but will fail in the region closer to the diffraction plane and for large diffraction angles. It is valid only when

$$z^3 >> \frac{\pi}{4\lambda}\left[(x-u)^2 + (y-v)^2\right]^2. \tag{2.38}$$

The Fresnel diffraction integral can be reduced even more to a Fourier transform operation by applying the so-called far-field approximation,

$$z >> \frac{k(u^2+v^2)}{2} \tag{2.39}$$

The approximation, alternatively referred to as the Fraunhofer approximation, holds validity solely for extremely large distances. At such distances, the radius of the spherical wavefront is considerable, allowing a portion of the spherical wavefront to be approximated as a planar wavefront [8, 37].

A diffraction-limited imaging system is characterized by its ability to convert a diverging spherical wave, originating from a point-source object, into a perfectly spherical wave once again [8, 37]. This newly formed spherical wave converges towards an ideal point within the image plane. The location of this ideal image point is directly related to the position of the original object point through a consistent scaling factor, known as magnification. For the system to be considered ideal, this magnification factor must remain uniform across all points within the image field of interest.

Hence, the key feature of a diffraction-limited imaging system is that a diverging spherical wave entering the entrance pupil is transformed by the system into a converging spherical wave at the exit pupil. Then, a diffraction-limited imaging system is the equivalent of a stigmatic optical system in geometrical optics.

Most of the optical imaging systems are radially symmetric, which implies that they have a circular shape aperture σ. As we have seen, the diffraction pattern depends on the shape of the aperture, as in Huygens integral. Therefore is it highly relevant the special case of the diffraction of a spherical wave converges towards an ideal point with a circular aperture. The diffraction of a diffraction-limited system with a circular aperture. thus results are well known in several textbooks. The outcome of the diffraction process, characterized by a prominent bright region at the center surrounded by a sequence of diminishing intensity concentric rings, is commonly referred to as the Airy disk. This diffraction pattern is described by a Bessel function if the Fraunhofer approximation is considered.

The Airy disk, the central lobe of the diffraction pattern, defines the smallest achievable focused spot produced by an optical system,

$$\emptyset_{\text{Airy disk}} = 2.44\frac{f\lambda}{D} \tag{2.40}$$

According to geometric optics, in the absence of aberrations (stigmatic), the focused spot size of a lens is theoretically infinitely small. But if we consider scalar diffraction then the spot size will be the Airy disk. The spot size can be even smaller for special cases considered by vector diffraction theory.

Figure 2.25 shows the diffraction pattern produced by an optical system. The Airy disk is the central lobe of the diffraction pattern. It the lobe presented in yellow and within the figure.

2.5 Polarization

We have mentioned before that diffraction can be analyzed with the scalar and vector diffraction theories. The vector theory is more accurate than the scalar since the electric field is a vector identity. We consider the electric field as a vector before discussing polarization.

Polarization of light refers to the orientation of the oscillations of the electric field component of an electromagnetic wave. In an unpolarized light wave, the electric field oscillates in all possible directions perpendicular to the

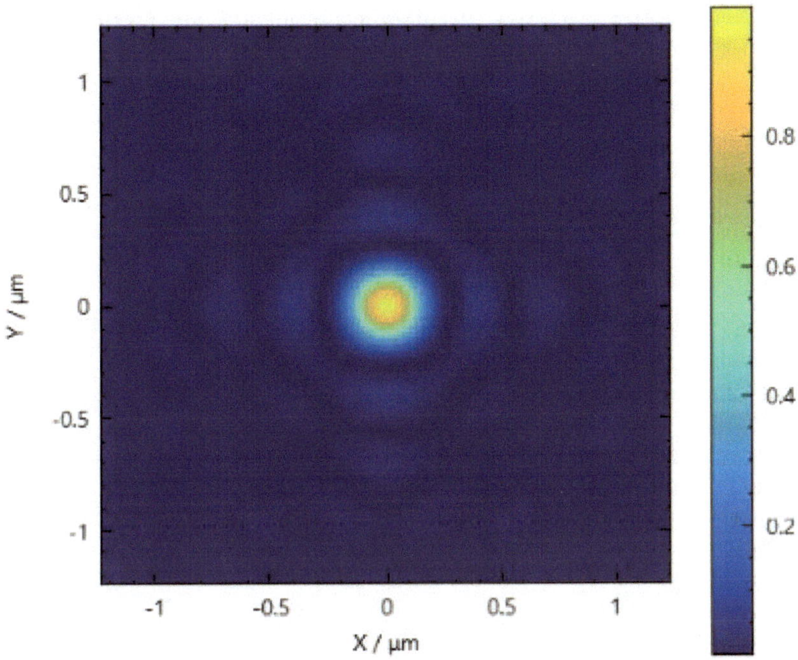

FIGURE 2.25
The Airy disk is the central lobe of the diffraction pattern. It the lobe presented in yellow and within the figure.

direction of propagation. When light is polarized, however, the electric field oscillates predominantly in a specific direction [38].

There are several ways in which light can be polarized, the most common cases are:

- **Linear polarization:** In linearly polarized light, the electric field oscillates in a single plane perpendicular to the direction of propagation. This can be achieved by passing unpolarized light through a polarizing filter, which selectively transmits light waves oscillating in a specific orientation while blocking others.
- **Circular polarization:** In circularly polarized light, the electric field vector rotates in a circular pattern as the wave propagates. Circular polarization can be achieved by superimposing two linearly polarized waves with perpendicular orientations and a phase difference of 90 degrees.
- **Elliptical polarization:** Elliptically polarized light is a combination of linear and circular polarization, where the electric field vector traces out an elliptical path as the wave propagates. This can be achieved by combining

two linearly polarized waves with different amplitudes and a phase difference between them.

Polarized light has numerous applications in various fields including photography, optical communications, 3D movie projection, scientific research, and polarization microscopy.

The polarization state of light changes at the interface between two different optical media. This occurs because the total amount of light that passes through the two different optical media is divided into the light that is transmitted and the light that is reflected.

Fresnel coefficients describe how light is reflected and refracted when it encounters such an interface, taking into account the differences in the refractive indices of the two media [38, 39].

There are typically two Fresnel coefficients associated with the reflection and transmission of light at an interface:

- **Reflection coefficient:** The reflection coefficient, often denoted as R, describes the fraction of incident light that is reflected back into the same medium at the interface. It is given by the ratio of the intensity of the reflected light to the intensity of the incident light. The reflection coefficient depends on the angle of incidence and the polarization state of the incident light.

- **Transmission coefficient:** The transmission coefficient, often denoted as T, describes the fraction of incident light that is transmitted into the second medium (i.e., refracted). It is given by the ratio of the intensity of the transmitted light to the intensity of the incident light. Similar to the reflection coefficient, the transmission coefficient depends on the angle of incidence and the polarization state of the incident light.

Both reflection and transmission coefficients are determined by the Fresnel equations, which are derived from Maxwell's equations describing the behavior of electromagnetic waves at boundaries between media with different refractive indices. These equations provide explicit formulas for calculating the reflection and transmission coefficients based on the angle of incidence, the refractive indices of the two media, and the polarization state of the incident light. The Fresnel equations are [5],

$$R_s(\theta_1) = \left[\frac{n_1 \cos\theta_1 - n_2 \cos\theta_2}{n_1 \cos\theta_1 + n_2 \cos\theta_2}\right]^2 \tag{2.41}$$

and

$$R_p(\theta_1) = \left[\frac{n_1 \cos\theta_2 - n_2 \cos\theta_1}{n_1 \cos\theta_2 + n_2 \cos\theta_1}\right]^2 \tag{2.42}$$

where θ_1 and θ_2 are the incident and refractive angles, respectively, $R_s(\theta_1)$ and $R_p(\theta_1)$ are the reflectance of light with polarization vertical and parallel

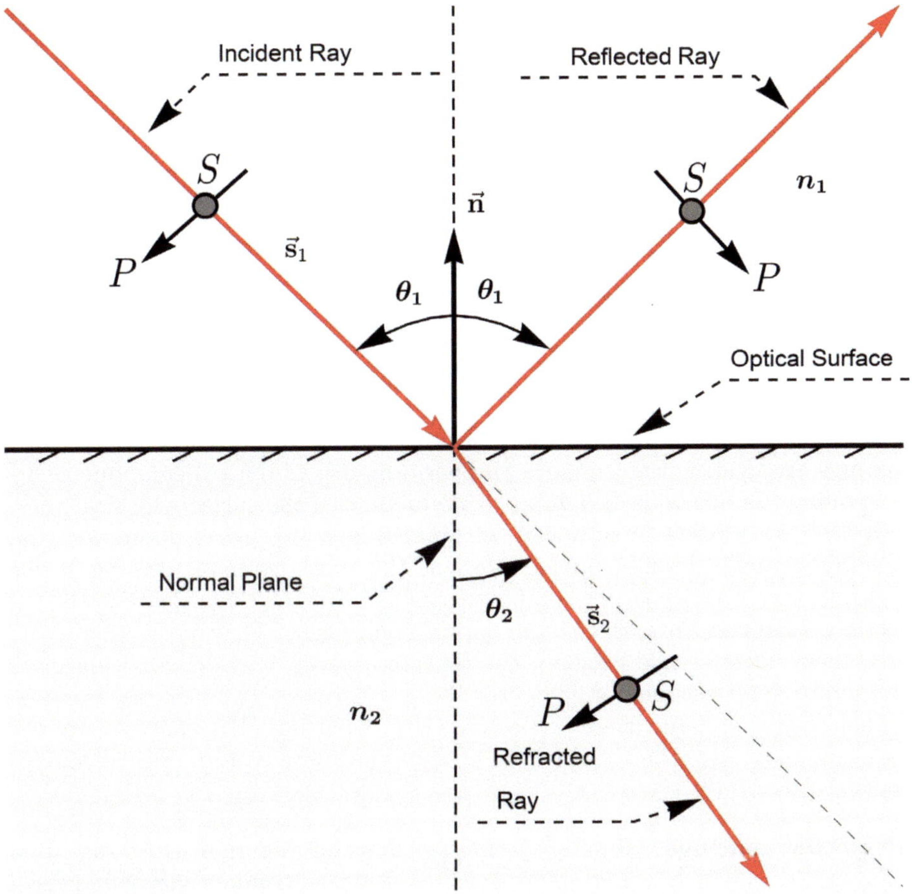

FIGURE 2.26
Polarized light incident on an optical interface. S and P are the direction vertical and parallel to the reflection plane, respectively. The S direction is going out of the page. θ_1 and θ_2 are the incident and refractive angles, respectively. $\vec{\mathbf{n}}$ is the normal vector of the surface. $\vec{\mathbf{s}}_1$ is a unit vector with the direction of the indecent ray and $\vec{\mathbf{s}}_2$ is a unit vector with the direction of the refracred ray. The refraction indexes are n_1 and n_2 for the first and second medium, respectively.

to the reflection plane, respectively, as illustrated in figure 2.26. In the figure, the refraction index of the first medium is n_1 the refraction index of the second medium is n_2. $\vec{\mathbf{n}}$ is the normal vector of the surface. $\vec{\mathbf{s}}_1$ is a unit vector with the direction of the indecent ray and $\vec{\mathbf{s}}_2$ is a unit vector with the direction of the refracted ray. S and P are the direction vertical and parallel to the reflection plane, respectively. The S direction is going out of the page.

Due to the principle of energy conservation, the transmitted amount of light can be determined by subtracting the portion of incident light that is reflected,

$$T_s(\theta_1) = 1 - R_s(\theta_1) \tag{2.43}$$

and

$$T_p(\theta_1) = 1 - R_p(\theta_1) \tag{2.44}$$

$T_s(\theta_1)$ and $T_p(\theta_1)$ are the reflectances of light with polarization vertical and parallel to the reflection plane.

2.5.1 Optical Coating

Optical coatings are thin layers of materials applied to optical surfaces to modify the transmission, reflection, or polarization properties of light. These coatings are designed to enhance the performance of optical components by controlling the behavior of light as it interacts with the surface. Optical coatings are widely used in a variety of applications in optics, photonics, and imaging systems. Usually in lens design, the most common coating is the anti-reflective (AR) coatings: AR coatings are designed to minimize the reflection of light at the surface of an optical component. They work by creating destructive interference between the reflected waves, thereby reducing glare and improving transmission [38].

On the other hand High Reflective (HR) Coatings: HR coatings are designed to maximize the reflection of light at the surface of an optical component.

- **Metallic coatings:** Metallic coatings, such as aluminum or silver, are used for applications where high reflectivity or electrical conductivity is required. They are often used in mirrors and optical components for infrared applications.

- **Polarizing coatings:** Polarizing coatings are designed to selectively transmit or reflect light of a specific polarization state. They are used in polarizing filters and beamsplitters.

- **Filters:** Optical filters are coatings designed to selectively transmit or block certain wavelengths of light. They are used in applications such as spectroscopy, fluorescence microscopy, and photography.

Overall, optical coatings play a crucial role in enhancing the performance and functionality of optical components across various particular applications. `Quadoa Optical CAD` can deal with several of these optical coatings, we are going to study them deeper in chapter 7.

2.6 Summary

In this chapter we have shown several useful concepts in optical design, starting from the fundamentals of the Fermat principle and Snell's law. Both are so important that we take the time to proof them step by step. We focus on also important concepts but not so transcendental like the Fermat principle and Snell's law. The concepts are just presented and explained but not proven but additional references for these topics are in the bibliography. The concepts are just introductory more concepts will be shown in the following chapters. The glossary of this chapter may be useful also as a good summary of it.

3

Optical Materials

> The diversity of the phenomena of nature is so great, and the treasures hidden in the heavens so rich, precisely in order that the human mind shall never be lacking in fresh nourishment.
>
> Johannes Kepler

3.1 Type of Optical Materials

Optical materials, possessing transparency, are utilized in the fabrication of optical lenses, prisms, windows, waveguides, and second-surface mirrors, serving as essential components in a wide array of optical instruments. Transparency, also known as pellucidity or diaphaneity, refers to the physical quality of permitting the passage of light through a substance without significant scattering [1].

3.1.1 Glasses

The most common optical material is glass. Glass is a non-crystalline solid known for its frequent transparency, brittleness, and chemical inertness. The main types of classes used in optical design are crown glasses and flint glasses [40]. Crown glasses typically exhibit lower refractive indices and larger `Abbe numbers`. Flint glasses generally feature higher refractive indices and smaller `Abbe numbers`.

The `Abbe number` serves as an approximate gauge of a material's dispersion, reflecting the change in the `Refractive Index` concerning wavelength [12]. Higher values of the `Abbe number` correlate with lower dispersion. Later in this chapter, we are going to explore the concept of `Abbe number`. For the moment we are going to show the most common optical materials [5].

DOI: 10.1201/9781003506355-3

3.1.2 Polymers

A polymer refers to a type of substance, whether natural or synthetic, comprised of extremely large molecules known as macromolecules. These macromolecules are composed of repeating simpler chemical units called monomers. Most optical polymers have a refraction index between 1.45 and 1.66 and their `Abbe number` can be around 20 and 50 [5].

3.1.3 UV Optical Materials

Most of the ultraviolet (UV) optical materials have a grade fused silica (down to 0.17 μm). Fused quartz, also known as fused silica or quartz glass, is a type of glass composed primarily of nearly pure silica (SiO2) in its non-crystalline or amorphous state. Other well-known materials for UV optical materials are magnesium fluoride and crystal quartz with (down to 0.15 μm) and (down to 0.16 μm), respectively [5].

3.1.4 IR Optical Materials

There is a greater variety of optical materials available for use in the infrared (IR) range compared to the UV range. Commonly utilized materials include calcium fluoride (usable up to 10 μm), IR-grade fused silica (up to 3.5 μm), gallium arsenide (up to 14 μm), germanium (up to 15 μm), sapphire (up to 5.5 μm), silicon (up to 11 μm), zinc selenide (up to 18 μm), and combinations of various IR materials tailored to achieve specific properties [5].

Also, it is important to mention that the `Refractive Index` of IR materials does not vary so dramatically, so IR materials don't have so much chromatic dispersion [5].

3.1.5 Mirrors

The essential characteristic sought in mirror substrates is low thermal expansion, as the primary function of mirrors is reflection. Therefore, parameters such as `Refractive Index`, dispersion, and transmission range of the materials become irrelevant in this context [24].

3.2 Dispersion

Now that we have exposed the main optical materials let's go more deeply into the main properties, dispersion, refraction index, `Abbe number`, transmission, and thermal properties, we start with dispersion [8].

Dispersion stands out as the paramount property among optical glasses. It's characterized by three key parameters: `Refractive Index` and `Abbe number`.

3.2.0.1 Refraction Index

The `Refractive Index` $n(\lambda)$ of any glass is a function dependent on wavelength λ. Various polynomials are commonly employed to fit the experimental data of the `Refractive Index` $n(\lambda)$. `Quadoa Optical CAD` uses several of these fitting polynomials [15, 18], `Sellmeier`, `Sellmeier Extended`, `Schott`, `Schott Extended`, `Conrady`, `Herzerberg`, etc.

We show some of the most used ones. We start with `Schott` formula,

$$n(\lambda)^2 = a_0 + a_1\lambda^2 + a_2\lambda^{-2} + a_3\lambda^{-4} + a_4\lambda^{-6} + a_4\lambda^{-6} + a_5\lambda^{-8} \tag{3.1}$$

where the a's are coefficients. The next formula is `Sellmeier`,

$$n(\lambda)^2 - 1 = \frac{K_1\lambda^2}{\lambda^2 - L_1} + \frac{K_2\lambda^2}{\lambda^2 - L_2} + \frac{K_3\lambda^2}{\lambda^2 - L_3} + \frac{K_4\lambda^2}{\lambda^2 - L_4} + \frac{K_5\lambda^2}{\lambda^2 - L_5} \tag{3.2}$$

where as you may guess $n(\lambda)$ is the refraction index given a wavelength, λ and the rest of the terms are coefficients. Finally we also include here `Herzberger` formula,

$$\begin{aligned} n(\lambda) = A + B\frac{1}{\lambda^2 - 0.028} + C\left(\frac{1}{\lambda^2 - 0.028}\right)^2 + D\left(\frac{1}{\lambda^2 - 0.028}\right)^2 \\ + D\left(\frac{1}{\lambda^2 - 0.028}\right)^4 + E\left(\frac{1}{\lambda^2 - 0.028}\right)^6, \end{aligned} \tag{3.3}$$

where A, B, C, D and E are coefficients.

3.2.1 Abbe Number

Each polynomial presented in the previous section offers a comprehensive portrayal of a glass's dispersion property. Nevertheless, these polynomials can be somewhat intricate to assess and can cost additional computational costs. To simplify, two parameters are often utilized to succinctly convey the most significant dispersion characteristic of a glass. The first parameter is the refraction index n_D at wavelength 0.589μm and the `Abbe number`. The `Abbe number` V_D is given by,

$$V_D = \frac{n_D - 1}{n_F - n_C} \tag{3.4}$$

where n_F and n_C correspond to the Fraunhofer F line and C line. These lines are at 0.486μm and 0.656μm, for F and C, respectively.

`Quadoa Optical CAD` in its `Materials Catalog` provides a map that plots the `Abbe number` and the refraction index for each the catalogs, as can be seen in figure 3.1 with `CDGM` catalog in `Quadoa Optical CAD`. This map is also called `Abbe diagram`.

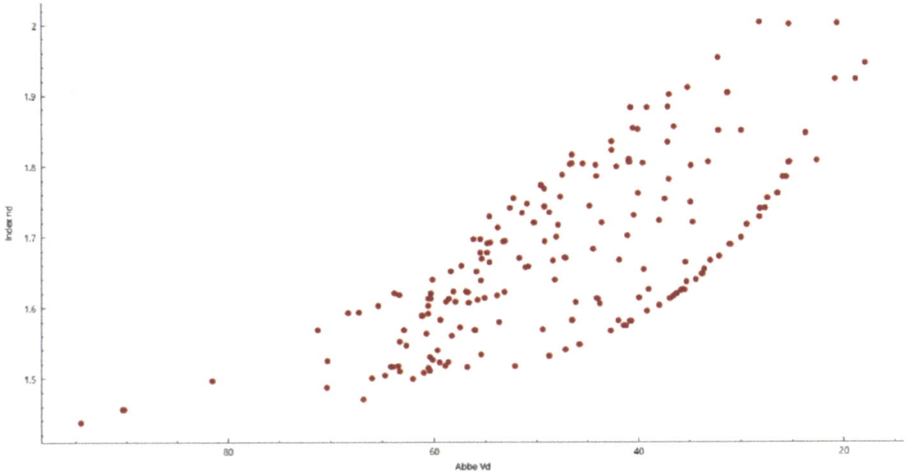

FIGURE 3.1
`Abbe diagram` of the `CDGM` catalog in `Quadoa Optical CAD`. The horizontal axis is the `Abbe number` and the vertical axis is the refraction index.

3.3 Transmission

The surface transmittance of a material refers to its ability to effectively transmit radiant energy [23]. It represents the portion of incident electromagnetic power that passes through a sample, as opposed to overall transmission. The transmission of a lens is determined by the combination of the transmittance of its two surfaces along with the absorption of the glass itself. If a proper AR coating is added to a lens the reflectance can be reduced to around 0.01 percent and the transmission can be around 0.99 percent [38]. Coatings will be studied in more detail in chapter 7.

3.4 Thermal Properties

As you may guess the refraction index changes with the temperature. There are also different fitting polynomials to describe the absolute thermal index variation $\Delta n_{Abs}(\lambda, t)$. In the case of `Quadoa Optical CAD` the following expression is used,

$$\Delta n_{Abs}(\lambda, t) = \frac{n(\lambda, t)^2 - 1}{2(\lambda, t)} \left[D_0(t - t_0) + D_1(t - t_0)^2 + D_2(t - t_0)^3 + \frac{E_0(t - t_0) + E_1(t_{t_0})^2}{\lambda^2 - \lambda_k^2} \right] \quad (3.5)$$

where D_0, D_1, D_2, E_0, E_1 and λ_k are coefficients. In the case of λ_k reference wavelength. t_0 is the temperature of 20° in degrees.

The catalogs provided by glass manufacturers in `Quadoa Optical CAD` typically include thermal data for the majority of glasses provided.

3.5 Quadoa Catalog

`The Material Catalog Dialog` serves as a tool for browsing materials, offering the capability to filter materials based on various criteria such as `Refractive Index`, `Abbe number`, or name. When opened from the material catalog icon of a lens material within the `Optical Design Editor` or from within a `Lens Wizard`, the selected material can be directly applied using the `Insert Material` button.

The material catalog offers two display options: a tree view and an `Abbe diagram` visualization. In the tree view, the `Refractive Index` and `Abbe number` are represented with small bar charts directly in the view. Alternatively, users can visualize materials via the `Abbe diagram`. Clicking on the chart in the `Abbe diagram` allows users to view the parameters of a material.

- **Material Catalog**: To specify the `Material`, click on the Catalog icon beside the `Material` name and navigate to the desired Catalog for the `Material`. Alternatively, you can directly enter the `Material` name into the Name parameter field. If only one `Material` with the specified name exists in any of the available `Material Catalog`, the `Material Catalog` parameter will be automatically inserted. However, if a `Material` with the same name exists in more than one `Material Catalog`, the user must specify the `Material Catalog` as well. If the `Material` is not specified, cannot be found in the specified `Material Catalog`, or does not exist, it will be highlighted in red, and an error message will be displayed in the `Status Info Bar`.

- **The model Material:** is a straightforward material option that allows the user to directly specify the `Refractive Index` and, optionally, the

Abbe number. The values for these parameters can be designated as variables during optimization. This feature is beneficial for determining a useful parameter range when searching for materials, such as for achromatization. The available options depend on the Abbe Type parameter. Also, Quadoa Optical CAD considers the option of a constant refraction index.

Also Quadoa Optical CAD can add materials to catalogs with several dispersion formulas, like Sellmeier, Sellmeier Extended, Schott, Schott Extended, Conrady, Herzerberg etc. To add a material just go to the Material Catalog and press add material.

3.6 Summary

In this chapter we have seen how Quadoa Optical CAD address the principal optical properties of optical materials, like refraction index, the Abbe number, the dispersion, the thermal properties and transmission. Also, we saw properties of the main optical materials and how to find them in Material Catalog .

4

Optical System Analysis

> The deep study of nature is the most fruitful source of mathematical discoveries. By offering to research a definite end, this study has the advantage of excluding vague questions and useless calculations; besides it is a sure means of forming analysis itself and of discovering the elements which it most concerns us to know, and which natural science ought always to conserve.
>
> Jean-Baptiste Joseph Fourier

4.1 Image Analysis

Image formation encompasses transforming a three-dimensional object into a two-dimensional representation on a surface. Ideally this process of forming the image for an optical system is captured by the sensor or the human eye. It is expected to concentrate a solitary object point into a single, or stigmatic, image point. So for a given field, the rays of the field should converge in a single point, ideally speaking. For several fields, they should converge into their respective points, the concatenation of these points forms what we call an image [1].

In practice, optical systems experience optical aberrations, causing rays to converge on a small area rather than a precise point, even in a finely tuned system. These imperfections, known as optical aberrations, manifest as deviations from an optical system's ideal image. They are categorized into two main types: monochromatic aberrations and chromatic aberrations [30, 31, 9]. In this chapter, we are going to explore both of them.

So, even in well-corrected systems designed to minimize aberrations, resolution is constrained by the wave nature of light, specifically by diffraction at the system's aperture. The Airy disk, studied in chapter 2, represents the diameter of maximum brightness in the diffraction pattern, highlighting this limitation. Diffraction occurs within an optical element as wavelets interfere,

DOI: 10.1201/9781003506355-4

resulting in an energy distribution around the focal region resembling a Bessel function shape. This distribution features a central peak of maximum brightness surrounded by rings of varying intensities [23, 25].

However, optical systems suffer from several aberrations, and there are many ways to express these aberrations, in this chapter, we are going to study them and analyze them in `Quadoa Optical CAD` [18].

4.2 Monochromatic Aberrations

Monochromatic aberrations represent image imperfections unaffected by the wavelength of light. They are directly associated with the aperture and field of an optical system. Larger apertures or fields lead to greater optical aberrations, more light comes into the system with higher angles and Snell's law shows more nonlinearity with higher angles [7, 24]. In such instances, rays deviate from the paraxial region and linearity is lost within the system.

Two methods are available for quantifying monochromatic aberrations: one can adopt either a ray or a wave perspective. Both are related to each other since at the end the rays are perpendicular to wavefronts [8, 17, 40].

The first approach is the ray optics, mostly based on geometrical optics. Rays are subsequently traced to measure the deviation of their focal point from the reference point. In the ray perspective, aberrations can be categorized into longitudinal and transverse types [41].

In figure 4.1, it is presented longitudinal aberration. The figure shows the image plane at the output of an optical system. In black are shown the reference ray (the rays should be) and in red are the actual rays. Longitudinal aberrations can be measured as the distance between a ray intersecting the optical axis and the reference point. Additionally, in specific instances where the image is at infinity, such as a collimated beam, aberrations can also be quantified based on the angle between a ray and the chief ray [3]. Transverse aberrations can be measured by the axial distance between the position where a ray intersects the image plane and the reference point [42] as shown in figure 4.1.

The wave perspective is based on a reference spherical wavefront (stigmatic) if the rays of an optical system at the exit pupil do not coincide with the rays of the reference spherical wavefront that means that the system suffers aberrations. Since the rays are perpendicular to the wavefronts, this deviation can be measured with the optical path length of each ray. Both perspectives are usually used to analyze an optical system and describe it with several measurements, plots, and tools that can be found in `Quadoa Optical CAD`.

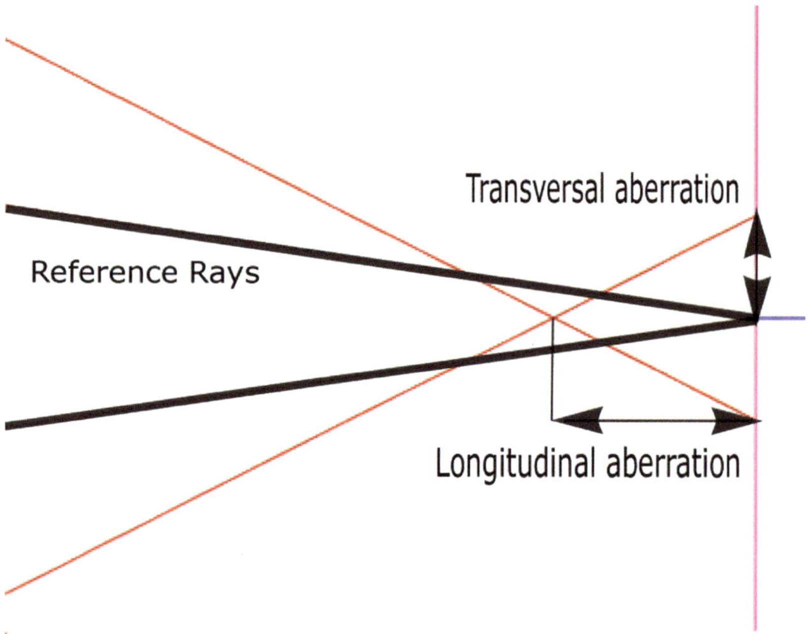

FIGURE 4.1
Longitudinal aberration and Transverse aberration.

4.2.1 Defocus and Tilt

Defocus occurs when the image plane is either closer to or farther from the image formation of an optical system, in other means the image plane is not placed where the rays do not converge [5, 42]. This issue can be rectified by adjusting either the position of the image plane or the optical system to align them. In figure 4.2, it is shown that the image plane (in pink) is not in the region where the fields converge, therefore the system is suffering from defocus.

From figure 4.2, we know that the image plane within a system experiences defocus. A system is considered to be in defocus if the image plane is not placed in the circle of least confusion. The circle of least confusion refers to the smallest circle encompassing all the reflected rays from an optical system. Its size and position can be readily determined by seeing where the caustic waist is smaller. A caustic represents the collective path of light rays that have been reflected or refracted by an optical system, or the projection of this path onto another surface. It's a curve or surface where each light ray is tangent, outlining a boundary of concentrated light rays [43, 44].

Tilt arises when the wavefront produced by an optical system is rotated in the tangential orientation, causing an angle mismatch, between the aberrated and the reference spherical wavefront. Figure 4.3, shows the optical system with tilt, the optical system is tilted concerning the image plane.

FIGURE 4.2
An optical system with defocus. The image, plane in color pink, is not where the fields converge it is a little bit away from it, where the fields start to diverge again.

FIGURE 4.3
Optical system suffering from tilt aberration.

4.2.2 Spherical Aberration

The pupil radius influences spherical aberration and is the primary on-axis monochromatic aberration. This phenomenon arises due to variations in the

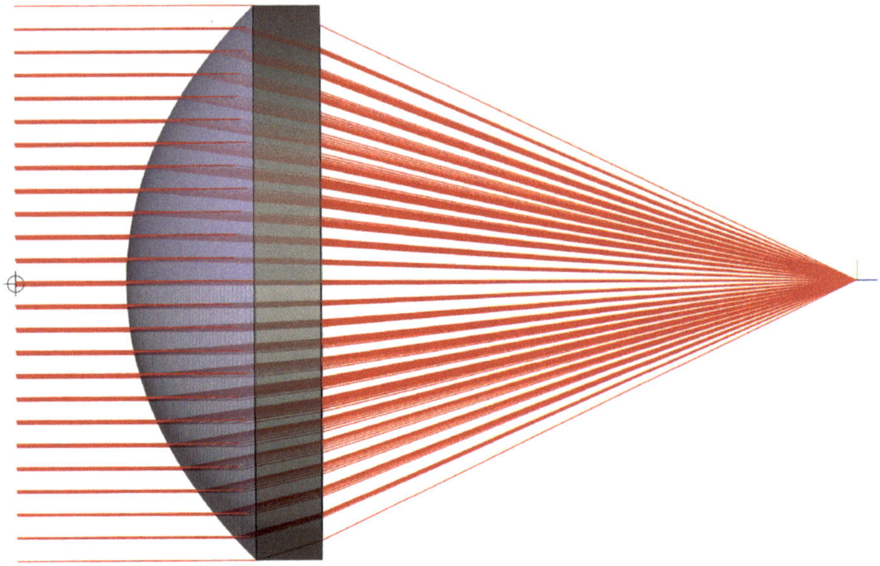

FIGURE 4.4
Raytring of a singlet lens with a single on-axis field.

optical power of a spherical surface, contingent upon the angle of the incoming ray about the surface's normal at various heights. Consequently, rays from different pupil radii converge at distinct positions along the optical axis [45].

In figure 4.4 and 4.5 we have the ray training of a single lens. In this instance, the rays seem to converge in figures 4.4 and 4.5. But if we look closer, as depicted in figure 4.6, the angle of a collimated bundle with the lens varies depending on the height of the ray where the surface is intersected. The central region exhibits only slight variation in these angles, thereby maintaining the validity of the paraxial approximation. Consequently, rays originating from the aperture zones in the central region converge relatively closely to each other along the optical axis. However, as the height of the rays progresses toward the edge of the surface, the angle between the rays and the surface's normal noticeably increases, rendering the paraxial approximation insufficiently accurate in those zones. By applying Snell's law, it becomes evident that the paraxial zone undergoes less refraction compared to the rays located farther from the optical axis. Consequently, the paraxial rays focus on the optical axis at a greater distance from the surface than the marginal rays. See figure 4.6.

Now let's introduce some colloquial nomenclature. When we say we have a fast system, that indicates it has a small f-number. For a fast system usually, the greater the spherical aberration it will exhibit. Conversely, a smaller radius of curvature results in higher curvature, leading to increased spherical aberration in the system. This is because there is a greater disparity in

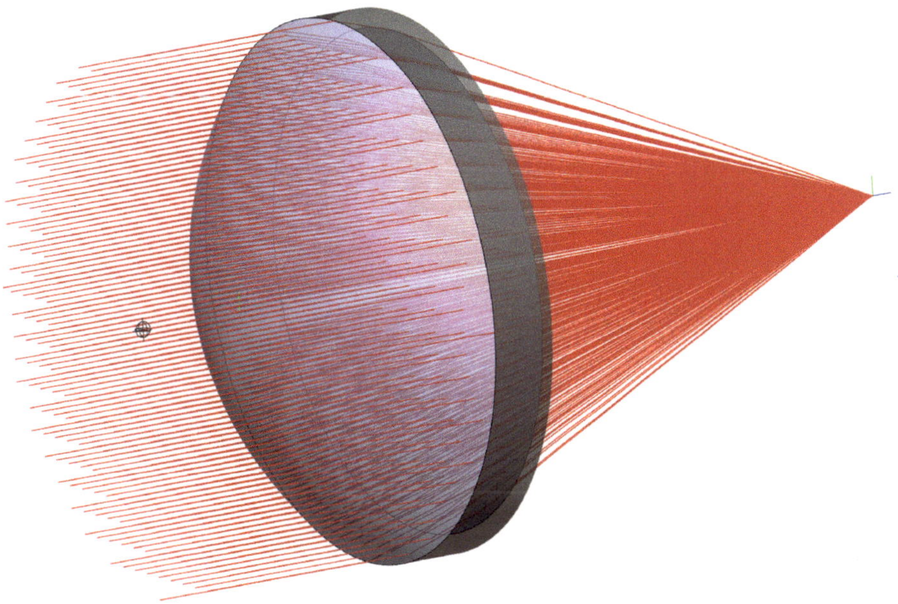

FIGURE 4.5
Side view of a singlet lens with a single on-axis field.

the angles between rays from different aperture zones and the normal to the surface.

Due to the varying positions of focal points along the optical axis for different aperture zones, selecting the proper image plane is crucial for minimizing spot size. The optimal position is a compromise between the paraxial and marginal rays, typically where the beam's waist is the smallest. This position is usually called the `Circle of Least Confusion`. In figure 4.7, it is graphically presented the `Circle of Least Confusion`. The curve formed by the outgoing beam's height concerning different focal lengths along the optical axis is known as the `Caustic Curve` [43, 44], illustrated in black in figure.

4.2.3 Coma

The symmetry of optical element surfaces is disrupted when an off-axis incident field interacts with them. Consequently, the angle at which the field intersects the surface varies across different annular zones of the surface. This discrepancy causes different aperture zones of the optical element to focus rays into different transversal positions, resulting in a comet-like image shape aberration known as coma. Similar to spherical aberration, marginal rays focus in a different position compared to paraxial rays. However, unlike spherical

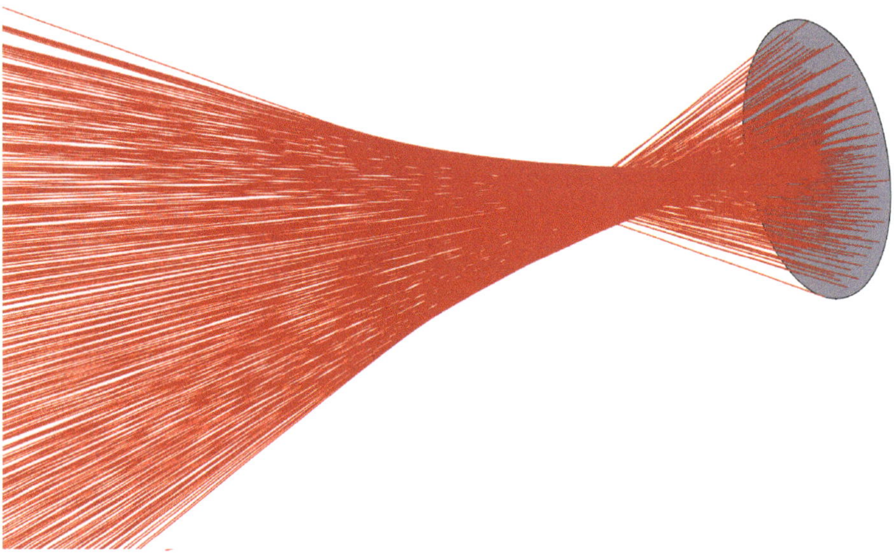

FIGURE 4.6
Side view of the zoom-up of the image plane. The rays do not converge in a single point due to the spherical aberration.

aberration, the presence of a coma is influenced not only by the aperture size but also by the angle of the oblique rays. Therefore, larger aperture sizes or field angles lead to greater coma in the system. Personally speaking, We can see coma aberration as an off-axis spherical aberration. Since spherical aberration is just for on-axis fields, and coma is just for off-axis fields, this is just a personal thought [5, 42].

In figure 4.8, there is a diagram of a lens, with several files, the on-axis field is in red. The two off-axis fields are in blue and green, respectively. If we made a zoom on the image plane and traced only the green off-axis field. We will see the coma aberration for the mentioned field as shown in figure 4.9.

4.2.4 Astigmatism

Off-axis rays are influenced not only by the aperture zone of the pupil but also by the plane in which they lie. When a ray bundle is situated in the tangential plane, the rays intersect the lens surface with a tilt in the sagittal plane. Consequently, due to the distinct paths of the sagittal and tangential rays, they exhibit different focal lengths. This variation in focal lengths results

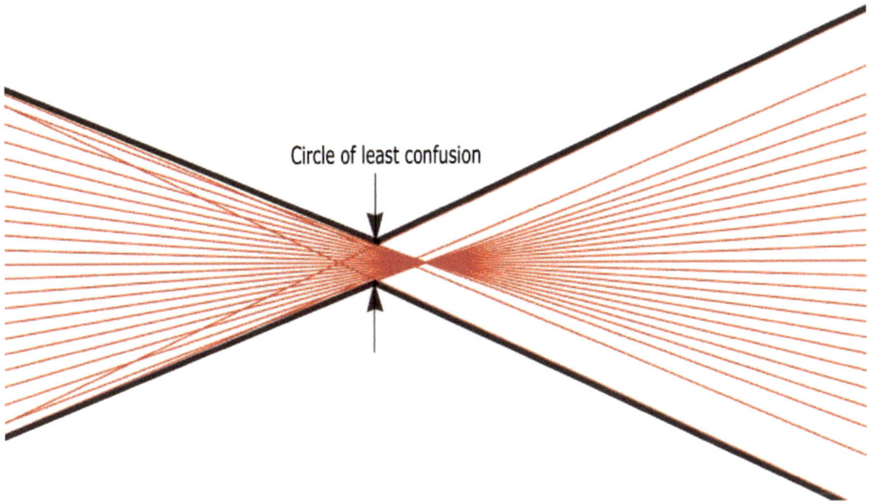

FIGURE 4.7
`Circle of Least Confusion` of an on-axis field.

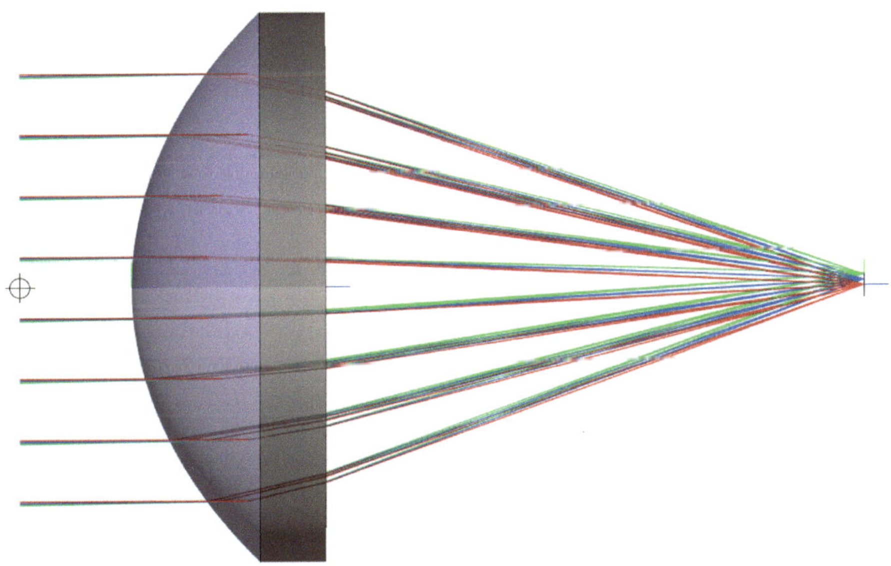

FIGURE 4.8
A singlet lens with one on-axis, and two off-axis fields.

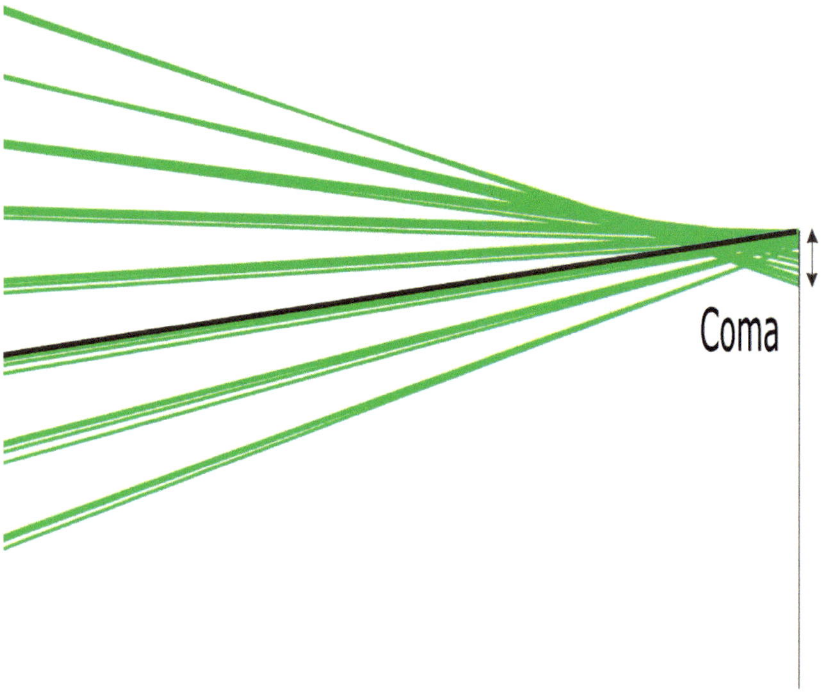

FIGURE 4.9
A zoom on the image plane of the singlet presented in figure 4.8. Here we just show the ray of the green off-axis field. The pattern formed by the rays, which does not converge in a single point, is called coma.

in the lens demonstrating different optical powers depending on the plane in which the rays lie [5, 42]. This aberration is referred to as astigmatism. Astigmatism is displayed in figure 4.10.

Figure 4.10 illustrates a bundle of rays in both sagittal and tangential orientations focusing on their respective focal points. These rays are the rays of the green off-axis field of the singlet lens presented at the bottom of figure 4.8. So notice that the circle of least confusion diameter changes in the sattignal and tangential plane, this is what we call astigmatism.

Rays refracted in the sagittal plane ultimately traverse a longer `Optical Path Length` within the lens compared to rays in the tangential plane. This effect is akin to the lens appearing thicker for the sagittal plane.

When the image plane is positioned at the focus of sagittal rays, the resulting image appears as a vertical line. This indicates that while the sagittal rays converge to a focal point at this position, the tangential rays either have

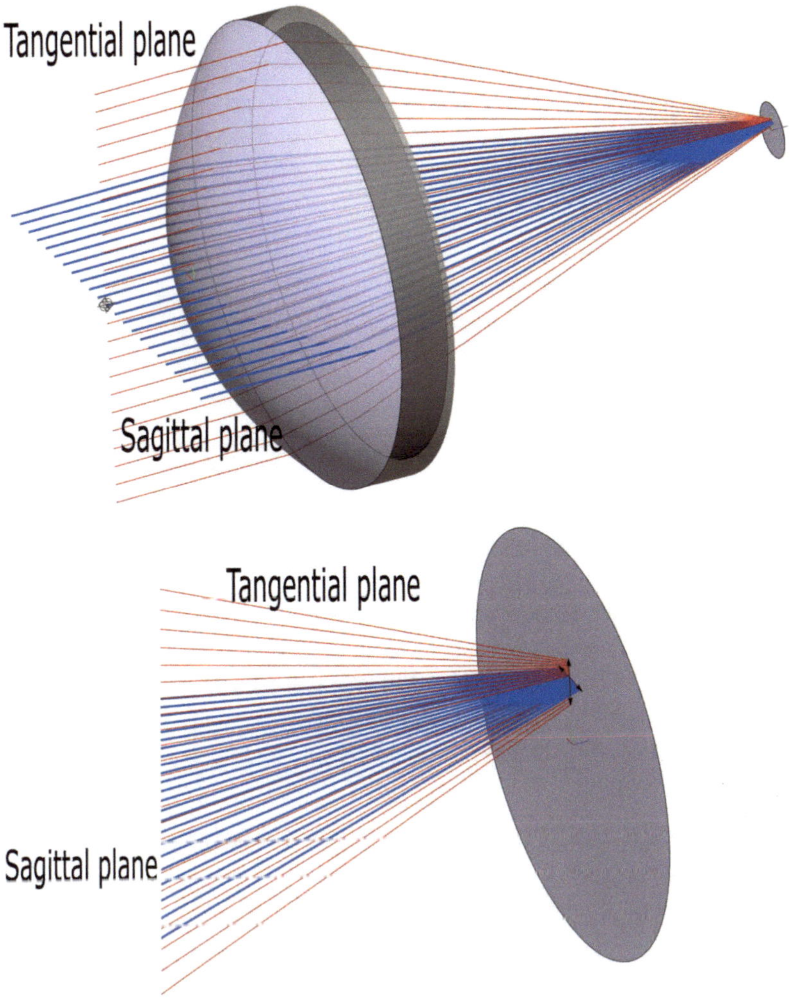

FIGURE 4.10
A singlet lens suffering from astigmatism.

not yet focused and are hence converging, or have already focused earlier and are diverging upon intersecting the image plane.

4.2.5 Field Curvature

The image formed by various field angles doesn't converge onto a single plane. With larger field angles, the focal point shifts further away from the image plane, as depicted in figure 4.11. Figure 4.11 shows a zoom-up of the three fields on the singlet lens of figure 4.8. Observer that the three circles of least

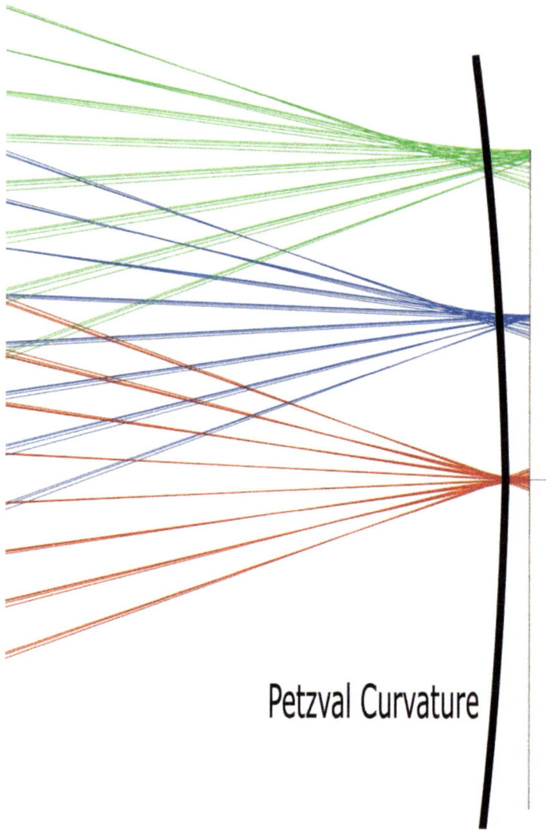

FIGURE 4.11
A zoom-up of the image plane of the singlet of figure 4.8. The points where the rays converge of the three fields are not aligned in a single plane. The surface that passes through all these converging points is the field curvature.

confusion of the respective three fields are not aligned in the same image plane. Consequently, the image forms on a curved surface, a phenomenon known as Field Curvature [5, 42]. While similar to astigmatism, Field Curvature is distinct in that it's independent of the azimuth angle.

Petzval curvature, another name for field curvature, refers to the curvature at which the image produced by an object is sharply focused. Just like in figure 4.11.

4.2.6 Distortion

Object points located at varying distances from the optical axis exhibit different transverse magnifications. Consequently, the geometric shape of the image undergoes alteration while maintaining sharpness. Even points situated farther from the paraxial region retain image quality, resulting in a sharp image. However, these points deviate from the ideal image point, producing a distortion effect. Hence, this aberration is appropriately termed distortion. Figure 4.12 shows the distortion of the lens from figure 4.8. The back rays are the predicted rays that focus on the ideal image. However, the off-axis fields do not focus on the same point. The percentage of this difference is what we call distortion [5, 42].

Mathematically speaking we can measure the distortion with eh following equation:

$$\mathrm{D} = \frac{\mathrm{AD} - \mathrm{PD}}{\mathrm{PD}} \times 100\% \tag{4.1}$$

where, D is the distortion, the actual distance is AD and the predicted distance is PD. The actual distance AD is where the actual rays lay in the image plane. The predicted distance PD is the predicted by the model. This one depends on the parading presented by the designer. There are several ways to express PD, but the two main used are given by

$$\mathrm{PD} = f \tan\theta \tag{4.2}$$

and

$$\mathrm{PD} = f\theta \tag{4.3}$$

where f is the effective focal length of the system and θ is the half field of view of the system. From the two given definitions of PD there is a vast discussion of when one should be used and when the other should be addressed. But in practice, usually, lens designers use $\mathrm{PD} = f \tan\theta$ for systems with half field of view around 45° or less, and $\mathrm{PD} = f\theta$ is used for systems with half field of view around 60° or more.

The shape of the image is influenced by the direction of the change in magnification. There are two types of distortion: pincushion distortion, also known as positive distortion, and barrel distortion, also known as negative distortion. Pincushion distortion occurs when the magnification increases further away from the optical axis, causing the lines to appear curved outwards with the borders extending beyond the central region. Conversely, barrel distortion arises when the magnification decreases, resulting in lines that appear curved inwards and create a barrel-like shape.

In practice, this aberration is not as detrimental to image quality as others, up to a certain extent, as it can be rectified using image processing techniques. Barrel distortion can be regarded as a desirable characteristic in certain optical systems for specialized applications. For instance, fish-eye lenses deliberately introduce significant distortion to expand the field of view.

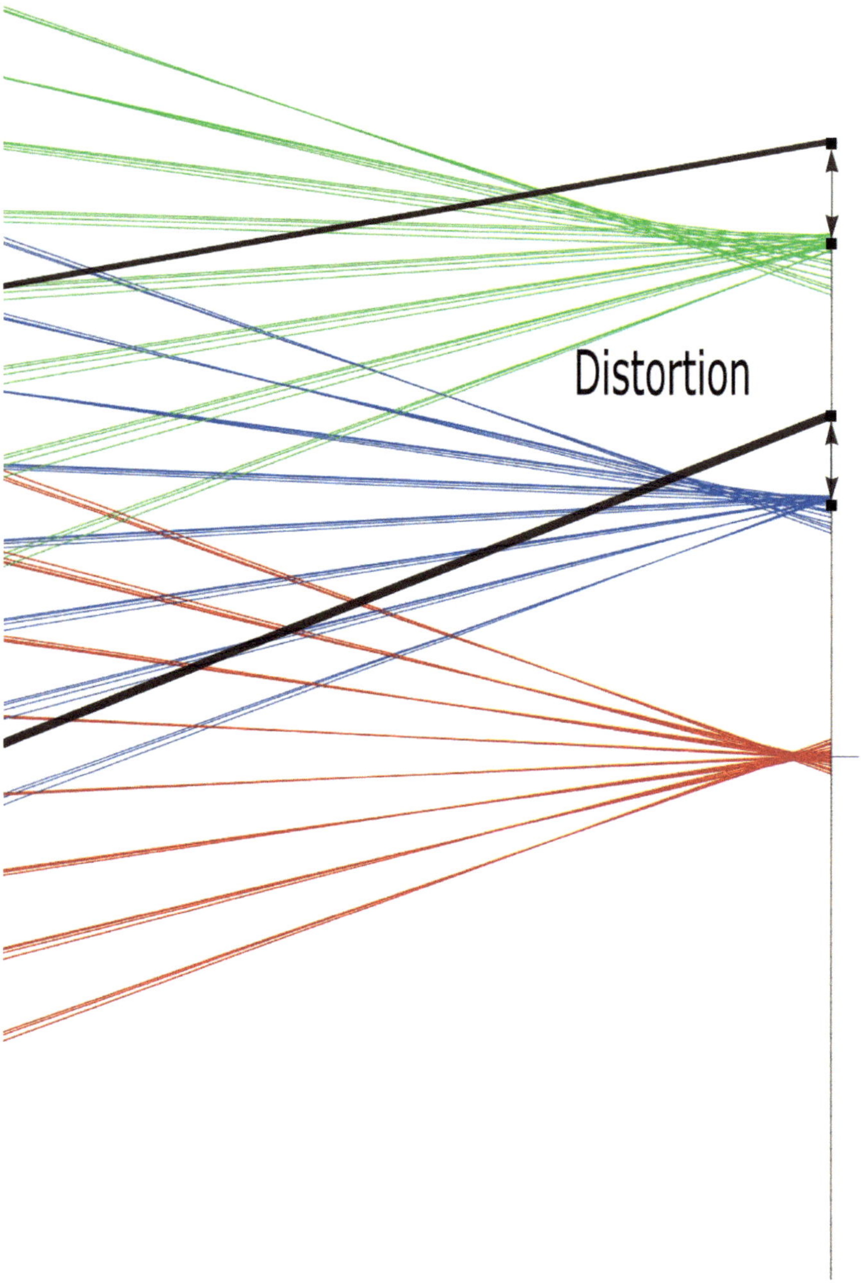

FIGURE 4.12
Distortion of the singlet of figure 4.8.

4.3 Chromatic Aberrations

Chromatic aberrations occur when an image is affected by different wavelengths failing to focus at the same point. This phenomenon arises due to the relationship between wavelength and refractive index, as seen in the previous chapter [5, 42].

The relationship between wavelength and refractive index implies that a refractive medium exhibits varying behavior for each wavelength. Consequently, a mirror, being a reflective surface, does not exhibit chromatic aberration. The refractive index is inversely proportional to the square of the wavelength. Consequently, two wavelengths equally spaced in the spectrum won't appear equally spaced in the image plane.

4.3.1 Longitudinal Chromatic Aberration

`Longitudinal Chromatic Aberration` manifests with on-axis rays. Rays of different wavelengths focus at distinct positions along the optical axis because they have different focal lengths resulting from varying optical powers across the spectrum. As the term suggests, this aberration, known as longitudinal chromatic aberration, exclusively affects the longitudinal dimension. Hence, if the image plane is positioned at the focal length of one wavelength, the others will be out of focus, either diverging or converging. Consequently, each wavelength exhibits a distinct `Back Focal Length`.

Figure 4.13 depicts the impact of `Longitudinal Chromatic Aberration` at the output of an optical system. The figure does not show the whole optical system but shows a zoom-up on the image plane. Here the three fields are displayed in three different colors, each color represents a different wavelength, but all the fields for this example are on-axis.

Observe that the system's image resembles defocus, albeit with variations across different wavelengths. When the image plane is positioned at the focal point of one wavelength, that specific wavelength focuses at the center of the image plane, while the others appear defocused. This difference in the position where the focal points are located is what we call `Longitudinal Chromatic Aberration`.

Due to variations in refractive indices and Abbe numbers among different materials, it is feasible to combine them to mitigate dispersion, thereby correcting `Longitudinal Chromatic Aberration`, particularly when pairing crown and flint glasses. To achieve this, a converging crown lens is paired with a diverging flint lens, forming a combination known as a doublet. However, this method solely corrects `Longitudinal Chromatic Aberration` for a pair of wavelengths positioned at opposite ends of the spectrum. Consequently, to correct additional wavelengths, it becomes necessary to incorporate more lenses into the system.

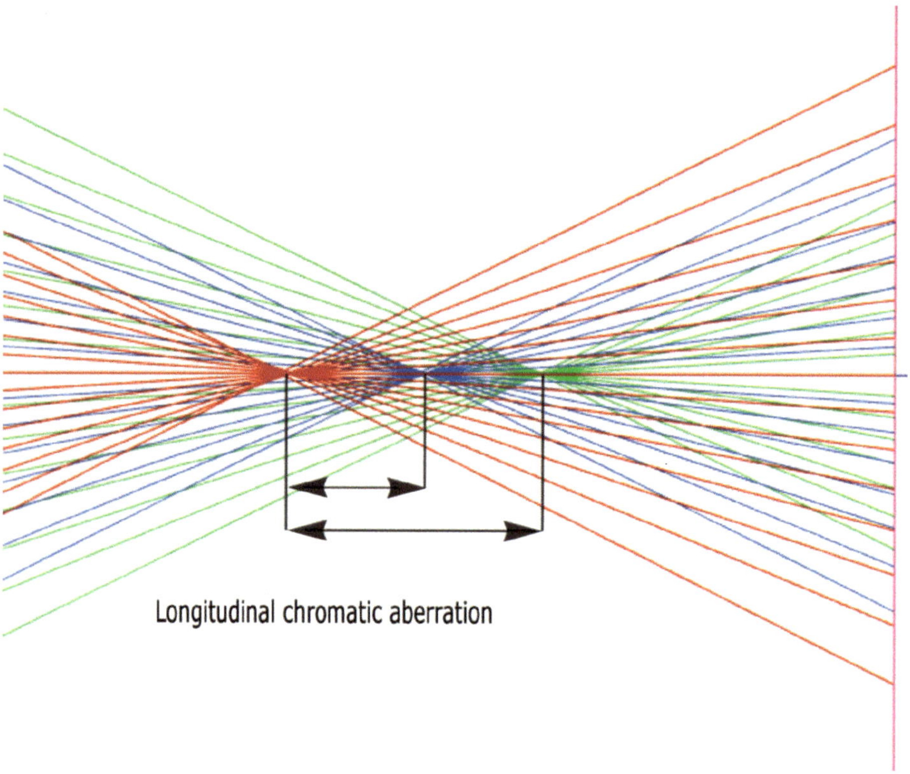

FIGURE 4.13
`Longitudinal Chromatic Aberration` at the output of an optical system. The figure does not show the whole optical system but shows a zoom-up on the image plane. Here the three fields are displayed in three different colors, each color represents a different wavelength, but all the fields for this example are on-axis. When the image plane is positioned at the focal point of one wavelength, that specific wavelength focuses at the center of the image plane, while the others appear defocused. This difference in the position where the focal points are located in what we call `Longitudinal Chromatic Aberration`.

4.3.2 Transverse Chromatic Aberration

The magnification of an optical system also varies with wavelength. The principal ray of off-axis rays occupies distinct positions along the image plane for each wavelength, resulting in transverse displacement [5, 42]. Consequently, the image size varies across different wavelengths. This displacement is referred to as `Transverse Chromatic Aberration` or lateral color.

As depicted in figure 4.14, these aberrations arise from the angular deviation. The figure shows the zoom-up of an image plane for an off-axis field,

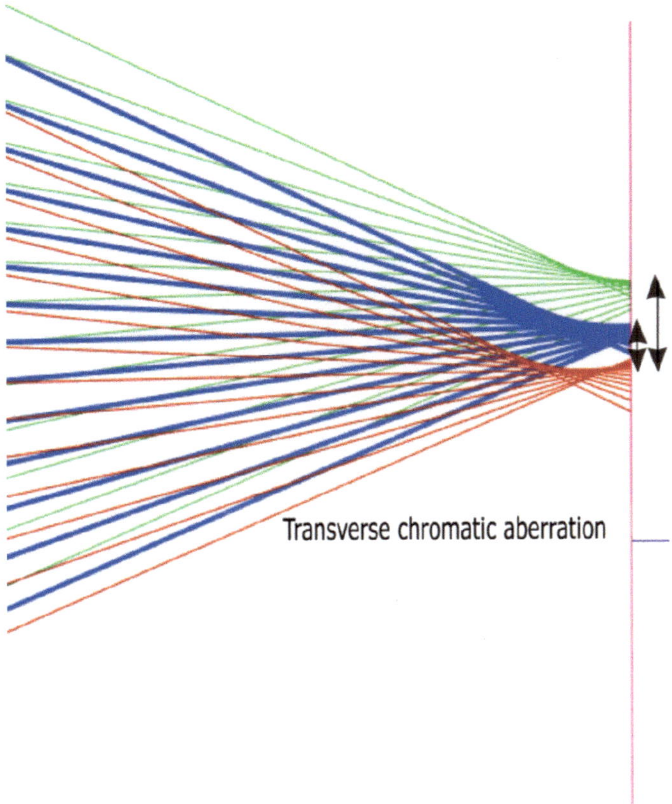

FIGURE 4.14
The magnification of an optical system also varies with wavelength. The principal ray of off-axis rays occupies distinct positions along the image plane for each wavelength, resulting in transverse displacement. Consequently, the image size varies across different wavelengths. This displacement is referred to as `Transverse Chromatic Aberration` or lateral color.

each color of the rays represents different wavelengths. Each wavelength has a different magnitude.

4.4 Geometrical Optical System Analysis

Optical designers have developed a variety of tools to geometrically analyze optical systems. In this section, we are going to explore them. All these tools are built-in `Quadoa Optical CAD`.

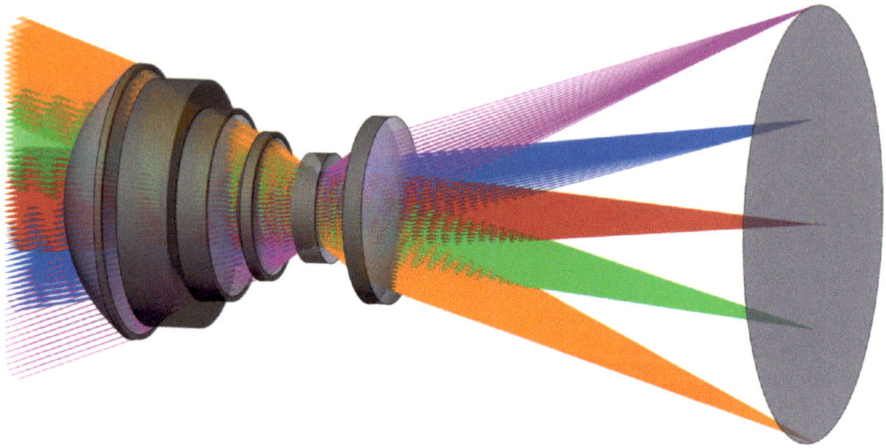

FIGURE 4.15
An example of an optical system. The file of this example is called `C:\Program Files\Quadoa\examples\Objectives\Camera_Lens` and can be found in the folder of the example presented with the installation of `Quadoa Optical CAD`.

We are going to learn how to analyze an optical system with the design example presented in figure 4.15. The file of this example is called `C:\Program Files\Quadoa\examples\Objectives\Camera_Lens` and it can be found in the examples files provided with the installation of `Quadoa Optical CAD`.

The design that is going to be discussed in this section is composed of six lenses. The materials of the lenses are `N-BK7` and `N-F2`. Both the materials can be found in the `SCHOTT Material Catalog` The system has five `Fields` that come from an object located at minus infinity. The angle of incidence of the field is the following 0°, 10°, 20°, −10°, and −20°. The stop is located at the first surface of the fifth lens. The back focal length is 79.33mm. This system is particularly slow, it has an f-number of 5.36. However, this system is perfect to analyze with the tool provided by `Quadoa Optical CAD`. So let us start!

4.4.1 Spot Diagram

Maybe the most popular tool to geometrically analyze an optical system is the `Spot Diagram`. The front view of the image plane, where the rays intersect, is termed as the `Spot Diagram` [5, 42]. This diagram serves as a useful tool for quickly identifying potential aberrations within an optical

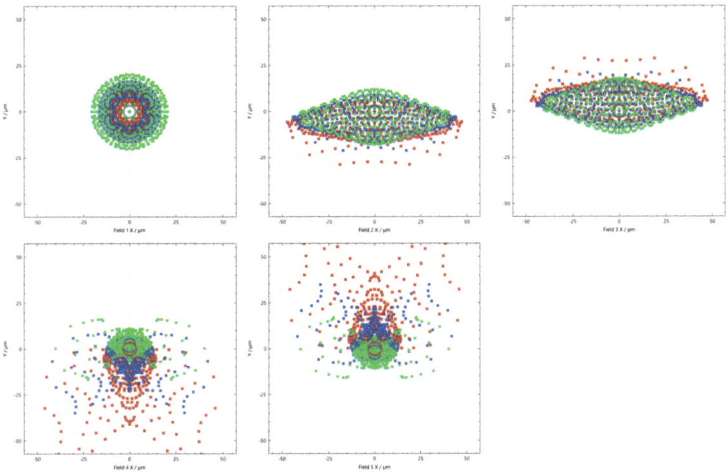

FIGURE 4.16
`Spot Diagram` of the file called `bla bla`. `Spot Diagram` shows a grind of spot diagrams, each diagram for each field. The color represents the wavelength. In this particular case..

system. While not as precise as other methods, the `Spot Diagram` offers a starting point for identifying aberrations since different aberrations produce distinct shapes within the diagram.

In `Quadoa Optical CAD`, the `Spot Diagram` visualizes the 2D distribution of rays on the image surface within a sequence. The spot diagram is visualized in figure 4.16, it is the spot diagram of the file called `C:\Program Files\Quadoa\examples\Objectives\Camera_Lens`. `Spot Diagram` shows a grind of spot diagrams, each diagram for each field. The color represents the wavelength. In this particular case..

Both defocus and spherical aberration produce spot diagrams characterized by rings surrounding the reference point. However, in defocus, there's a consistent energy distribution across the evenly spaced rings. So we can see at the top=left of figure 4.16 that the field has `C:\Program Files\Quadoa\examples\Objectives\Camera_Lens` suffers from defocus and spherical aberration. It makes sense since it is the on-axis field.

In instances of off-axis aberrations, the reference point may not align with the optical axis, and instead, the centroid of the spot diagram can serve as the reference. The coordinates of the centroid can be determined using equations (4.4) and (4.5).

$$\Delta x_c = \frac{1}{N}\sum_j \Delta x_j \tag{4.4}$$

and

$$\Delta y_c = \frac{1}{N}\sum_j \Delta y_j \tag{4.5}$$

In these equations, N represents the total number of rays being traced, and j serves as the index for each ray. The Root-Mean-Square (RMS), also known as the Gaussian Moment, utilizes the centroid as a reference point to compute the deviation of the rays in the image plane. The RMS radius is calculated using equation (4.6).

$$\Delta r_{RMS} = \sqrt{\frac{1}{N}\sum_j [(\Delta x_c - \Delta x_j)^2 + (\Delta y_c - \Delta y_j)^2]} \tag{4.6}$$

Sagittal coma refers to the disparity between the reference point and the nearest point of the circle delineated by the largest annular zone of the pupil. Tangential coma, on the other hand, denotes the discrepancy between the reference point and the furthest point of the circle formed by the same aperture zone.

Astigmatism manifests with a distinctive ellipsoidal shape when it occurs outside of the previously mentioned planes. In fact, many optical systems afflicted with astigmatism exhibit this shape, whether in horizontal or vertical orientations. Figure 4.16 illustrates a field with astigmatism at the top-center and top-right. From the same figure, we can see that in the bottom diagrams, the fields have large coma aberration, this can be easily determined since its spot diagram looks like a comment.

4.4.2 Field vs Wavelength Plot

Since the `Spot Diagram` overlaps the spots of several wavelengths in a single diagram, `Quadoa Optical CAD` has another tool called `The Field vs. Wavelength Spot Diagram Plot` to avoid the overlapping.

`The Field vs. Wavelength Spot Diagram Plot` depicts the 2D ray distribution of rays on the image surface within a sequence. It organizes the rays into a grid, where each subplot represents the rays for one field/wavelength combination. The wavelengths are arranged from left to right, and the fields are arranged from top to bottom, following the order specified in the sequence. `The Field vs. Wavelength Spot Diagram Plot` of the design of figure 4.15 is presented in figure 4.17.

4.4.3 Footprint Diagram

On many occasions, it is very useful to see where the rays strike in the image plane or any other surface. Therefore `Quadoa Optical CAD` has `The Footprint Diagram Plot` tool. `The Footprint Diagram Plot` displays the points where the rays strike a surface. The x and y coordinates are plotted

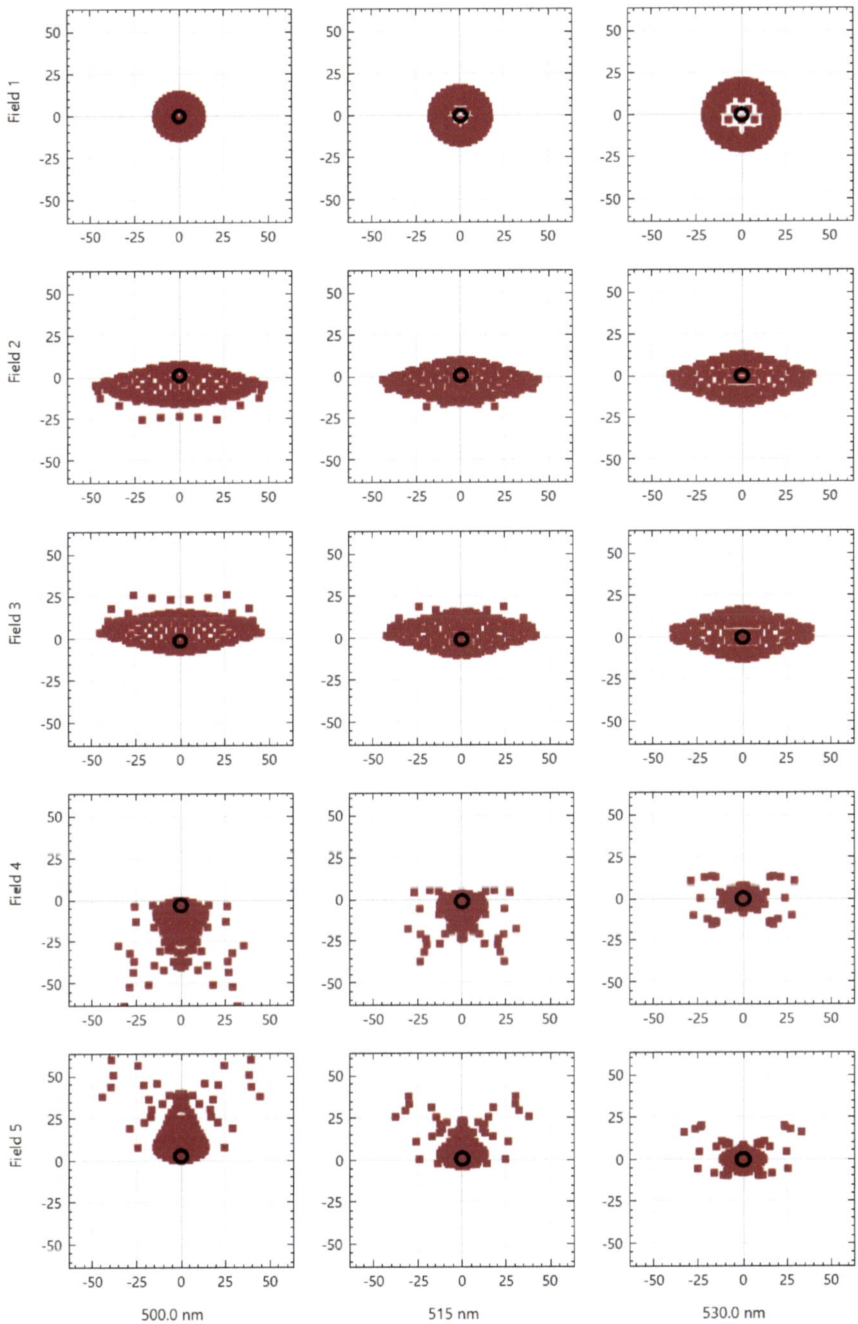

FIGURE 4.17
`The Field vs. Wavelength Spot Diagram Plot` of the design presented in figure 4.15.

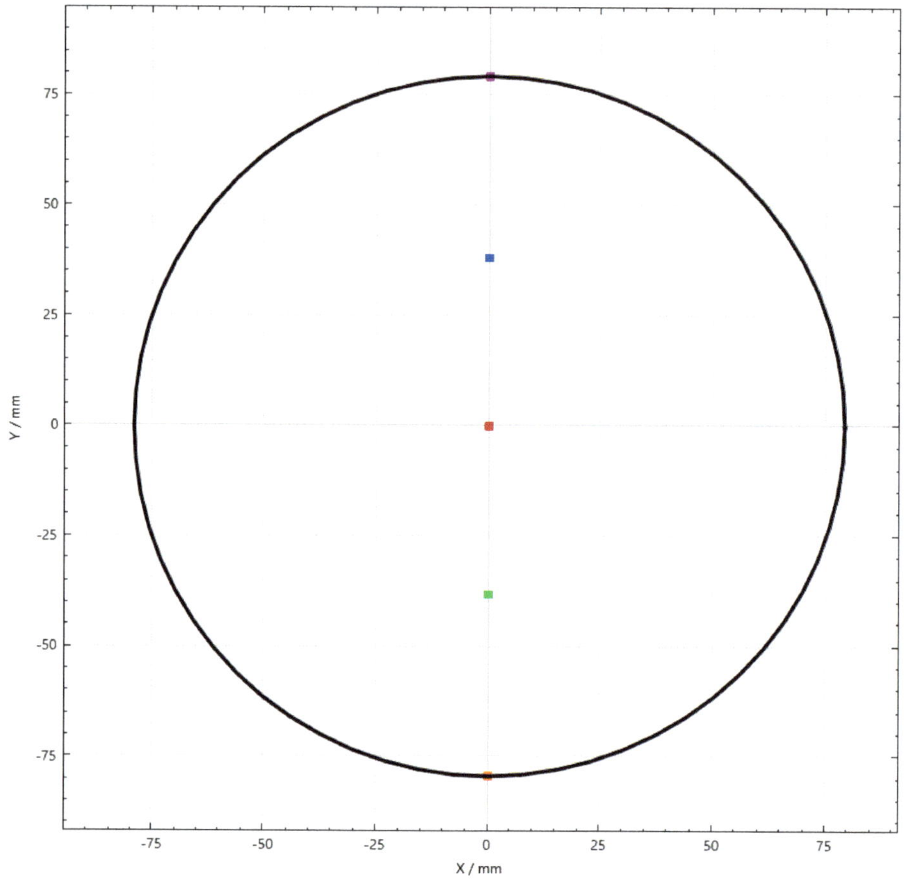

FIGURE 4.18
`Footprint Diagram Plot` of the filed called . The diagram not just only shows where the rays strike but also shows the aperture of the chosen surface. In this case, the surface is the image plane.

within the local surface coordinate system. Additionally, the plot includes a visual representation of the surface's aperture for reference. The `Footprint Diagram Plot` is very similar to the spot diagram but it not only considers the region around the spot but all regions of the surface itself. In figure 4.18 shows the `Footprint Diagram Plot` of the design of the figure 4.15.

4.4.4 OPD Fan Plot

In the analysis of an optical system is relevant to see the optical path difference between the fields in a sequence. `Quadoa Optical CAD` compute

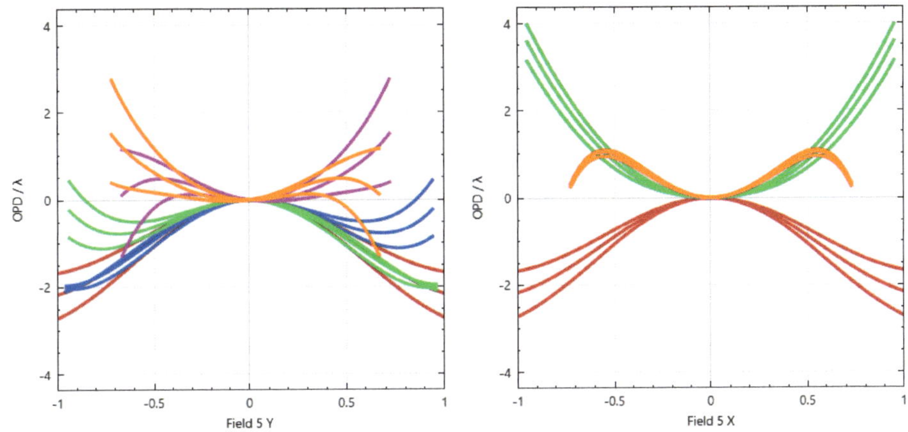

FIGURE 4.19
`OPD Fan Plot` of the design of figure 4.15. The colors of the plots are associated with the fields in figure 4.15 in a color code. The field with the color orange has a respective OPD plot line in orange and so on. On the right of the figure is the OPD along the y dimension and in the left is the OPD along the x dimension.

the OPD and displays it in the `OPD Fan Plot`. The `OPD Fan Plot` illustrates the Optical Path Difference along the x and y axes across the pupil of the sequence. The specifics of the plot are contingent upon the Sequence parameters, including OPD Reference and OPD Chief Reference. This means which optical path is to be considered the one for reference. In figure 4.19 is the `OPD Fan Plot` of the design of figure 4.15. The colors of the plots are associated with the fields in figure 4.15. Notice that both fields and plots have the same color, they are associated. On the right side is the OPD along the x dimension and on the left is the OPD along the y dimension. From the figure, it is clear that the OPD is zero with the ray at the center of the pupil since it is the reference ray and the OPD starts to increase from that point.

4.4.5 Transversal Ray Fan Plot

Transveres aberrations can be computed in `Quadoa Optical CAD` with `Transversal Ray Fan Plot`. The `Transversal Ray Fan Plot` displays the transversal aberration along the x and y axes across the pupil of the sequence. Remember that transverse aberrations are defined by the axial distance between the position where a ray intersects the image plane and the reference point. Figure 4.20 shows the transverse aberrations of the design in figure 4.15. For each wavelength in the design of figure 4.15 there is a plot in

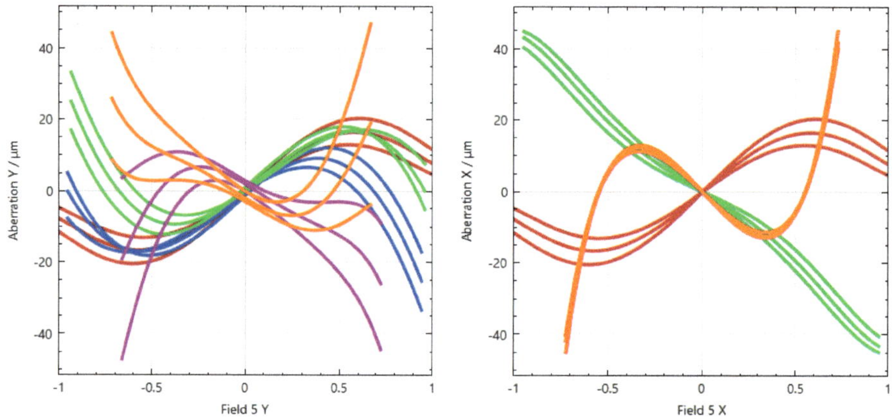

FIGURE 4.20
`Transversal Ray Fan Plot` of the design of figure 4.15. The colors of the plots are associated with the fields in figure 4.15 in a color code. On the right of the figure is the `Transversal Ray Fan Plot` along the x dimension and on the left is the `Transversal Ray Fan Plot` along the y dimension.

figure 4.20, the correspondence is given by color-code. In other means, they share the same color.

4.4.6 Longitudinal Aberration Plot

Just like transverse aberration can be computed and plotted in `Quadoa Optical CAD` the longitudinal aberration can be computed too. The `Longitudinal Aberration Plot` depicts the focal shift of a lens plotted against the wavelength. This calculation utilizes the on-axis field with field coordinates (0, 0). The range of the wavelength axis is determined by the minimum and maximum wavelength in the sequence. Figure 4.21 show `Longitudinal Aberration Plot` of the design presented in figure 4.15.

4.4.7 Longitudinal Pupil Aberration Plot

Longitudinal aberration can be also spotted across the pupil. `Quadoa Optical CAD` computes the longitudinal pupil aberration with a built-in function called `Longitudinal Pupil Aberration Plot`. The `Longitudinal Pupil Aberration Plot` illustrates the focal shift of a lens plotted across the pupil for each wavelength in the sequence. This calculation employs the on-axis field with field coordinates (0, 0). Figure 4.22 shows `Longitudinal Pupil Aberration Plot` of the design of the file called `C:\Program Files\Quadoa\examples\Objectives\Camera_Lens`. Each plot

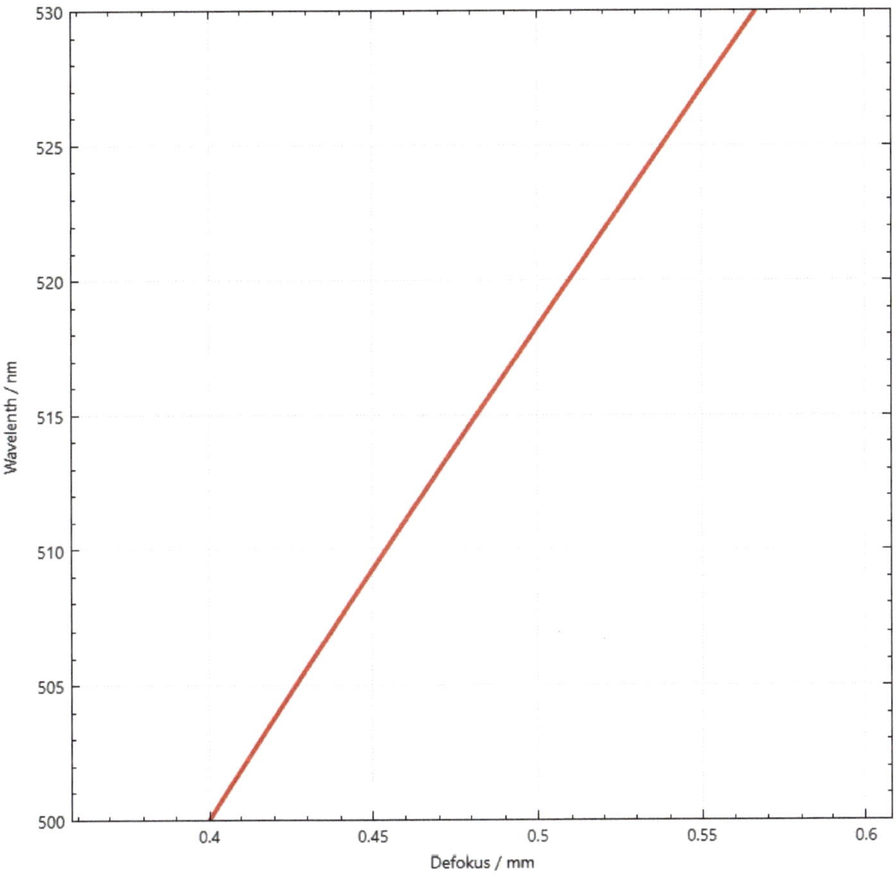

FIGURE 4.21
`Longitudinal Aberration Plot` of the design of the figure 4.15. The range of the wavelength axis is determined by the minimum and maximum wavelength in the sequence.

represents a different wavelength, for this case the wavelengths are.. Observe that in `Longitudinal Pupil Aberration Plot` the pupil is normalized to 1.

4.4.8 Lateral Chromatic Aberration Plot

When we consider the lateral aberration across several wavelengths we shall take a look to the `Lateral Chromatic Aberration Plot`, a tool that `Quadoa Optical CAD` offers for this topic. The `Lateral Chromatic`

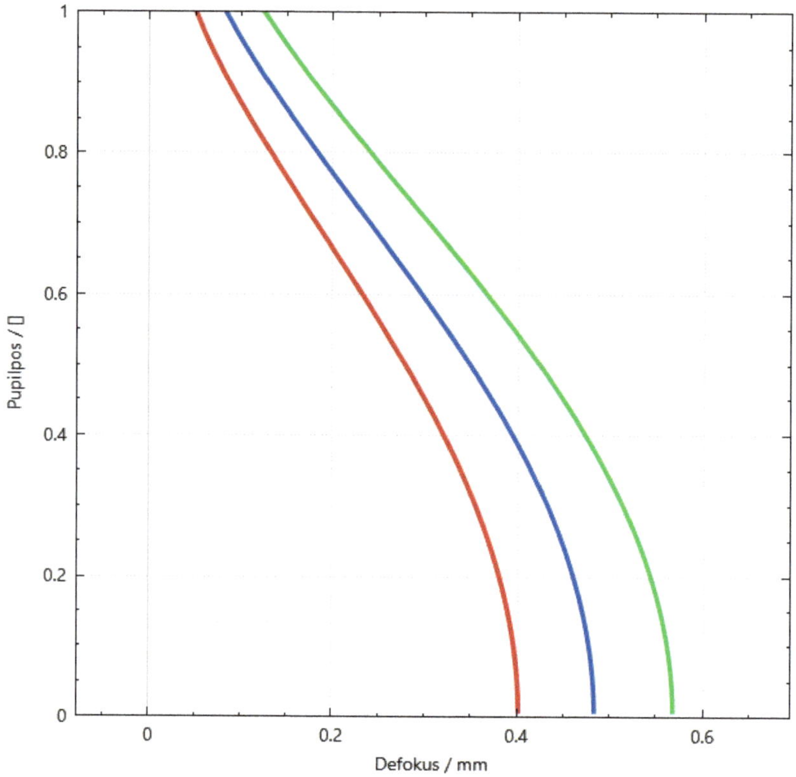

FIGURE 4.22
`Longitudinal Pupil Aberration Plot` of the design of the design in figure 4.15. Each plot represents a different wavelength defined in the sequence in figure 4.15. For this example, the wavelengths are red, blue, and green, respectively.

`Aberration Plot` displays the lateral chromatic aberration of a lens plotted across the field for each wavelength in the sequence. We show a `Lateral Chromatic Aberration Plot` in figure 4.23 of the design in figure 4.15. Just like the `Longitudinal Pupil Aberration Plot`, each plot represents a different wavelength. For this example, the wavelengths are red, blue, and green, respectively. Notice that the vertical axis in the plot shows the field of view in degrees (left) and normalized field of view (right). The horizontal axis is in mm is the later displacement. In this example, the reference wavelength in the sequence is in red. Therefore we have a vertical line. The other lines show the difference in lateral aberration with respect to the main wavelength.

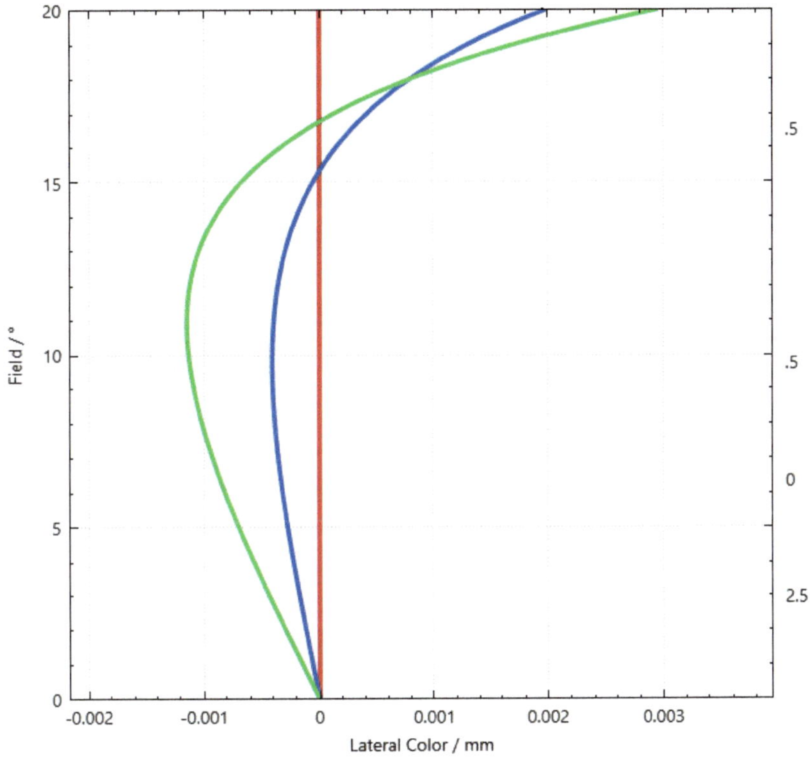

FIGURE 4.23
`Lateral Chromatic Aberration Plot` of the design at figure 4.15

4.4.9 Petzval Curvature Plot 1D

The `Petzval Curvature Plot 1D` illustrates the field curvature of a lens across the field coordinates in X and y plane. In this plot, the defocus is calculated by tracing a set of paraxial rays. Figure 4.24 shows `Petzval Curvature Plot 1D` of the design of the figure 4.15.

4.4.10 Petzval Curvature Plot 2D

The `Petzval Curvature Plot 2D` illustrates the field curvature of a lens across the field. In this plot, the defocus is calculated by tracing a set of paraxial rays. The plot can be computed by any of the wavelengths of the system but by default is the primary wavelength. For this example the primary wavelength is 500nm. The axes of the plot are given by the image height in x and y dimensions respectively. The deviation of the focal plane is in the color plot, where for this case, yellow represents zero deviation. For this particular

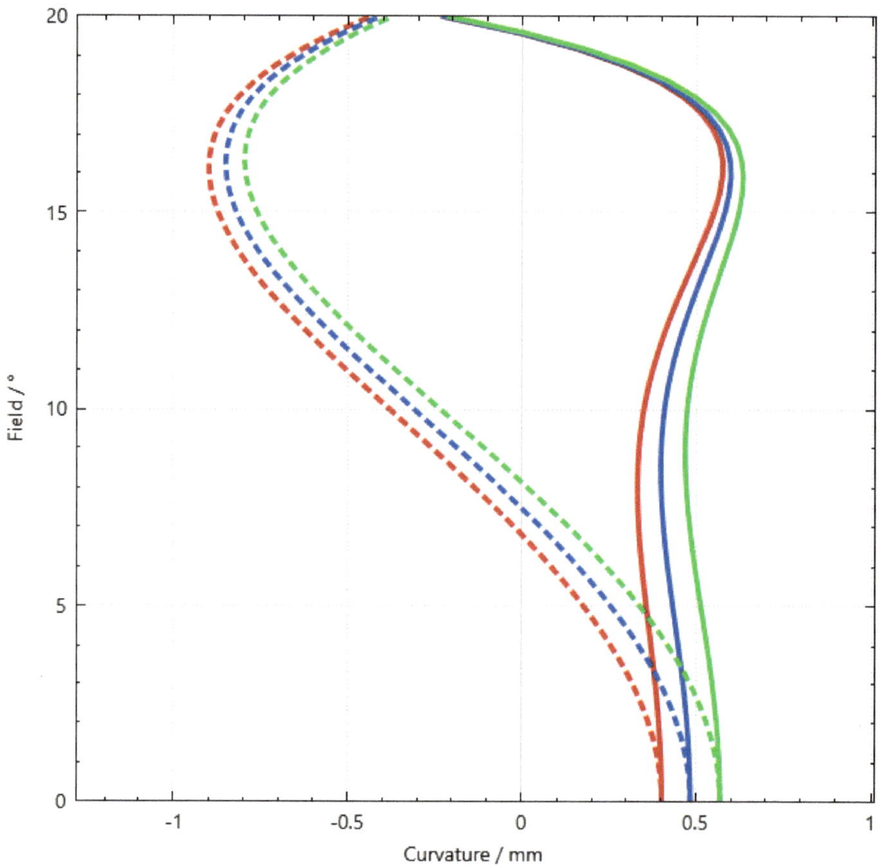

FIGURE 4.24
`Petzval Curvature 1D Plot` of a design of figure 4.15. The continuous lines correspond to the y plane and the dashed lines to the x plane. The three different colors, red, blue, and green, represent the wavelengths, which are 500 nm 515 nm, and 53 nm respectively.

example, we can see that the system of figure 4.15 does not suffer too much from `Petzval Curvature`, since the plot presented in figure 4.25 has mostly a homogenous color. Only in the corners, the `Petzval Curvature` increase.

4.4.11 Distortion

Distortion is one of the most important aberrations in an optical system, not because it affects the quality of the image but because it affects where the image points are displayed. Therefore `Quadoa Optical CAD` has two

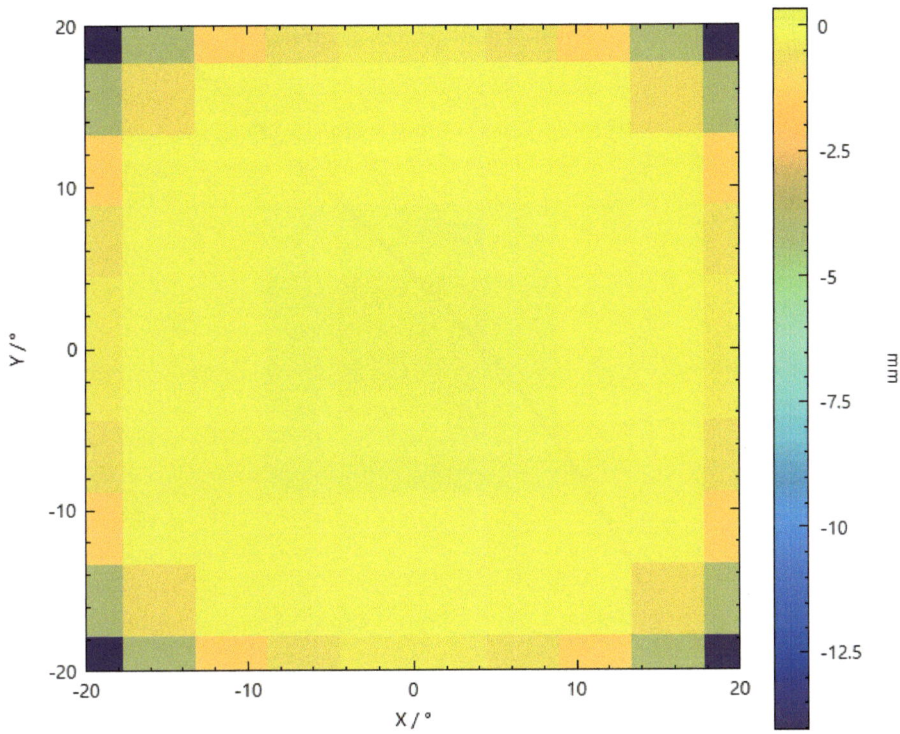

FIGURE 4.25
`Petzval Curvature Plot 2D` of a design of figure 4.15. The primary wavelength is 500nm. The axes of the plot are given by the image height in x and y dimensions respectively. The deviation of the focal plane is in the color plot.

main ways to compute it. The first one is a `Distortion 1D Plot`, which illustrates the distortion of a lens along a fan of field points. Which means $\pm x$ and $\pm y$ dimension in the image plane. This tool is very useful for radially symmetric optical systems. In figure 4.26 we show the `Distortion 1D Plot` of our example in figure 4.15. Depending on the optical system the distortion can be tolerated. For a cellphone camera can be less than 2%. But for sure it highly depends on each system and the field of view.

The `Distortion 2D Plot` illustrates the distortion exhibited by a lens across a grid spanning the field coordinates. It is also very useful for radially symmetric systems but also non-radially symmetric systems. In figure 4.27 we show the `Distortion 2D Plot` of the design of figure 4.15. The axes of the plot are given by the image height in x and y dimensions respectively. For this particular example, we can see that the system of figure 4.15 does not suffer too much from distortion, since the image is very similar to a rectangular grid

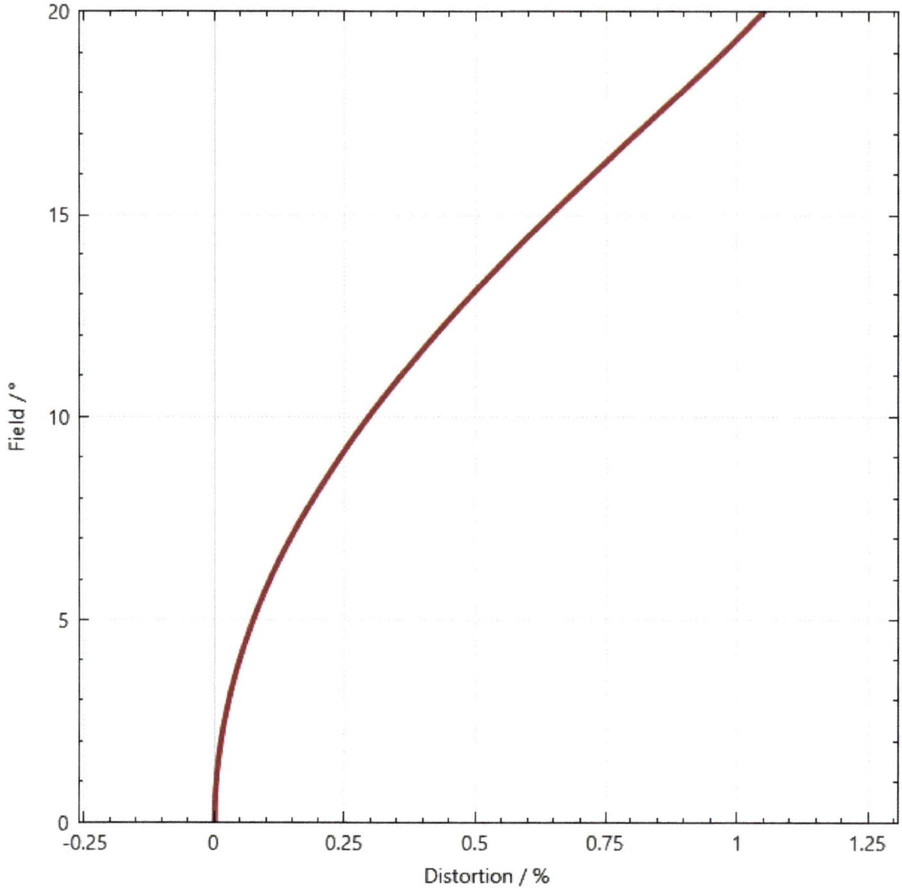

FIGURE 4.26
`Distortion 1D Plot` of a design of figure 4.15.

as should be. Only in the corners, we can see more distortion. Which is widely common, fields with bigger incident angles tend to have larger distortion.

4.5 Wave Optical System Analysis

Also, optical designers have developed a variety of tools to analyze optical systems with wave-optics. In this section, we are going to describe them using `Quadoa Optical CAD`.

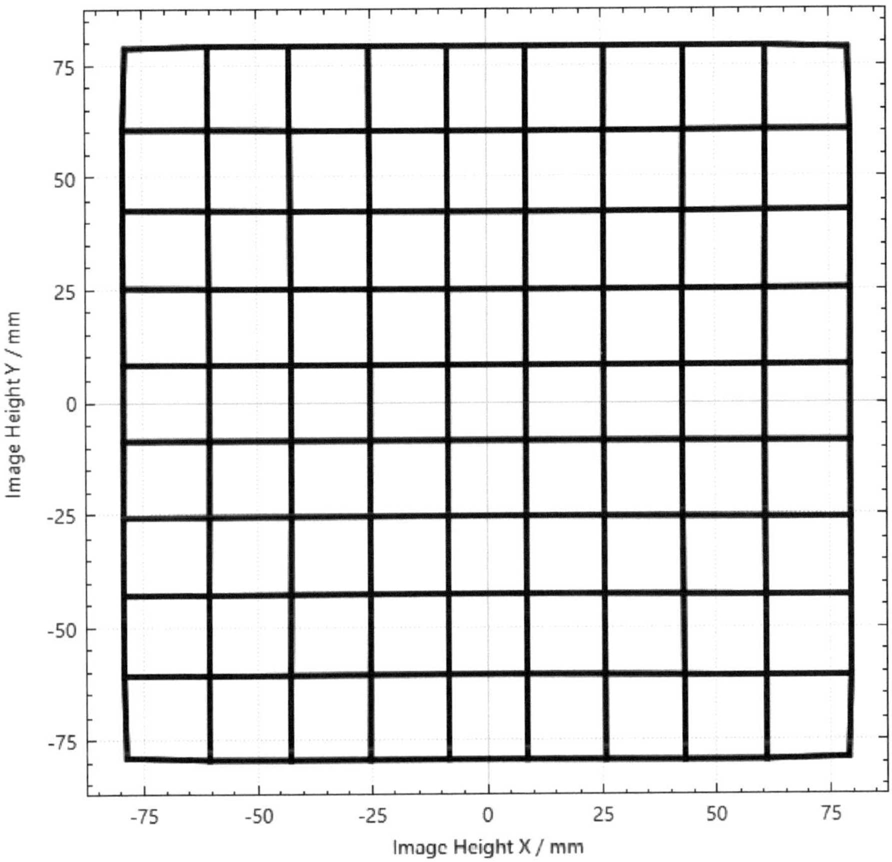

FIGURE 4.27
`Distortion 2D Plot` of the example presented in figure 4.15. The axes of the plots are given by the image height in x and y dimensions respectively.

4.5.1 Aberration Coefficients

The wavefront at the exit pupil of an optical system can be described in several mathematical ways. The most common is to express it as a polynomial expansion. Each element of the polynomial is related to a particular aberration. Each element is multiplied by a respective coefficient which describes the magnitude of the aberration suffered by the optical system. The polynomial expansion can be extended to infinite terms, and then the optical designer proposes to call the aberration by a certain order. For example, there is a spherical aberration of the third order, fifth order, and so on. In practice, just a few terms are needed it to express the aberrations of the system.

`Quadoa Optical CAD` considers this polynomial expansion in a finite number of terms. The main polynomials are `Seidel Coefficients` and `Zernike Coefficients`. Both can be found in `Quadoa Optical CAD`.

4.5.2 Seidel Coefficients

The Seidel coefficients comprise paraxial approximations. As demonstrated in previous sections, wavefront aberrations can be transformed into transverse ray aberrations through differentiation. But it is important to mention that Seidel aberrations yield meaningful values solely for classical systems employing spherical lenses, devoid of tilted or decentered elements.

Seidel polynomial expansion is not complete an orthogonal polynomial. This means that the terms of the Seidel polynomial are not independent from each other in the wavefront that want to describe. So, the fitting may differ a lot from by the number of terms implemented. Since the terms are not independent of each other, the adding or removing terms may affect the overall fitting of the exit pupil wavefront. This crucial fact, we are not going to spend too much time describing them, but in the bibliography of this chapter there are some good books about them.

So let go back to `Quadoa Optical CAD`, with the `Seidel Bar plot` . The `Seidel Bar plot` visualizes the `Seidel Aberration Surface` contribution along with the `Seidel Sum` for a specified Sequence in the form of a bar chart. The bars are organized into groups, where each group represents the surface contribution of one surface. The `Seidel Sum` is depicted in the last column of the chart.

Figure 4.28 shows the `Seidel Bar plot` of the design presented in figure 4.15. The horizontal axis shows the number of surfaces. The vertical axis shows the magnitude of the coefficients. Each surface has a group of bars. Each bar has a color-code that represents an aberration. The color code can be changed by the user, but for this particular example, the colors blue, green, red, violet, orange, and light green represent, spherical aberration, coma, astigmatism, field curvature, and distortion, respectively.

The `Seidel Report` in `Quadoa Optical CAD` presents the Seidel aberration surface contribution along with the `Seidel Sum` for a specified sequence in the form of a table. See figure 4.29. The magnitude of values are presented in SII or wave notation. For more details in this representation of the Seidel coefficients see book [42, 46].

4.5.3 Zernike Coefficients and Report

The `Zernike polynomials` are a complete set of continuous functions orthogonal over a unit circle. When the definition says complete, it means that they can express all the possible waterfronts at the exit pupil, When the definition says orthogonal, it means that each term of the polynomial expansion means that the terms are independent from each other. In other words, each term

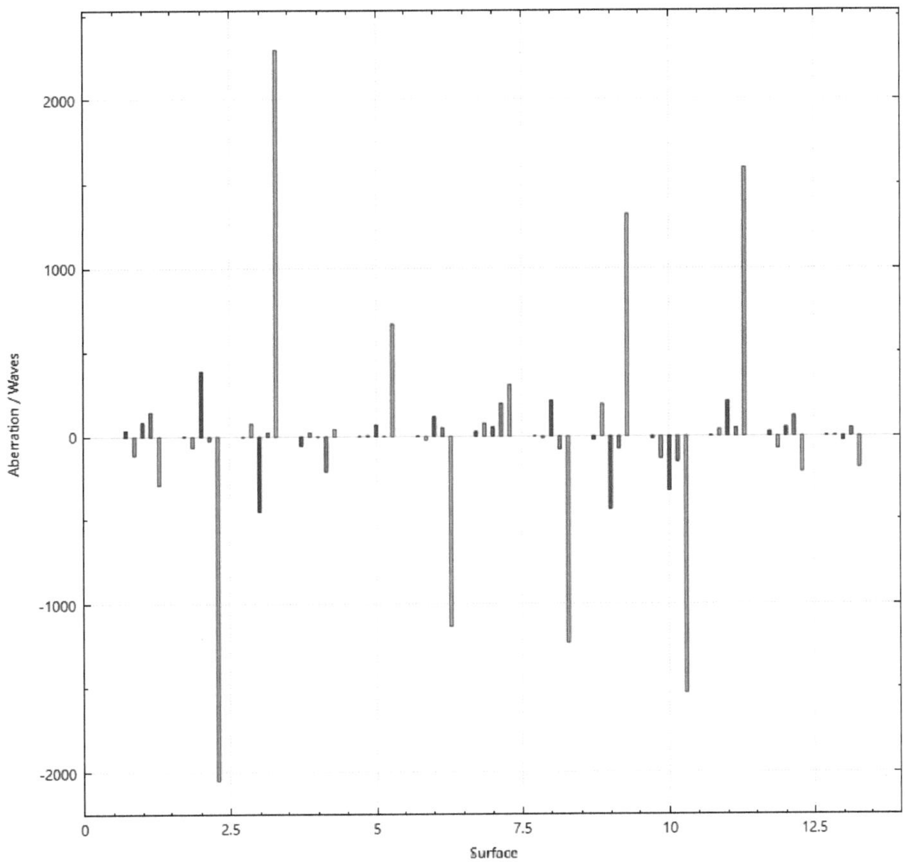

FIGURE 4.28
`Seidel Bar plot` of the design presented in figure 4.15. The horizontal axis shows the number of surfaces. The vertical axis shows the magnitude of the coefficients. Each surface has a group of bars. Each bar has a color code that represents an aberration. For this example, the colors blue, green, red, violet, orange, and light green represent spherical aberration, coma, astigmatism, field curvature, and distortion, respectively.

carries specific information that does not share with any other term. In practice, this better fits the `Zernike polynomial` and the wavefront at the exit pupil [47, 48, 49].

In `Quadoa Optical CAD`, various types of `Zernike Polynomials` are implemented, distinguished primarily by their sorting based on radial and angular modes. Additionally, the norming prefactor may vary among these polynomials.

Seidel Aberration Report

Surface	SI Sph.	SII Coma	SIII Asti.	SIV Pzvl.	SV Dist.
1	0.14332	-0.11054	0.085264	0.28771	-0.28768
2	0.010411	-0.063371	0.38574	-0.049512	-2.0466
3	-0.013446	0.07752	-0.44693	0.047949	2.3002
4	-0.21859	0.022343	-0.0022838	-0.41736	0.042893
5	0.00066553	0.0066674	0.066795	5.6331e-07	0.66917
6	0.0043235	-0.022424	0.11631	0.10109	-1.1276
7	0.11024	0.076444	0.05301	0.38885	0.30641
8	0.0004903	-0.010205	0.21242	-0.15342	-1.2281
9	-0.083516	0.19045	-0.43432	-0.14588	1.3231
10	-0.053572	-0.13128	-0.32173	-0.30058	-1.5251
11	0.0074654	0.039337	0.20728	0.095813	1.5971
12	0.10122	-0.072592	0.052063	0.24166	-0.21066
Sum	0.0090005	0.0023457	-0.026388	0.096323	-0.18671

Surface	W040 Sph.	W131 Coma	W222 Asti.	W220 Pzvl.	W311 Dist.
1	35.829	-110.54	85.264	143.86	-287.68
2	2.6027	-63.371	385.74	-24.756	-2046.6
3	-3.3615	77.52	-446.93	23.974	2300.2
4	-54.648	22.343	-2.2838	-208.68	42.893
5	0.16638	6.6674	66.795	0.00028165	669.17
6	1.0809	-22.424	116.31	50.544	-1127.6
7	27.559	76.444	53.01	194.42	306.41
8	0.12258	-10.205	212.42	-76.709	-1228.1
9	-20.879	190.45	-434.32	-72.939	1323.1
10	-13.393	-131.28	-321.73	-150.29	-1525.1
11	1.8664	39.337	207.28	47.906	1597.1
12	25.304	-72.592	52.063	120.83	-210.66
Sum	2.2501	2.3457	-26.388	48.161	-186.71

FIGURE 4.29
`Seidel Report` of the design is presented in figure 4.15. `Seidel Report` displays the Seidel aberration surface contribution along with the Seidel Sum for a specified Sequence in the form of a table. The data is displayed in SII or Woh notation.

The even `Symmetric Zernike Polynomials` are given by,

$$Z_n^m(\rho, \phi) = A \cdot R_n^m(\rho)\cos(m\phi), \tag{4.7}$$

and the odd `Symmetric Zernike Polynomials` are express as,

$$Z_n^{-m}(\rho, \phi) = A \cdot R_n^m(\rho)\sin(m\phi) \tag{4.8}$$

where $R_n^m(\rho)$ is the radial part of the `Symmetric Zernike Polynomials`,

$$R_n^m(\rho) = \sum_{k=0}^{(n-m)/2} \frac{(-1)^k (n-k)!}{k![(n+m)/2-k]![(n-m)/2-k]!}\rho^{n-2k} \tag{4.9}$$

The above expression is valid if $n-m$ is even. But, If $n-m$ is odd $R_n^m(\rho)$ is reduce to,

$$R_n^m(\rho) = 0 \tag{4.10}$$

The coefficient A is determined to satisfy the following normalization condition,

$$\int_0^{2\pi}\int_0^1 Z_j^2 \rho \mathrm{d}\rho \mathrm{d}\phi = \pi \tag{4.11}$$

The `Symmetric Zernike Polynomials` utilized in `Quadoa Optical CAD` align with those outlined in the `ANSI Z80.28/ISO 24157 Standard`. They are normalized to satisfy the norming condition specified in equation (4.11). The sorting of the `Symmetric Zernike Polynomials` follows the arrangement depicted in figure 4.30. In the `Optical Design Editor`, the index of the Zernike coefficients are counted row by row, commencing from the top with `Z0` for the piston, and progressing from left to right in the figure.

`Symmetric Zernike Polynomials` according to `ANSI Z80.28/ISO 24157 Standard` in table 4.1.

`Quadoa Optical CAD` also considers `Zernike Fringe Polynomials`. The `Zernike Fringe Polynomials` set is a subset of the `Symmetric Zernike Polynomials`. The mapping from `Symmetric Zernike Polynomials` to `Zernike Fringe Polynomials`, $Z_n^m \rightarrow Z_j$

$$j = \left(1 + \frac{n+|m|}{2}\right)^2 - 2|m| + \frac{1-\mathrm{sgn}(m)}{2} \tag{4.12}$$

where $\mathrm{sgn}(\cdot)$ is the sign function, and its outcome is the sign of its argument.

The plot of the first terms of `Zernike Fringe Polynomial` is shown in figure 4.31. Starting with $Z1$ which is the piston at the top, and counting from top to bottom and left to right.

Table 4.2 shows the coefficients of `Zernike Fringe Polynomials`.

The `Zernike Report` in `Quadoa Optical CAD` lists the wavefront aberrations in the form of coefficients of Zernike polynomials fitted to the wavefront. The specifics of this report depend on the Sequence parameters, including OPD Reference and OPD Chief Reference. The `Zernike Report` of the example presented in figure 4.15 is in figure 4.32. The `Zernike Report` in `Quadoa Optical CAD` can display the results in terms of `Symmetric Zernike Polynomials` and `Zernike Fringe Polynomials`.

FIGURE 4.30
`Symmetric Zernike Polynomials` according to `ANSI Z80.28/ISO 24157 Standard`. The index of the Zernike coefficients is counted row by row, commencing from the top with `Z0` for the piston, and progressing from left to right in the figure.

4.5.4 Point Spread Function

 The spot diagram is useful to calculate the spot size, but it is based only on the geometrical optics concepts. The wave behavior of light is ignored, which means interference and intensity of the spot are ignored, and the diffraction that forms the spot is ignored too. So the spot diagram is very limited in that way.

Remember that we mention some diffraction integrals in chapter two, the diffraction pattern obtained by these integrals, if is a converging optical

TABLE 4.1
Formula of Zernike Polynomials with ANSI Z80.28/ISO 24157 sorting.

Coefficient	Formula
Z0	1
Z1	$2x$
Z2	$2y$
Z3	$\sqrt{6}(2xy)$
Z4	$\sqrt{3}(-1+2y^2+2x^2)$
Z5	$\sqrt{6}(-1y^2+1x^2)$
Z6	$\sqrt{8}(-1y^3+3x^2y)$
Z7	$\sqrt{8}(-2y+3y^3+3x^2y)$
Z8	$\sqrt{8}(-2x+3xy^2+3x^3)$
Z9	$\sqrt{8}(-3xy^2+1x^3)$
Z10	$\sqrt{10}(-4xy^3+4x^3y)$
Z11	$\sqrt{10}(-6xy+8xy^3+8x^3y)$
Z12	$\sqrt{5}(1-6y^2+6y^4-6x^2+12x^2y^2+6x^4)$
Z13	$\sqrt{10}(3y^2-4y^4-3x^2+4x^4)$
Z14	$\sqrt{10}(1y^4-6x^2y^2+1x^4)$
Z15	$\sqrt{12}(1y^5-10x^2y^3+5x^4y)$
Z16	$\sqrt{12}(4y^3-5y^5-12x^2y+10x^2y^3+15x^4y)$
Z17	$\sqrt{12}(3y-12y^3+10y^5-12x^2y+20x^2y^3+10x^4y)$
Z18	$\sqrt{12}(3x-12xy^2+10xy^4-12x^3+20x^3y^2+10x^5)$
Z19	$\sqrt{12}(12xy^2-15xy^4-4x^3-10x^3y^2+5x^5)$
Z20	$\sqrt{12}(5xy^4-10x^3y^2+1x^5)$
Z21	$\sqrt{14}(6xy^5-20x^3y^3+6x^5y)$
Z22	$\sqrt{14}(20xy^3-24xy^5-20x^3y+24x^5y)$
Z23	$\sqrt{14}(12xy-40xy^3+30xy^5-40x^3y+60x^3y^3+30x^5y)$
Z24	$\sqrt{7}(-1+12y^2-29y^4+20y^6+12x^2-59x^2y^2+59x^2y^4)$ $+\sqrt{7}(-29x^4+59x^4y^2+20x^6)$
Z25	$\sqrt{14}(-6y^2+20y^4-15y^6+6x^2-15x^2y^4-20x^4+15x^4y^2$ $+15x^6)$
Z26	$\sqrt{14}(-5y^4+6y^6+30x^2y^2-30x^2y^4-5x^4-30x^4y^2+6x^6)$
Z27	$\sqrt{14}(-1y^6+15x^2y^4-15x^4y^2+1x^6)$

system, is also called point spread function (PSF) the magnitude value of the diffraction pattern. Let's analyze the name of the PSF, it means how the point (focus, spot) spreads as a function of the local coordinates in the local plane of the image. Then, we can see that the PSF is the spot according to the paradigm provided by wave-optics [50].

Since there are several diffraction integrals, there are several ways to compute the PSF [18]. `Quadoa Optical CAD` considers the following three

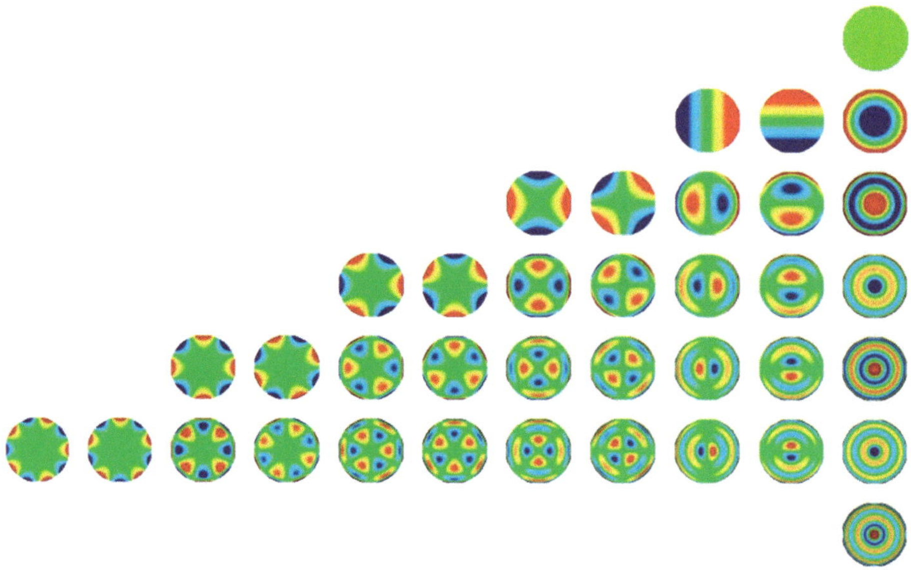

FIGURE 4.31
The Zernike Fringe Polynomials set is a subset of the Symmetric Zernike Polynomials. The mapping is given by equation (4.12). The following procedure gives the index of Zernike Fringe Polynomial: starting with $Z1$ which is the piston at the top, and counting from top to bottom and left to right.

options, Fast Fourier Transform PSF, Huygens PSF and Geometric PSF.

The three methods compute the PSF for a single field in a given sequence. The default field is the on-axis field, but the PSF can be computed also for off-axis fields. The option can easily be accessed by the user interface, which is equal for the three methods.

The intensity shown in the PSF for all the methods is normalized and it is displayed with a color plot. The coordinates of each pixel represented a coordinate in the image space. The color levels the intensity at that point.

Before we focus on computing the PSF with these three different methods in Quadoa Optical CAD. Let's review some concepts of the Fast Fourier Transform PSF Fast Fourier Transform PSF.

The Fast Fourier Transform PSF Fast Fourier Transform PSF is computed using the Franfouer approximation of the Huygens integral.

Remember that under this approximation the Huygens integral becomes the integral known as the Fourier transform. The Fourier transform is a

TABLE 4.2
Formula of Zernike Fringe Polynomials with

Coeficient	Formula
Z1	1
Z2	$2x$
Z3	$2y$
Z4	$-1 + 2y^2 + 2x^2$
Z5	$-1y^2 + 1x^2$
Z6	$2xy$
Z7	$-2x + 3xy^2 + 3x^3$
Z8	$-2y + 3y^3 + 3x^2y$
Z9	$1 - 6y^2 + 6y^4 - 6x^2 + 12x^2y^2 + 6x^4$
Z10	$-3xy^2 + 1x^3$
Z11	$-1y^3 + 3x^2y$
Z12	$3y^2 - 4y^4 - 3x^2 + 4x^4$
Z13	$-6xy + 8xy^3 + 8x^3y$
Z14	$3x - 12xy^2 + 10xy^4 - 12x^3 + 20x^3y^2 + 10x^5$
Z15	$3y - 12y^3 + 10y^5 - 12x^2y + 20x^2y^3 + 10x^4y$
Z16	$-1 + 12y^2 - 30y^4 + 20y^6 + 12x^2 - 60x^2y^2 + 60x^2y^4$
	$-30x^4 + 60x^4y^2 + 20x^6$
Z17	$1y^4 - 6x^2y^2 + 1x^4)$
Z18	$-4xy^3 + 4x^3y$
Z19	$12xy^2 - 15xy^4 - 4x^3 - 10x^3y^2 + 5x^5$
Z20	$4y^3 - 5y^5 - 12x^2y + 10x^2y^3 + 15x^4y$
Z21	$-6y^2 + 20y^4 - 15y^6 + 6x^2 - 15x^2y^4 - 20x^4 + 15x^4y^2 + 15x^6$
Z22	$12xy - 40xy^3 + 30xy^5 - 40x^3y + 60x^3y^3 + 30x^5y$
Z23	$-4x + 30xy^2 - 60xy^4 + 35xy^6 + 30x^3 - 120x^3y^2$
	$+105x^3y^4 - 60x^5 + 105x^5y^2 + 35x^7$
Z24	$-4y + 30y^3 - 60y^5 + 35y^7 + 30x^2y - 120x^2y^3 + 105x^2y^5$
	$-60x^4y + 105x^4y^3 + 35x^6y$

super very well-known integral. In many cases, this integral can be computed analytically but in practice in lens design, it is always computed numerically because the wavefront at the exit pupil is represented very complex polynomial. Which is not easy to compute analytically. When it is computed discretely/numerically the algorithm to compute it is called fast Fourier Transform. Particularly this algorithm, called the fast Fourier Transform algorithm, is just the most computationally cost-effective way to compute the Fourier transform for a discrete set.

4.5.5 Fast Fourier Transform PSF

The `Fast Fourier Transform PSF Plot` displays the point spread function of a Sequence. The PSF is calculated by directing a set of rays to the

Quadoa 24.02.1

Zernike Wavefront Report

Sequence:	1 Reflex 1
Field:	1 (X:0.0, Y:0.0) °
Wavelength:	1 (500.0 nm)
Config:	Current

Type: Zernike ANSI Z80.28 / ISO 24157 Fit

Fit Radius Type:	auto
Fit Radius:	17.614 mm
Fit Order:	37
RMS Fit Error:	8.1532e-08
Unit:	Waves
Ray Grid:	65

Zern	Value	Formula
Z0	6.849e+05	1
Z1	1.0062e-10	$2x$
Z2	-6.3466e-12	$2y$
Z3	3.1071e-13	$6^{1/2}(2xy)$
Z4	0.46759	$3^{1/2}(-1 + 2y^2 + 2x^2)$
Z5	-2.0764e-11	$6^{1/2}(-1y^2 + 1x^2)$
Z6	5.5101e-11	$8^{1/2}(-1y^3 + 3x^2y)$
Z7	-9.6962e-11	$8^{1/2}(-2y + 3y^3 + 3x^2y)$
Z8	-4.7063e-12	$8^{1/2}(-2x + 3xy^2 + 3x^3)$
Z9	-8.1592e-13	$8^{1/2}(-3xy^2 + 1x^3)$
Z10	3.1362e-12	$10^{1/2}(-4xy^3 + 4x^3y)$
Z11	-3.4866e-12	$10^{1/2}(-6xy + 8xy^3 + 8x^3y)$
Z12	-0.11385	$5^{1/2}(1 - 6y^2 + 6y^4 - 6x^2 + 12x^2y^2 + 6x^4)$
Z13	-1.2996e-11	$10^{1/2}(3y^2 - 4y^4 - 3x^2 + 4x^4)$
Z14	-5.8405e-06	$10^{1/2}(1y^4 - 6x^2y^2 + 1x^4)$
Z15	-6.2376e-12	$12^{1/2}(1y^5 - 10x^2y^3 + 5x^4y)$
Z16	1.3762e-12	$12^{1/2}(4y^3 - 5y^5 - 12x^2y + 10x^2y^3 + 15x^4y)$
Z17	1.3829e-11	$12^{1/2}(3y - 12y^3 + 10y^5 - 12x^2y + 20x^2y^3 + 10x^4y)$
Z18	2.4332e-12	$12^{1/2}(3x - 12xy^2 + 10xy^4 - 12x^3 + 20x^3y^2 + 10x^5)$
Z19	1.7051e-12	$12^{1/2}(12xy^2 - 15xy^4 - 4x^3 - 10x^3y^2 + 5x^5)$
Z20	2.7761e-12	$12^{1/2}(5xy^4 - 10x^3y^2 + 1x^5)$
Z21	3.9883e-12	$14^{1/2}(6xy^5 - 20x^3y^3 + 6x^5y)$
Z22	8.6585e-13	$14^{1/2}(20xy^3 - 24xy^5 - 20x^3y + 24x^5y)$
Z23	4.8574e-14	$14^{1/2}(12xy - 40xy^3 + 30xy^5 - 40x^3y + 60x^3y^3 + 30x^5y)$
Z24	0.010423	$7^{1/2}(-1 + 12y^2 - 29y^4 + 20y^6 + 12x^2 - 59x^2y^2 + 59x^2y^4 - 29x^4 + 59x^4y^2 + 20x^6)$
Z25	-8.1681e-11	$14^{1/2}(-6y^2 + 20y^4 - 15y^6 + 6x^2 - 15x^2y^4 - 20x^4 + 15x^4y^2 + 15x^6)$
Z26	-6.5223e-06	$14^{1/2}(-5y^4 + 6y^6 + 30x^2y^2 - 30x^2y^4 - 5x^4 - 30x^4y^2 + 6x^6)$
Z27	2.8081e-11	$14^{1/2}(-1y^6 + 15x^2y^4 - 15x^4y^2 + 1x^6)$
Z28	-1.1064e-11	$4(-1y^7 + 21x^2y^5 - 35x^4y^3 + 7x^6y)$
Z29	8.3776e-13	$4(-6y^5 + 7y^7 + 60x^2y^3 - 63x^2y^5 - 30x^4y - 35x^4y^3 + 35x^6y)$
Z30	4.1794e-12	$4(-10y^3 + 30y^5 - 21y^7 + 30x^2y - 60x^2y^3 + 21x^2y^5 - 90x^4y + 105x^4y^3 + 63x^6y)$
Z31	1.3157e-11	$4(-4y + 30y^3 - 60y^5 + 35y^7 + 30x^2y - 120x^2y^3 + 105x^2y^5 - 60x^4y + 105x^4y^3 + 35x^6y)$
Z32	6.9762e-13	$4(-4x + 30xy^2 - 60xy^4 + 35xy^6 + 30x^3 - 120x^3y^2 + 105x^3y^4 - 60x^5 + 105x^5y^2 + 35x^7)$
Z33	-1.4611e-13	$4(-30xy^2 + 90xy^4 - 63xy^6 + 10x^3 + 60x^3y^2 - 105x^3y^4 - 30x^5 - 21x^5y^2 + 21x^7)$
Z34	1.0603e-12	$4(-30xy^4 + 35xy^6 + 60x^3y^2 - 35x^3y^4 - 6x^5 - 63x^5y^2 + 7x^7)$
Z35	-4.0838e-12	$4(-7xy^6 + 35x^3y^4 - 21x^5y^2 + 1x^7)$
Z36	7.8452e-13	$18^{1/2}(-8xy^7 + 56x^3y^5 - 56x^5y^3 + 8x^7y)$

FIGURE 4.32
`Zernike Report` of the design presented in figure 4.15.

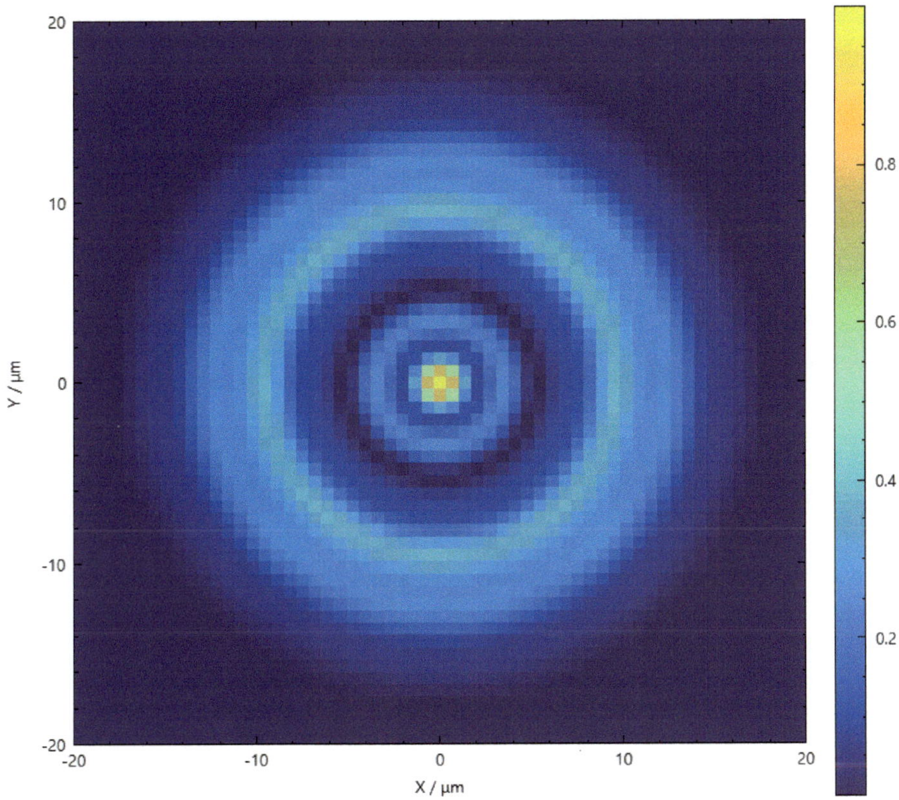

FIGURE 4.33
`Fast Fourier Transform PSF Plot` of the on-axis field of the example presented in figure 4.15. The intensity shown in the `Fast Fourier Transform PSF Plot` is normalized. It is displayed with a color plot. The coordinates of each pixel represented a coordinate in the image space. The color levels the intensity at that point.

exit pupil of the system, resulting in a map of the complex amplitude at the exit pupil's location. Figure 4.33 shows `Fast Fourier Transform PSF Plot` of the on-axis field of the example presented in figure 4.15.

4.5.6 Huygens PSF

The Huygens PSF is computed directly using the Huygens integral. To compute the Huygens PSF, a set of rays is initially directed towards the exit pupil of the system. In the subsequent step, from each ray within the pupil, a 2D fan of rays (one ray per image space pixel) is launched towards the image surface. This process corresponds to the emission of a grid of spherical

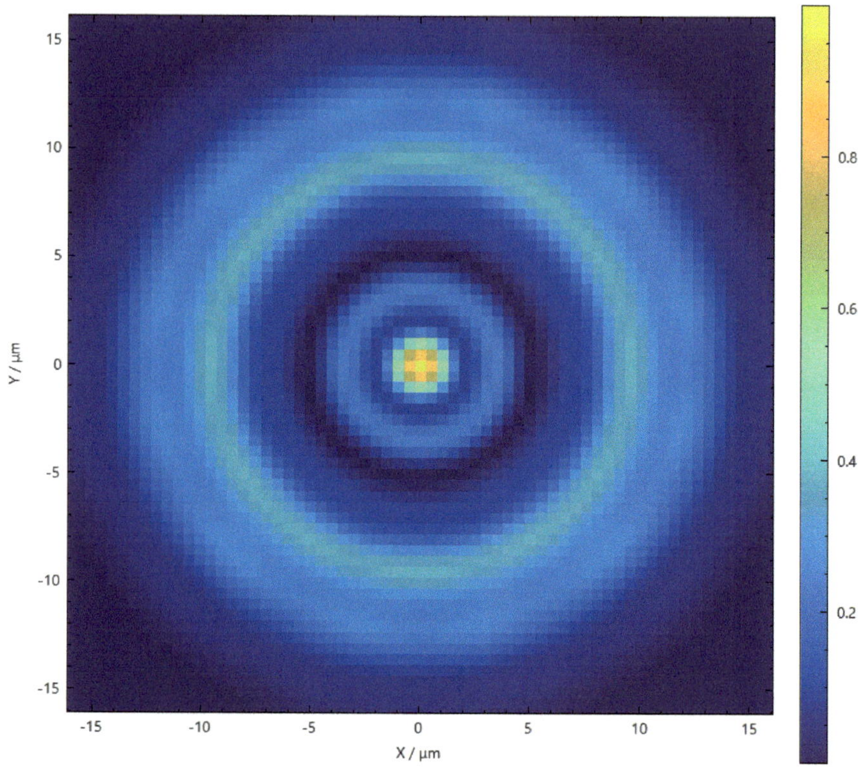

FIGURE 4.34
`Huygens PSF Plot` of the on-axis field of the example presented in figure 4.15.

wavefronts based on Huygens' principle. Upon reaching the image surface, all the rays converging at a particular pixel interfere with each other, resulting in the accumulation of complex amplitudes. It is important to remark that since the Franfouer approximation is a direct approximation of the Huygens integral. In general the Huygens PSF is more accurate than the Fast Fourier Transform PSF. But since the Franfouer approximation was done to simplify the math of the Huygens PSF, in general it takes less time to compute the Fast Fourier Transform PSF than Huygens PSF. The `Huygens PSF Plot` of the on-axis field of the example presented in figure 4.15 is in figure 4.34.

4.5.7 Geometric PSF

The `Geometric PSF Plot` depicts the density of the ray distribution from a ray trace on the image surface. It disregards diffraction effects, making it suitable for systems with significant aberrations. In such cases, where

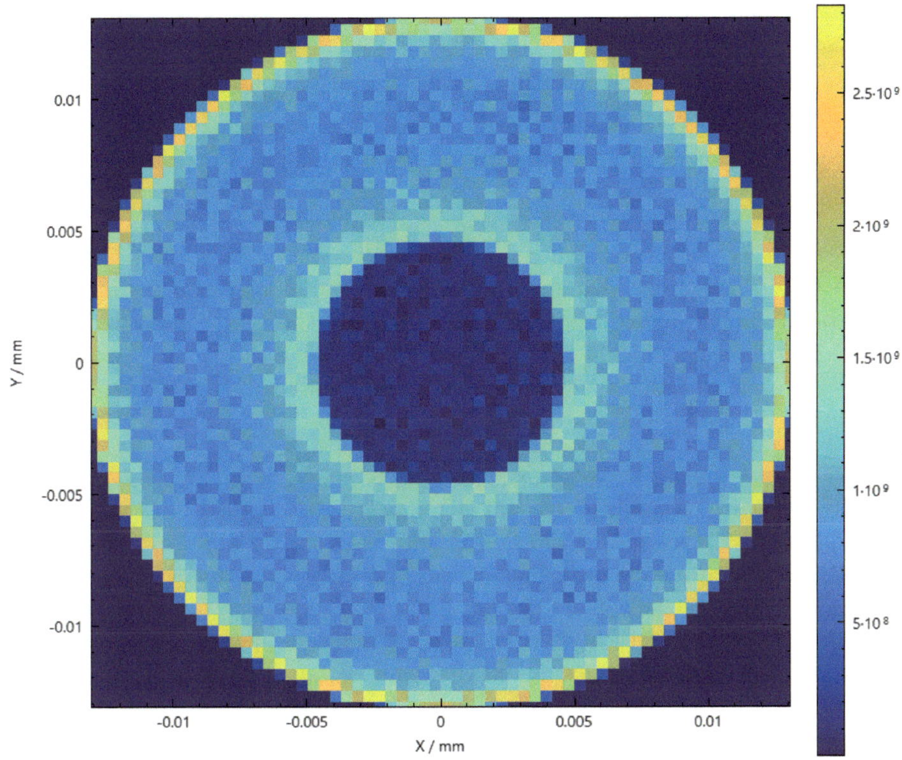

FIGURE 4.35
`Geometric PSF Plot` of the on-axis field of the example presented in figure 4.15.

aberrations outweigh diffraction effects, the `Geometric PSF Plot` is preferable over the `FFT PSF Plot` and `Huygens PSF Plot`, which may become less accurate. Figure 4.35 shows the `Geometric PSF Plot` of the example of figure 4.15.

4.6 Modulation Transfer Function

We mentioned in the past that Franfouer approximation is very useful, to simplify math computation since the diffraction integral is reduced to the Fourier transform. We also mention that the Fourier transform is super well-know by scientists and engineers in many areas. The fundamental reason is that the Fourier transform is an integral transform that converts a function into a form that describes the frequencies present in the original function.

The description of the frequencies present in the original function can be extremely useful in several areas of science and engineering. For our case, optical design, we can compute the Fourier transform of the PSF, to study its behavior in the frequency space. This computation is called optical transfer function (OTF) [50],

$$\mathrm{OTF}(v_x, v_y) = \int\int \mathrm{PSF}(x, y) e^{2\pi i v_x x} e^{2\pi i v_y y} \mathrm{d}x\mathrm{d}y \tag{4.13}$$

In the provided context, x and y represent the two orthogonal spatial coordinates of the image plane, while v_x and v_y denote the two spatial frequencies along the x and y directions, respectively.

$\mathrm{OTF}(v_x, v_y)$ characterizes how various spatial frequencies are captured or transmitted by an optical system in the image plane. It serves as a vital tool for optical engineers to elucidate how optics project light from objects or scenes onto the image plane. Remember that the $PSF(x, y)$ is in the image plane.

The optical transfer function (OTF) is inherently complex-valued (imaginary numbers) due to its nature as a Fourier transform there is an alternative function, the modulation transfer function (MTF). It disregards phase effects but is often equivalent to the OTF in many scenarios. The modulation transfer function (MTF) is formally expressed as the magnitude (absolute value) of the complex OTF [50],

$$\mathrm{MTF}(v_x, v_y) = |\mathrm{OTF}(v_x, v_y)| \tag{4.14}$$

The relative contrast is given by MTF. When a sinusoidal target is utilized, the resulting image formed by an optical system is referred to as the MTF. The MTF value is also determined by the contrast. The values of the modulation transfer function (MTF) signify the extent to which the contrast of the object is retained in the image relative to spatial frequency. Typically, the MTF diminishes as spatial frequency rises, transitioning from 1 to 0, ultimately reaching its minimum at the diffraction limit. Therefore, if the period of a normalized sinusoidal target significantly exceeds the image spot size, the MTF value virtually equals 1. However, if the period of the sinusoidal target closely matches the image spot size, the MTF value approaches 0, resulting in the image of the target becoming uniformly illuminated. Given a specific lens and field, with the image spot size remaining constant, the MTF value decreases as the spatial frequency of the target increases.

The impulse response of a well-focused optical system exhibits a three-dimensional intensity distribution, peaking at the focal plane. This distribution could be assessed by capturing a series of images while shifting the detector axially. Consequently, the three-dimensional optical transfer function is defined as the Fourier transform of this impulse response across three dimensions. Although often only a one-dimensional or two-dimensional section is utilized. `Quadoa Optical CAD` computes two one-dimensional MTF, in the x and y, directions of the image plane, respectively.

The one-dimensional MTF can be calculated as the discrete Fourier transform of the line spread function. This data is graphed against the spatial frequency data. The line spread function is the x or y directions of the image plane of the PSF.

For the case of x,

$$\mathrm{MTF}(v_x, 0) = |\int\int \mathrm{PSF}(x, 0)e^{2\pi i v_x x}\mathrm{d}x|, \tag{4.15}$$

on y,

$$\mathrm{MTF}(0, v_y) = |\int\int \mathrm{PSF}(0, y)e^{2\pi i v_y y}\mathrm{d}y|. \tag{4.16}$$

Since there are several ways to compute PSF, then there are several ways to compute the MTF. `Quadoa Optical CAD` considers the following three options, Fast Fourier Transform MTF, Huygens MTF and Geometric MTF.

- **Sequence** choose the sequence in which the MTF will be computed.
- **Filed** The MTF can be computed for all fields or a single specific field.
- **Wavelength** The MTF can be computed polychromatic or a single specific wavelength.
- **Config** Choose which configuration of the optical system. A configuration in optical design can be a set of parameters that compose an optical system for example the position of lenses, the field of view of the system, the position of an object/image, etc. So when an optical system has several configurations means that this set of parameters differs from at least one value with other configurations. More details in chapter 11.
- **Color by** Defined color of the plots.
- **Pupil sampling** The number of points the pupil sampling.
- **Image sampling** The number of points the image sampling.

4.6.1 Fast Fourier Transform MTF

The `Fast Fourier Transform Modulation Transfer Function`, or `FFT MTF Plot`, illustrates the contrast of an image plotted against the spatial frequency of the image in the image plane. The computation of the PSF employs the method of the `FFT PSF`. Which is displayed at figure 4.36.

4.6.2 Huygens MTF

The `Huygens Modulation Transfer Function` or also called `Huygens MTF` illustrates the contrast of an image plotted against the spatial frequency of the image in the image plane. The computation of the `PSF` utilizes the method of the `Huygens PSF`, which is shown in figure 4.37.

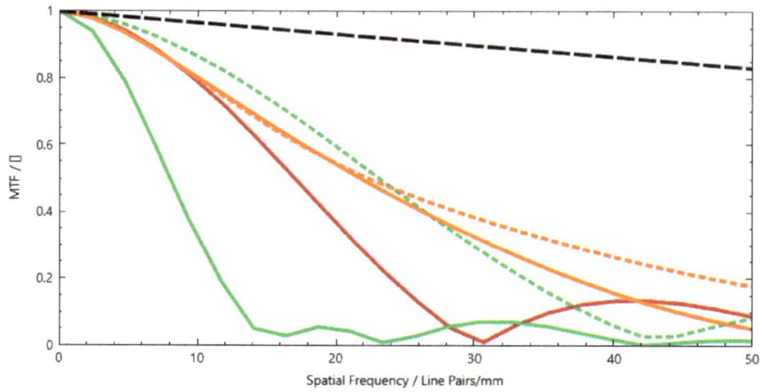

FIGURE 4.36
`FFT MTF Plot` of the on-axis field of the example presented in figure 4.15.

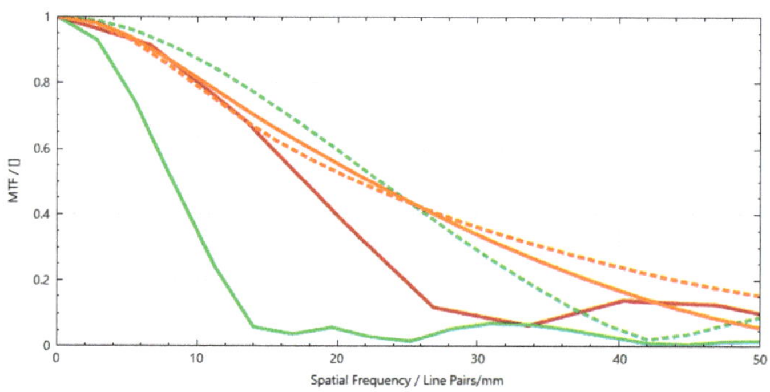

FIGURE 4.37
`Huygens MTF Plot` of the on-axis field of the example presented in figure 4.15.

4.6.3 Geometric MTF

The `Geometric Modulation Transfer Function` displays the contrast of an image plotted against the spatial frequency of the image in the image plane. The computation of the PSF employs the method of the `Geometric PSF`. It is displayed in figure 4.38.

4.6.4 Sampling of MTF in `Quadoa Optical CAD`

For the three MTF that can be computed in `Quadoa Optical CAD` the sampling of the described set in which the Fourier transform will be computed is relevant to the outcome computation. Therefore, it's important to note that the MTF may vary significantly depending on the sampling parameters, especially if the sampling is too low. Therefore, it's recommended to first examine

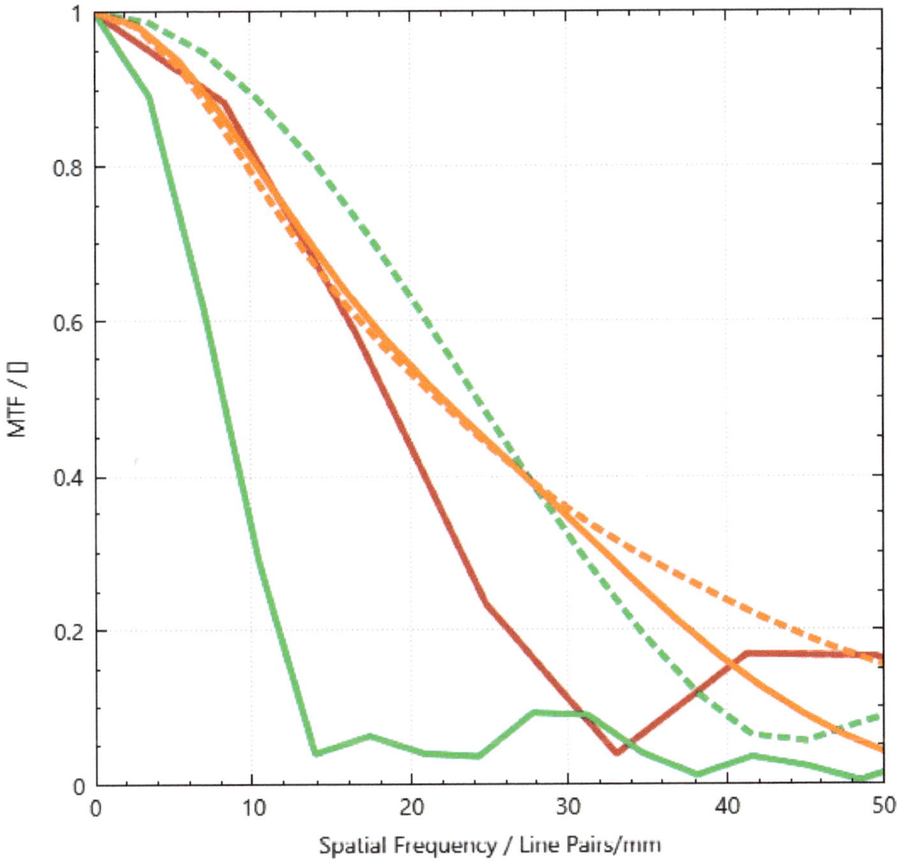

FIGURE 4.38
`Geometric MTF Plot` of the on-axis field of the example presented in figure 4.15.

the PSF for various wavelengths and fields to ensure adequate sampling for a given system with its aberrations. Only rely on the MTF result if all the PSF plots exhibit reasonable outcomes. If the sampling works for the PSF, the sampling parameters should be adjusted accordingly for the MTF plot.

4.7 Summary

In this chapter, we have studied the main form of analysis and optical system from the ray and wave perspectives. We review the main aberrations on an optical system and how to measure them in `Quadoa Optical CAD`.

Part II

Optimization of an Optical System

5

Optimization

> I do not know what I may appear to the world, but to myself I seem to have been only like a boy playing on the seashore, and diverting myself in now and then finding a smoother pebble or a prettier shell than ordinary, whilst the great ocean of truth lay all undiscovered before me.
>
> Isaac Newton

5.1 Design Requirements

Design requirements and independent variables are key components in the lens design process, each serving distinct purposes [1].

Design requirements typically encompass the performance specifications and constraints that the optical system must satisfy. These requirements may include parameters such as focal length, aperture size, image quality metrics (e.g., MTF, distortion), weight, cost, and tolerances. Design requirements serve as the targets or goals that the optical design process aims to achieve [9, 30, 31].

Independent variables, on the other hand, are the parameters that the optical designer has control over and can adjust during the design process to meet the design requirements. These variables may include the number of lens elements, curvature of lens surfaces, lens material selection, lens thicknesses, surface curvatures, and aspheric coefficients. By adjusting these independent variables, the designer can optimize the optical system to achieve the desired design requirements [5, 12, 13].

In summary, design requirements outline the performance criteria that the optical system must meet, while independent variables represent the adjustable parameters that the designer can manipulate to achieve those requirements [11, 16, 51].

DOI: 10.1201/9781003506355-5

Therefore, when working with lens data, various characteristics of the lens can be assessed using optical theory. The typical independent variables in lens design include [11, 16, 51],

- Total surface number
- Material between surfaces
- Curvature of surface
- Distance between surfaces
- Clear aperture of surface
- Surface number to become aspheric
- Shape of asphere
- Stop surface number
- Stop surface diameter, and so on.

On the other hand, the typical dependent variables in lens design include [51, 11, 16],

- Effective focal length
- Back focal length
- Distance from the object surface to the image surface
- Imaging magnification
- Ray aberration
- Root mean square optical path difference (RMS OPD)
- Modulation transfer function (MTF)
- Distortion
- Relative illumination
- Ghost image intensity
- Total track length
- Maximum surface height
- Glass weight
- Transmittance
- Glass cost
- Tolerance sensitivity, and so on.

A significant aspect of lens design involves optimization, which is defined as the process of achieving the values of independent variables that realize the target values of dependent variables.

5.2 Least Squares

The method of least squares is a parameter estimation technique in regression analysis. It involves minimizing the sum of the squares of the residuals, where a residual is defined as the difference between an observed value and the fitted value provided by a model, for each equation in the dataset. The method of least squares is the most used in the optical design to optimize a system, and `Quadoa Optical CAD` is not the exception [15, 18]. It also used the method of least squares in the built-in optimization functions. But there are other methods that will not be considered in this book [52, 53, 54, 55]

In general in optical design, the number of dependent variables exceeds the number of independent variables, and it becomes impossible to simultaneously achieve the target values of all dependent variables. In such cases, the problem typically requires the method of least squares approach to find the best fit solution. Therefore a merit function is defined. A merit function quantifies the agreement between data and the fitting model for a specific set of parameters. Typically, the merit function is minimized when the agreement between data and model is optimal. Therefore, lens designers need to optimize numerically some merit functions to meet some specifications [56, 57, 58].

Mathematically speaking, the merit function is defined,usually as,

$$\Phi = \sum_{i=1}^{n} [w_i(f_i - t_i)]^2, \tag{5.1}$$

let f_i denote the value of the i-th dependent variable, t_i represent the i-th target value, w_i signify the weight, and n represent the number of dependent variables.

The initial step in optimization involves obtaining the differential coefficients of the dependent variables with respect to the independent variables. These coefficients are commonly estimated using finite differences. In other means, we want to express the dependent variable as a linear function of the independent variable.

Thus, we use a Taylor series expansion for the dependent, using just the first terms for each independent variable,

$$f_i = f_{0i} + \sum_{j=1}^{m} a_{ij} \Delta x_j, \tag{5.2}$$

where f_{0i} is the current value of the i-th dependent f_i, Δx_j is the variation of the j-th independent x_j, and a_{ij} is the derivative of f_i respect to x_j,

$$a_{ij} = \frac{\partial f_i}{\partial x_j} \approx \frac{f_i(x_j + \delta x_j) - f_i(x_j)}{\delta x_j}. \tag{5.3}$$

Replacing equation (5.2) in (5.1),

$$\Phi = \sum_{i=1}^{n} \left[w_i \left(f_{0i} - t_i + \sum_{j=1}^{m} a_{ij} \Delta x_j \right) \right]^2, \tag{5.4}$$

Then from calculus, we know that we can find the minima, by deviating the previous expression and equalizing to zero,

$$\frac{\partial \phi}{\partial x_k} = \sum_{i=1}^{n} 2w_i^2 \left(f_i - t_i + \sum_{j=1}^{m} a_{ij} \Delta x_j \right) a_{ik} \tag{5.5}$$

reordering terms,

$$\frac{\partial \phi}{\partial x_k} = 2\sum_{i=1}^{n} w_i^2 (f_i - t_i) a_{ik} + 2\sum_{j=1}^{m} \Delta x_j \sum_{i=1}^{n} w_i^2 a_{ij} a_{ik} = 0. \tag{5.6}$$

Notice that we have an expression like the previous one for each independent variable. Therefore we can group all these equation in a more convent way, a matricidal form give by,

$$A^T \vec{y} + A^T A \Delta \vec{x} = 0, \tag{5.7}$$

where A is a matrix, which is coefficients are given by,

$$A_{ij} = w_i a_{ij}. \tag{5.8}$$

A^T stands for the transpose of A. The elements of the column vector $\vec{y}$ are,

$$y_i = w_i (f_{0i} - t_i). \tag{5.9}$$

The column vector $\Delta \vec{x}$ is the unknown of our system. It is the change we need to apply to our independent variables in order to decrease the value of Φ. It can be obtained with some matricidal algebra applied to equation (5.7),

$$\Delta \vec{x} = -(A^T A)^{-1} A^T \vec{y} \tag{5.10}$$

Since we reduce our dependent variables to the first terms of the Taylor series the result provided by the last equation is not fully accurate. For that, we need to add all the elements of the Taylor series but we cannot since the system becomes non-linear. Thus, what is usually done is that the process explained in this section is repeated in a loop until Φ converges. The number of iterations needed for the convergence highly depends on the system itself [51].

5.3 Merit Function

In Quadoa Optical CAD, The Merit Function serves as the framework for defining all parameters required for optimization. Multiple Merit Functions can be defined as needed. They are mostly based on dependent variables (Goals and Constraints) and independent variables (Quadoa Optical CAD just called variables).

In the next subsection, we are going to present the most common dependent variables and targets in Quadoa Optical CAD.

5.3.1 Ray Trace Goals and Constraints

Several Ray Trace Goals and Constraints can be set in the merit function in Quadoa Optical CAD. In this section we enlist the most common, to see all the goals and constraints referred to in the manual. In the following list for each goal and constraint, there is a small explanation of it.

- **Spot Radius RMS:** The Root Mean Square (RMS) distance of all ray positions from their respective centroids on the image surface is calculated. Each centroid is computed independently for every field. In the case of a curved image surface, the rays' z-coordinate is disregarded. It's important to note that this target cannot be utilized as a constraint.

- **Spot Radius PV:** The maximum distance of all ray positions from their centroid on the image surface is determined. The centroid is calculated individually for each field, and in the case of a curved image surface, the rays' z-coordinate is not considered.

- **Wavefront RMS:** The Root Mean Square of all rays' optical path lengths at the optical path difference reference surface is computed after subtracting their respective means for each Field/Wavelength. It's important to note that this target cannot be employed as a constraint.

- **Wavefront PV:** The Peak-to-Valley (PV) of all rays' optical path lengths at the optical path difference reference surface is measured.

- **Collimation RMS:** The collimation of all rays at the image space surface is assessed. Collimation is characterized by the angular deviation of the rays about the chief ray.

- **Lateral Chromatic Aberration:** It computes the maximum lateral distance of spots across all wavelengths in the Ray Trace on the image surface.

- **Longitudinal Chromatic Aberration:** It determines the maximum z-distance of spots across all wavelengths in the `Ray Trace` on the image surface.

- **Field Curvature:** It computes the maximum z-distance of spots for all fields in the `Ray Trace` for the selected wavelength on the image surface.

- **Distortion:** The distortion is calculated by comparing the position or angle of the chief ray for the fields defined in the `Ray Trace` with their ideal counterparts. The options available to compute the distortion are $f \tan\theta$ or $f\theta$. In practice, $f\theta$ is used for a system with a wide field of view.

- **Zernike:** It conducts a Zernike polynomial fit on the chosen wavefront at the optical path difference reference surface and establishes a single coefficient as the target value.

- **Tilt:** It conducts a Zernike polynomial fit on the selected wavefront at the optical path difference reference surface and sets a single tilt coefficient as the target value.

- **Defocus:** Performs the coefficient of Zernike polynomial for defocus.

- **Astigmatism:** Performs the coefficient of Zernike polynomial for astigmatism.

- **Coma:** Computes the coefficient of Zernike polynomial for coma.

- **Spherical Aberration:** Calculates the coefficient of Zernike polynomial for spherical aberration.

- **FFT MTF:** It determines the value of the `Modulation Transfer Function` (MTF) computed using the `Fast Fourier Transform Point Spread Function` at a specified spatial frequency.

- **Huygens MTF:** It computes the value of the `Modulation Transfer Function` (MTF) calculated using the `Huygens Point Spread Function` at a specified spatial frequency.

- **Geometric MTF:** It computes the value of the `Modulation Transfer Function` (MTF) calculated using the `Geometric Point Spread Function` at a specified spatial frequency.

5.3.2 Optical Properties

In `Quadoa Optical CAD`, multiple optical properties can be configured within the merit function. We refer to optical properties as the properties that optically describe the system. In this section, geometrical considerations or aberrations goals are not considered.

Below, we provide a list of the most common ones, along with brief explanations for each:

- **Image Space NA:** It computes the working numerical aperture of the incident rays on the image surface. This is determined by utilizing the maximum angle between the rays and their respective chief ray. It's important to note that this target always employs the on-axis field.

- **Object Space NA:** It computes the working numerical aperture of the incident rays on the object/source surface. This is determined by utilizing the maximum angle between the rays and their respective chief ray. Notice that this target always employs the on-axis field.

- **Image Space f-number:** It computes the working f-number of the incident rays on the image surface. This is determined by utilizing the maximum angle between the rays and their respective chief ray. It only works with the on-axis field.

- **Object Space f-number:** It computes the working f-number of the incident rays on the object/source surface. This is determined by utilizing the maximum angle between the rays and their respective chief ray. It only works with the on-axis field.

- **Effective f-number:** It computes the working f-number of the incident rays on the image surface for an object at infinity. This is determined by utilizing the maximum angle between the rays and their respective chief ray.

- **Effective Focal Length:** Determines the effective focal length for the sequence, assuming an object at infinity. The on-axis field is used in this computation.

- **Entrance Pupil Position:** Calculates the entrance pupil position for the sequence, defined as the z-distance from the source surface. The on-axis field is used in this computation.

- **Entrance Pupil Radius:** Calculates the entrance pupil radius for the sequence using the on-axis field.

- **Exit Pupil Position:** Calculates the exit pupil position for the sequence, defined as the z-distance from the source surface using the on-axis field.

- **Exit Pupil Radius:** Calculates the exit pupil radius for the sequence using the on-axis field.

- **Magnification:** Calculates the magnification for the sequence with on-axis field.

- **Angular Magnification:** Computes the magnification for the sequence with on-axis field.

- **Object Space Telecentricity:** Calculates the deviation from telecentricity in object space. This is determined by calculating the angle at which the chief ray of all fields defined in the `Ray Trace` is launched at the object surface. This target does not have any additional parameters.

- **Imgage Space Telecentricity:** Calculates the deviation from telecentricity in image space. This is determined by calculating the angle at which the chief ray of all fields defined in the `Ray Trace` hits the image surface. This target does not have any additional parameters.

These optical properties are commonly utilized in optimizing optical systems within `Quadoa Optical CAD`. For a comprehensive list of all goals and constraints referenced in the manual, please refer to the documentation.

5.3.3 Ray Data Targets

In `Quadoa Optical CAD`, various ray data targets can be configured within the merit function. `Ray Data Targets` refer to data associated with the ray in a given optical system. This section functions to get all the data associated with the ray, but it has the most common.

- **Ray Position:** Computes the position of the ray in the local surface coordinate system.

- **Ray Position Global:** Determines the position of the ray within the global coordinate system.

- **Ray Incident Angle:** Calculates the incident ray angle to the surface normal of a given surface.

- **Ray Exit Angle:** Calculates the exit ray angle to the surface normal of the selected surface.

- **Ray Refraction Angle:** Calculates the angle between the incoming and outgoing rays at a surface.

- **Incoming Ray Angle:** Gets the angle between the incoming ray and the $x - z$ or $y - z$ plane of the surface coordinate system.

- **Outgoing Ray Angle:** Computes the angle between the outgoing ray and the $x - z$ or $y - z$ plane of the surface coordinate system.

- **Incoming Ray Angle Global:** Calculates the angle between the incoming ray and the $x - z$ or $y - z$ plane of the global coordinate system.

- **Outgoing Ray Angle Global:** Gets the angle between the outgoing ray and the $x - z$ or $y - z$ plane of the global coordinate system.

- **OPL:** Determinate the absolute optical path length of a ray.

5.3.4 Global Goals and Constraints

All goals and constraints that are not specific to a single Sequence are segregated from the ray trace goals and constraints. These pertain to parameters associated with element geometry and materials.

- **Center Thickness:** Gets the central thickens of a given lens.

- **Edge Thickness:** Compute the edge thickness of a selected surface.

- **Center Air Gap:** Computes the air gap between two consecutive surfaces.

- **Edge Air Gap:** Computes the edge air gap between two consecutive surfaces.

- **Global Position:** Calculates the position of a surface vertex inside the global coordinate system.

- **General Distance 3D:** Compute a general distance in the optical system.

- **General Distance 1D:** compute a general distance in the optical system in one dimension.

- **Aperture Radius:** Determines the physical `Aperture` of the selected surface.

- **Surface Sag:** Determines the maximum sag of the selected surface.

- **Angle at Edge:** Calculates the angle of the surface relative to the $y-$axis of the local surface coordinate system at the edge of the aperture in the positive y direction.

- **Refractive Index:** Gets the refractive index of an element material.

- **Abbe Number:** Gets the Abbe number of an element material.

5.3.5 Math Expression

The optimization goals and constraints listed previously encompass the most common scenarios necessary for setting up a merit function. However, for certain specialized cases, it may be essential to establish a merit function that necessitates targets not currently available. For that exists `Math Expression Target`.

- **Math Expression Target:** Enables the insertion of a user-defined mathematical expression capable of computing an arbitrary equation based on the values of other targets.

5.3.6 Variables in `Quadoa Optical CAD`

Many parameters within the optimization process can be designated as `Variable`, indicating they are independent variables. See the list of independent variables in section 1 of this chapter.

The independent variables are like x_j in the second section of this chapter. In other words, they are the ones we will change to improve the `Merit Function`.

To set a parameter as a `Variable`, simply click the variable icon located next to it. Conversely, you can revert a parameter to a fixed state by clicking the icon again. When a parameter is designated as `Variable`, the icon next to it will turn green, and it will appear within the `Merit Functions Variable List` in the `System Setup Editor`.

Constraints regarding the range of Variables are specified within this `Variable List`. If multiple `Merit Functions` are defined, the activation state of a `Variable` can be adjusted independently for each function. To select the current `Merit Function` being worked on, you can utilize the `Combo-Box` positioned above the `Optical Design Editor`.

However, certain parameters, such as the radius of an aperture, cannot function as `Variables` since they do not influence the paths of rays. Consequently, the variable icon is not accessible for such parameters.

5.4 Types of Optimization in `Quadoa Optical CAD`

`Quadoa Optical CAD` offers a variety of `Optimizers` tailored for different stages of the design process. Depending on the current stage, one may find a particular optimizer more suitable for the task at hand. Typically, `Global Optimization` is employed when exploring the design space for potential solutions. The outcome of `Global Optimization` serves as an initial reference point for the design process. However, if there already exists a preliminary lens design, `Global Optimization` may not be necessary. In such cases, the `Local` and `Extended Optimizers` are often preferred choices.

The `Local Optimizer` rigorously seeks the local minimum and is known for its efficiency. Conversely, the `Extended Optimizer` not only targets the local minimum but also explores the surrounding area of the current solution. This broader exploration may involve traversing neighboring valleys in the potential landscape and engaging in non-continuous operations like substituting materials. Upon discovering a satisfactory solution, it is advisable to experiment with the `Extended Optimizer` for a period, as it frequently unlocks further performance improvements by delving into nearby areas inaccessible to the strictly local optimizer.

5.4.1 Local Optimizer

The `Local Optimizer` employs one of the `Internal Solver Types` specified in the Settings of the `Merit Function` to search for a local minimum within the design space. Its primary advantage lies in its speed, and due to its strict focus on optimizing the local minimum, it avoids unintended *jumps* to undesired solutions. Consequently, when the lens design is already finalized and requires minor adjustments without significant alterations, the `Local Optimizer` typically stands out as the optimal choice. Moreover, for interactive lens design involving iterative manual tuning and optimization, the `Local Optimizer` is often the most suitable option.

5.4.2 Extended Optimization

The `Extended Optimizer` extends its search beyond the current solution's vicinity, aiming to find a minimum within the surrounding area. Unlike the `Local Optimizer`, it doesn't confine its search strictly to the local minimum of the `Merit Function`. Instead, it can explore neighboring valleys within the design space, akin to 'climbing up the hill'. Additionally, it isn't constrained to strictly continuous changes and can perform operations like `Material Substitution`. The `Internal Solver Types` compatible with the Extended Optimizer can be specified in the Merit Function's Settings.

Typically, the Extended Optimizer is employed to run for a period when a promising solution exists for the design specification, yet the desired performance level hasn't been attained, or the user seeks to uncover an even better solution

5.4.3 Global Optimization

The Global Optimizer seeks an arbitrary design that constitutes a local minimum of the Merit Function. Given the typically high-dimensional search space for optical designs, finding a satisfactory solution may entail a considerable amount of time. Moreover, there's no assurance that the best solution uncovered represents the global optimum. Due to the sheer scale of the design space, exhaustive exploration within a practical time frame is infeasible. However, the Global Optimizer proves beneficial in identifying a promising starting point for an optical design.

5.5 Substitute Material

One of the hardest independent variable to change is the material of a lens. Since the change if not smooth if we are considering a real material.

Quadoa Optical CAD can manage substitution of material in a optical system. Any catalog material utilized for an element in the Optical Design Editor can be designated as Substitute . This designation implies that the material may potentially be substituted with another material chosen from the substitute candidates specified in the Substitute List.

It's important to note that attempts to substitute a material will only occur when using the Extended or Global Optimizer . The Local Optimizer strictly seeks the local minimum of the Merit Function and does not execute any *jumps* as would occur when switching materials. However, to optimize materials within the Local Optimizer , a Model Material with variable index of refraction and/or dispersion can be employed.

5.6 Example of a Merit Function

Now, we are going to see an example of a Merit Function defined in Quadoa Optical CAD . The Merit Function is going to be applied to

FIGURE 5.1
Ray-tracing of `Camera_Lens`.

FIGURE 5.2
`Merit Function` of `Camera_Lens`.

the design presented in figure 5.1. The path of this file is: `C:\Program Files\Quadoa\examples\Objectives\Camera_Lens`

Here we are going to call this design as `Camera_Lens`. `Camera_Lens` is composed of six lenses and five `Fields` with the following incident angles 0°, 10°, −10°, 20°, and −20°.

The `Merit Function` is presented in figure 5.2. Its main components are presented in `Trace Goals` and `Global Constraints`.

In the `Trace Goals` are four parameters. `Spot Radius RMS` to reduce the spot size. `Astigmatism` to control the mentioned aberration. `Img. Space Na` and `Magnification` whose target is not placed yet at still by default to zero. But can be set by the user in the column called `Target`. In the column called `Weight` is the weight, the constant coefficient that will be multiplied by the different of the `Value` and the `Target`.

In the `Global Constraints` are two parameters to control the `Center Thickness` of the first two lenses.

In the next chapter, we are going to go deeper into the `Merit Function` concept by optimizing a design from scratch.

5.7 Summary

In this chapter, we saw the main properties and ideas behind a `Merit Function` in lens design but also in `Quadoa Optical CAD`. The most common version of a `Merit Function` is the sum of the difference between the `Values` and `Targets`, multiplied their respective `Weigth`. This function is minimized by the least-squares algorithm.

In the next chapter, we explore more about the `Merit Function` concept by optimizing an on-axis optical design from scratch.

6

Lens Design Example 1: On-Axis Optical System

All science requires mathematics. The knowledge of mathematical things is almost innate in us. This is the easiest of sciences, a fact which is obvious in that no one's brain rejects it; for laymen and people who are utterly illiterate know how to count and reckon.

Roger Bacon

6.1 Purpose of the Chapter

The purpose of this chapter is to show how an on-axis optical system can be optimized and analyzed using `Quadoa Optical CAD`. There are several ways to do it and highly depends on the goals and constraints of the optical system but also on the experience of the designer [5, 56, 57, 58, 51].

But before we start directly with the example we are going yo to recall and introduce several concepts of how `Quadoa Optical CAD` deals and optical design, its hierarchical structure, its coordinate systems, and for sure its ray tracing.

6.2 Coordinate Systems and Hierarchical Structure

As you may have noticed at this point, in general, `Quadoa Optical CAD` deals with several coordinate systems. The most important is the global coordinate system which is the coordinate system which deals with the optical system itself, it has the global position of each element [18]. Each element is

DOI: 10.1201/9781003506355-6

made of surfaces, a singlet lens has two surfaces a doublet has three surfaces, and so on. Each surface has a local coordinate system. The center of the local coordinate system is the center of the surface if this one has no offsets.

Therefore, the primary data structure governing the optical system setup in `Quadoa Optical CAD` is organized in a hierarchical tree-like format. Typically, each element within this structure can have zero to several child items. The fundamental elements include `Surfaces`, `Lenses`, `Doublets`, `Triplets`, `Custom Elements`, and `Assemblies`, etc.

Additionally, property items are utilized to specify the characteristics of the parent item. For elements and lenses, the material item defines the material composition. In the case of kitted lenses, multiple materials are employed to define the element. As for `Surfaces`, a diverse array of property items is available.

These items encompass attributes such as form, apertures, phase function, coatings, polarization properties, or any other characteristic pertinent to the `Surfaces` they are applied. These property items can often be combined to articulate more intricate surface properties. These attributes are going to be studied in the next chapter with more detail.

6.2.1 Position Parameters

The elements in `Quadoa Optical CAD` that share position properties are `Surfaces`, `Lenses`, `Doublets`, `Triplets`, `Custom Elements`, and `Assemblies`.

Every item with positioning capabilities, like those mentioned above, functions similarly to any other item that can be positioned within the `Optical Design Editor`.

The coordinate system follows a right-handed convention, with the z-axis aligned along the optical axis, the x-axis extending towards the monitor, and the y-axis oriented vertically upwards. So, as we have mentioned before, in `Quadoa Optical CAD`, each coordinate system adheres to the right-handed convention. The z-axis, typically denoting the optical axis, extends to the right. Meanwhile, the x-axis radiates outward from the user towards the monitor, and the y-axis ascends vertically upwards.

- **Pos Z:** Specifies the position of the object along the z-direction, measured in length units,typically along the optical axis, within the local coordinate system.

- **Reference:** Determines the position type, with options including `Absolute` and `Relative`. Absolute means that the element is positioned in global coordinate space. Relative means that the element is positioned relative to another element of the optical system.

- **Off Axis:** Toggles visibility of all position parameters except Pos z. If any position parameter has a non-zero value, Off Axis is automatically activated, and parameters cannot be hidden.

- **Pos X:** Defines the object's position along the x-direction, measured in length units within the local coordinate system.

- **Pos Y:** Specifies the object's position along the y-direction, measured in length units within the local coordinate system.

- **Rot X, Y, Z:** Describes the rotation of the object around the x-axis, y-axis, and z-axis of the local coordinate system, respectively, measured in angular units.

- **Rot Center Y, Z:** Defines the center of rotation along the x-direction, y-direction, z-direction relative to the object's origin.

It is important to notice then that every item within the tree structure, representing a physical object such as a Surfaces, Lenses, Doublets, Triplets, Custom Elements, and Assemblies, possesses its individual Local Coordinate System. The position of each object is consistently described within the Local Coordinate System established by its parent object. In cases where an element lacks a parent element, its position is determined within the Global Coordinate System.

6.2.2 Absolute and Relative Positions

As we see before in the position parameters subsection, the position of any object can be specified either in absolute or relative coordinates using the Reference parameter. Quadoa Optical CAD has several options for absolute and relative Positions but here we just address the two most used by lens designers.

- **Absolute:** When coordinates are designated as Absolute, the item is positioned within the Local Coordinate System of its parent item at the specified coordinates. If the item lacks a parent, it is positioned within the global coordinate space.

- **Relative To Preceding:** If the position is defined as Relative To Preceding, the item's position is relative to the preceding item in the tree, typically the last surface of the preceding lens. The preceding item refers to the one encountered when moving upward in the tree view of the Optical Design Editor, and it doesn't necessarily have to be part of the same Element or Assembly. However, the first surface of an Element cannot be set to relative. This setting mirrors the relative positioning of surfaces in traditional spreadsheet-based optical design software. Observer that

Relative To Preceding that is very similar in its structure to other commercial software based on spreadsheets instead of CAD elements like Quadoa Optical CAD.

6.3 Sequential Raytracing

As we have mentioned before, in Quadoa Optical CAD, the process of tracing light along a sequential path is distinctly segregated from the optical system description. This is achieved through the use of the Sequence Object to define sequential raytraces. With this tool, users can establish multiple sequences within a single model, offering versatility for tasks such as ghost analysis, and modeling systems with multiple optical paths, for example, an interferometer. The incremental setup and design of complex systems, and various other applications.

6.3.1 Sequence

The Sequence serves as the fundamental object used to establish a sequential ray trace within the system. It outlines the sequential trajectory for the rays traced through the system. Additionally, properties of the rays such as Fields, Wavelengths λ, Aperture Type, and other ray characteristics are defined through the Source Object of each Sequence. The Sequence encompasses several parameters but optically speaking the most fundamental are Aperture Type, Field Type, Stop Surface, OPD Reference, and OPD Chief Reference.

6.3.1.1 Aperture Type

The Aperture Type parameter is utilized to specify the method by which the Aperture of the Sequence is defined.

- **Object Space NA**: This option defines the numerical aperture of the system in object space. The chief ray of each field typically passes through the center of the Stop Surface, unless the center for a Field is shifted using Vignetting Factors. This Aperture Type is recommended for situations where low f-numbers are crucial like microscopes.

- **Entrance Pupil**: This Option defines the radius of the entrance pupil. Similar to the **Object Space NA** option, the chief ray of each field typically passes through the center of the Stop Surface, unless the center for a Field is shifted using Vignetting Factors. This Aperture Type is recommended when getting a low f-number is paramount.

- **From Stop**: This option generates the ray distribution to precisely cover the aperture of the stop surface. It automatically calculates the numerical aperture in object space to ensure that the rays fill the entire aperture of the stop `Surface`. Consequently, it's not feasible to have a stop surface without a fixed aperture surface defined.

- **None**: This option indicates that the rays are traced completely independently of the `Stop Surface`. When this option is chosen, the `Stop Surface` parameter becomes unavailable. In this mode, users have the freedom to define all parameters such as tilt and decenter for every field independently. This `Aperture Type` is primarily intended for illumination applications rather than classical imaging optics. Therefore we are not going to study it in detail in this treatise.

6.3.1.2 Field Type

The `Field Type` parameter is employed to specify how the field is constituted.

- **Object height**: This field type defines the field based on the position of the rays on the first `Surface` within the `Sequence`. This option is exclusively available for point-like sources. We are going to explore the source types later in this chapter.

- **Angle**: This field type is determined by the angle, measured in angular units, at the first `Surface` within the `Sequence`. It is by far the most common field type.

- **Image height**: This field type is determined by the position of the chief ray on the last surface of the `Sequence` usually the image plane. Then it may be used for focal systems with specific image plane sizes, for example, the sensor size of a camera.

6.3.1.3 Stop Surface

`Stop Surface` parameter is utilized to designate which surface within the `Sequence` serves as the `Stop Surface`. Typically, the chief ray of each field passes through the center of the `Stop Surface`, unless the central position for a `Field` is adjusted using `Vignetting Factors`.

This parameter becomes unavailable if the `Aperture Type` is set to **none**, as this type of source operates independently of a stop surface definition.

6.3.1.4 OPD Reference

The `OPD Reference` parameter specifies the method used to compute the `Optical Path Length` and `Optical Path Difference` of a system. For the purpose of this book, the most important are:

- **Exit Pupil:** This setting is typically preferred for imaging lenses or similar systems where rays converge sharply onto the image surface. When `Exit`

`Pupil` is chosen, the rays are traced to the `Exit Pupil`, and the `Optical Path Difference` is represented by the position of the pupil. Here, the `Optical Path Difference` reflects the deviation of the optical path from a perfect spherical wavefront within the exit pupil, aiming for an aberration-free focus.

- **Abs over Img**: This setting is suitable for simulating wavefront sensors, interference, or other systems where the primary focus is to emulate the wavefront at the last surface. In the plots, the lateral coordinates (x, y) represent the position of the ray on this surface and the `Optical Path Difference` is calculated as the absolute `Optical Path Length` of the ray minus the `Optical Path Length` of the chief ray. However, if the system is focal or there exists a caustic zone in the vicinity of the last surface, this may result in self-intersecting `Optical Path Difference Plots` or `Wave Front Plots`. This occurs because, in the presence of caustics, the definition of a wavefront is no longer physically meaningful.

6.3.1.5 OPD Chief Reference

The `Optical Path Difference Chief Reference` parameter is utilized to specify the reference ray that defines the location of the exit pupil.

- **Primary:** When this option is chosen, the chief ray of the primary wavelength is employed to determine the location of the exit pupil. The primary wavelength is the wavelength in which `Quadoa Optical CAD` will make most of its computations to describe the properties of the optical system.
- **Individual:** When this feature is pressed, the chief ray of each wavelength is utilized to establish the location of the exit pupil. This results in the computation of a separate exit pupil for each wavelength.

6.3.2 Auto Sequence

The `Auto Sequence` represents the default type of `Sequence`. It's particularly suitable for various classical optical systems such as objective lenses, imaging optics, and beam shaping systems. In scenarios where only a single optical path of the model is of primary interest and there are no beam splitters or double passes through lenses, the `Auto Sequence` is typically the preferred choice.

The `Auto Sequence` traces rays starting from the first `Surface Surface` in the `Optical Design Editor`, proceeding from top to bottom until reaching the last `Surface Surface`. At each `Surface`:

- If no `Coating` is presented, the default action is transmission. We are going to study `Coatings` later in this book.

- If the `Coating` has a reflectance exceeding 50%, the default action is reflection.
- If the `Surface` includes `Phase Property Objects`, the first order of diffraction is utilized.
- To modify the default behavior, the `Auto Sequence` needs to be converted to a `User Sequence` .

6.3.3 User Sequence

The `User Sequence` serves to define a sequential ray trace. However, unlike the `Auto Sequence`, it offers the additional capability to the user to specify the order of surfaces through which the rays are traced in the `Surface List`.

In a `User Sequence`, the `Surface List` is employed to determine the `Sequence` in which the `Surfaces` are traversed during the sequential ray trace. Additionally, it allows for the specification of actions to be taken at each `Surface Surface`, such as whether the rays are transmitted or reflected, which diffraction order or polarization element is applied, and so on. The rays are traced from the top to the bottom of the `Surface List`. To include a `Surface` in the `Surface List`, you can add it using the appropriate function or interface within the software.

Note in `Stop Surface`: For `User Sequence`, the `Stop Surface` can alternatively be directly specified in the `Surface List`. To modify the `Stop Surface`, right-click on the `Surface` you wish to designate as the `Stop Surface` and select `Make Stop`.

6.3.4 Source

The source: Within every `Sequence`, the `Source object` is the initial child item. It serves the purpose of delineating the properties of the traced rays. These properties encompass `Wavelengths`, `Fields`, `Polarization`, `Apodization`, as well as the `Distribution` of the rays.

- **Point Source**: In this scenario, all rays originate from a single point located at the first `Surface`. This option is commonly employed in the design of imaging systems, particularly when the object is situated at a finite distance, or when simulating a perfect point source.
- **Plane WF**: This option is suitable for imaging systems that capture an object located at an infinite distance, or more generally, to simulate a point source that is infinitely far away. In such cases, the resulting wavefront is rendered as a plane, this will collimated rays.

- **WF from Surface**: This option allows for the simulation of a wavefront with an arbitrary shape, where the form of the wavefront is determined by the Form of the first `Surface` in the `Sequence`. It is important to remember that due to the Fermat principle, the rays are orthogonal to the wavefront.

6.3.5 Distribution

The distribution and quantity of the traced rays are determined by the `Distribution Object`. This object is consistently positioned as the first child of the `Source object` within every `Sequence`. This book only deals with imaging systems therefore we will only focus on the main `Available Ray Distributions`.

- **Single Ray**: Generates a single ray, and the coordinates of the ray at the `Stop Surface` can be defined by the parameters pupil x and pupil y.
- **Y-Fan** and **X-Fan**: Generates a fan of rays spread out in the y-direction and x-direction, respectively.
- **XY-Fan**: Generates two fans of rays spread out in both the x- and y-directions.
- **Ring**: Generates a single ring of rays that lie on a circle at the border of the aperture.
- **Equal Rings**: Generates a set of concentric rings, with the spacing of the rays on each ring equal to the distance between the rings.
- **Hex Grid**: Generates a hexagonal honeycomb pattern.
- **Rect Grid**: Generates an equally spaced rectangular grid in both the x- and y-directions.
- **Random**: Generates random rays evenly distributed over the entire aperture.

The main parameter for all the options is for sure, the number of rays.

6.3.6 Apodization

`Apodization` is utilized to define the intensity distribution over the pupil by adjusting the flux of the rays based on their positions within the pupil. It affects any computation involving the flux of a ray, such as `Point Spread Functions`, and `MTF Analysis`.

The types of apodization can be divided into main types: constant and Gaussian-related.

- **Constant** type generates each ray with the same flux.
- **The Gaussian-related apodizations** express the apodization in terms of Gaussian distribution. The reader is invited to read the manual for a complete description of the Gaussian-related apodizations. In the example of this book, we are going to consider only constant apodization.

6.3.7 Wavelength

λ The `Wavelength Container` is the parent item that holds the individual wavelength definitions of a `Source Object`. In this container, the primary wavelength can be defined. Recalling, the primary wavelength definition. It is the wavelength in which `Quadoa Optical CAD` will make most of the computation to get the optical properties of the optical system.

6.3.8 Fields

The `Field object` is utilized to define a field coordinate of the rays. A field object has three main parameters, tilts, decenter, weight, and `Vignetting Factors`.

- The tilt parameters, `Tilt X` and `Tilt Y`, are employed to define the tilt of the `Field` in degrees when the `Field Type` is set to `Angle` These parameters are not accessible for other `Field Types`.
- The decenter parameters, `Decenter X` and `Decenter Y`, are utilized to specify the object height of the Field when the `Field Type` is `Obj. Height`, or the image height when the `Field Type` is `Img. Height`, measured in length units. These parameters are not available for other `Field Types`.
- The `Weight` parameter determines the relative importance of a `Field` during system optimization. However, out of optimization procedures, this parameter does not influence the system.

Vignetting manifests when light rays of a `Sequence` fail to traverse the entire lens system to reach the sensor, often obstructed by the edges of individual lens elements or mechanical stops. This obstruction of rays can be deliberate or inadvertent, and in certain scenarios, it may be inevitable. Vignetting is commonly observed in or around lower f-numbers, short focal length lenses, or lenses aimed at achieving higher resolutions at a lower cost.

The `Vignetting Factors` are so important and common in many situations that we are going to describe them in more detail outside the previous point list.

The `Vignetting Factors` are instrumental in adjusting the ray distribution for each `Field` independently, which holds significant importance for optimization purposes. In instances where a `Field` encounters vignetting from certain apertures within the system, the absence of `Vignetting Factors` would limit the number of rays that can be traced. Consequently, this results in inadequate sampling of the pupil, leading to subpar optimization outcomes. By utilizing the `Vignetting Factors`, it becomes feasible to compress and/or shift the ray distribution, thereby enhancing the sampling of the vignetted pupil within the `Field`.

The `Vignetting Factors` are:

- `Vig. X and Vig. Y`: Adjustment of the ray distribution in the x-direction and the y-direction, respectively. By default, the value is set to 1.0, having no impact on the ray distribution. `Vig. X` parameter alters the x-coordinate of the ray, scaling it in the x-direction and `Vig. Y` parameter alters the y-coordinate of the ray, scaling it in the y-direction.

- `Vig. Dec. X and Vig. Y`: The default value of this parameter is 0.0, resulting in no effect. Vig. `Dec. X` parameter shifts the chief ray on the `Stop Surface` in the x-direction by the specified value. `Vig. Dec. Y` parameter shifts the chief ray on the `Stop surface` in the y-direction by the chosen value.

6.4 Polarization Raytracing

In addition to the fundamental `Sequential Raytracing` capabilities, `Quadoa Optical CAD` possesses the advanced ability to conduct `Polarization Raytracing`. With `Polarization Raytracing`, users can specify the polarization state of the rays using the `Polarization Parameters` within the `Source object` for each `Sequence`. The initial polarization state can be defined using either a `Jones Vector` or a `Stokes Vector`.

The `Jones Vector` is a mathematical representation used to describe the polarization state of light. It's particularly useful in analyzing and manipulating polarized light in various optical systems. The `Jones vector` is typically represented as a column matrix, vector [38, 39],

$$\vec{\mathbf{J}} = \begin{pmatrix} E_x \\ E_y \end{pmatrix} \tag{6.1}$$

where E_x and E_y represent the complex amplitudes of the electric field components of the light wave along two orthogonal axes, usually chosen to be the x and y directions. These complex amplitudes represent both the magnitude and phase of the electric field components.

The `Jones Vector` can undergo transformations when interacting with optical elements such as polarizers, waveplates, and birefringent materials. These transformations are often represented by Jones matrices, which are used to calculate the resulting Jones vector after the interaction [38, 39].

The `Stokes Vector` is another mathematical representation used in optics to describe the polarization state of light. It provides a more comprehensive description compared to the Jones vector, as it takes into account not only the amplitudes but also the phases and relative orientations of the electric field components [38, 39].

The Stokes vector is typically represented as a column matrix

$$\vec{\mathbf{S}} = \begin{pmatrix} S_0 \\ S_1 \\ S_2 \\ S_3 \end{pmatrix}, \tag{6.2}$$

where S_0, S_1, S_2, and S_3 are the four Stokes parameters. These parameters represent different aspects of the polarization state of light.

S_0 represents the total intensity of the light. S_1 represents the difference in intensity between horizontally and vertically polarized light. S_2represents the difference in intensity between +45° and -45° linearly polarized light and S_3 represents the difference in intensity between right-handed and left-handed circularly polarized light.

The Stokes parameters provide a physically meaningful description of the polarization state of light. They quantify various aspects of polarization, such as the degree of polarization, the orientation of the polarization ellipse, and the presence of elliptical or circular polarization [38, 39].

In `Quadoa Optical CAD`, when a polarized source is defined, all intensity data for analysis is derived from the electrical field of the rays. Any interactions between rays and surfaces are computed while considering the relevant polarization mechanisms. By default, the Fresnel equations are applied to compute these interactions. However, if a `Coating` or a `Polarization Element` is specified for the `Surface` instead of the default Fresnel equations for interfaces like air/glass or glass/glass, the specified `Polarization Element` or `Coating` dictates the mechanism for computation.

6.4.1 Polarization Coordinate Systems

All operations in Jones calculus are conducted within the plane of polarization of the rays, perpendicular to the direction of propagation $\vec{\mathbf{k}}$. As all Jones operations within a `Sequence` are determined per `Surface` (whether it's the source, interaction, or analysis surface), the local coordinate system of a ray $(\vec{\mathbf{x}}_j, \vec{\mathbf{y}}_j, \vec{\mathbf{z}}_j)$ must be derived from the coordinate system of the surfaces given by $(\vec{\mathbf{x}}, \vec{\mathbf{y}}, \vec{\mathbf{z}})$. Various options are available to accomplish this,

- **X-Axis Reference:** The local x-axis of the ray resides in the plane formed by the vectors $\vec{\mathbf{x}}$ and $\vec{\mathbf{k}}$, perpendicular to $\vec{\mathbf{k}}$.

$$\vec{\mathbf{y}}_j = \vec{\mathbf{k}} \times \vec{\mathbf{x}}, \quad \vec{\mathbf{x}}_j = \vec{\mathbf{y}}_j \times \vec{\mathbf{k}}, \quad \vec{\mathbf{z}}_j = \vec{\mathbf{k}} \tag{6.3}$$

- **Y-Axis Reference:** The local y-axis of the ray resides in the plane formed by the vectors $\vec{\mathbf{y}}$ and $\vec{\mathbf{k}}$, perpendicular to $\vec{\mathbf{k}}$.

$$\vec{\mathbf{x}}_j = \vec{\mathbf{y}} \times \vec{\mathbf{k}}, \quad \vec{\mathbf{y}}_j = \vec{\mathbf{k}}_j \times \vec{\mathbf{x}}_j, \quad \vec{\mathbf{z}}_j = \vec{\mathbf{k}} \tag{6.4}$$

- **S/P:** The local x-axis of the ray is perpendicular to the plane of incidence, which is defined by the surface normal $\vec{\mathbf{n}}$ and the ray direction $\vec{\mathbf{k}}$. Meanwhile, the local y-axis of the ray lies within the plane of incidence.

$$\vec{\mathbf{x}}_j = \vec{\mathbf{k}} \times \vec{\mathbf{n}}, \quad \vec{\mathbf{y}}_j = \vec{\mathbf{k}} \times \vec{\mathbf{n}}, \quad \vec{\mathbf{z}}_j = \vec{\mathbf{k}} \tag{6.5}$$

6.4.2 Polarized Source

As we mentioned before A polarized source can be described using either a `Jones Vector` or a `Stokes Vector`. These vectors are applied in a coordinate system perpendicular to the ray's direction of propagation $\vec{\mathbf{k}}$ The coordinate system for each ray is derived from the local surface coordinate system $(\vec{\mathbf{x}}, \vec{\mathbf{y}}, \vec{\mathbf{z}})$. `Quadoa Optical CAD` provides various options for defining these local ray coordinates.

6.4.2.1 Jones Vector

Multiple options exist for defining the polarization of a source using a normalized `Jones Vector` $|\vec{\mathbf{J}}| = 1$.

- **Linear:** Linear polarization with orientation determined by an angle α relative to the x-axis.

$$\vec{\mathbf{J}} = \begin{pmatrix} \cos\alpha \\ \sin\alpha \end{pmatrix} \tag{6.6}$$

- **Circular:** Circular polarization left-handed

$$\vec{\mathbf{J}} = \frac{1}{\sqrt{2}} \begin{pmatrix} 1 \\ i \end{pmatrix} \tag{6.7}$$

and circular polarization right-handed

$$\vec{\mathbf{J}} = \frac{1}{\sqrt{2}} \begin{pmatrix} 1 \\ -i \end{pmatrix} \tag{6.8}$$

- **Jones Complex:** The real and imaginary components of the Jones vector are input directly,

$$\vec{\mathbf{J}} = \begin{pmatrix} E_x + iE_{c_x} \\ E_y + iE_{c_y} \end{pmatrix}, \tag{6.9}$$

where E_x and E_y are the real part, and E_{c_x} and E_{c_y} are the complex part.

- **Jones Phase:** Jones vector is defined by the amplitude and phase of the components,

$$\vec{\mathbf{J}} = \begin{pmatrix} A_x e^{i2\pi p_x} \\ A_y e^{i2\pi p_x} \end{pmatrix}, \tag{6.10}$$

where A_x and A_y are amplitudes, and p_x and p_y are phases.

- **Random:** The polarization state for each ray is randomly generated. This results in a general elliptical polarization where the ellipticity, the orientation of the ellipse, as well as the handedness are completely random for each ray.

6.4.2.2 Stokes Vector

`Quadoa Optical CAD` can also define a polarized source with the `Stokes Vector`,

$$\vec{\mathbf{S}} = \begin{pmatrix} S_0 \\ S_1 \\ S_2 \\ S_3 \end{pmatrix} = \begin{pmatrix} I \\ Q \\ U \\ V \end{pmatrix}, \tag{6.11}$$

where I, Q, U, and V are the four Stokes parameters. I denotes the total intensity of the light. Q indicates the disparity in intensity between horizontally and vertically polarized light. U signifies the difference in intensity between light polarized at +45° and -45°, and V denotes the discrepancy in intensity between right-handed and left-handed circularly polarized light [38, 39] .

Therefore, the special case of the unpolarized source is given by,

$$\vec{\mathbf{S}} = \begin{pmatrix} 1 \\ 0 \\ 0 \\ 0 \end{pmatrix}. \tag{6.12}$$

6.5 Example of Design from Scratch

Let's start with our first example of an optimization in `Quadoa Optical CAD`. This example can be found in the example file presented once `Quadoa Optical CAD` is successfully installed on your computer. The path file is:

`C:\Program Files\Quadoa\examples\Optimization\Optimize_5Let`

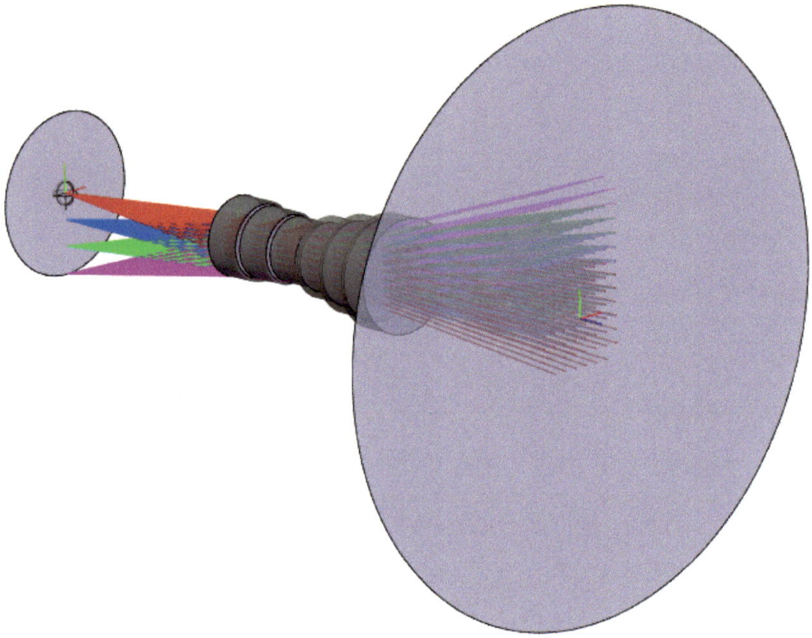

FIGURE 6.1
Ray-tracing of `Optimize_5Let`. The source has three different wavelengths, 500nm, 600nm, and 700nm. The lenses are made by glass N-F2.

From now on we will call this design as `Optimize_5Let`. This file contains a design made by five flat lenses. The lenses are made by glass called `N-F2` which can be found in the `SCHOTT Catalog`.

The source in this file contains three different wavelengths, which are 500nm, 600nm, and 700nm. Also, it has four `Fields` with angles of incident given by 0°, 5°, 10°, 15°, respectively.

The ray-tracing of `Optimize_5Let` is presented in figure 6.1 and vertical cut in figure 6.2. The file has already a built-in merit function. The merit function is divided in `Trace Goals` and `Global Constrains`. In `Trace Goals`, there are only two parameters. The first one is to reduce the RMS spot radius and an additional to control distortion. The `Global Constrains` are shown in figure 6.3. Please take a look, and spend the necessary time to understand each of the parameters that control the edge thickness, center air gap, and edge air gap. These operands have the function that the lenses do not cross or touch each other the values of tolerance between the distance of edges or gaps are

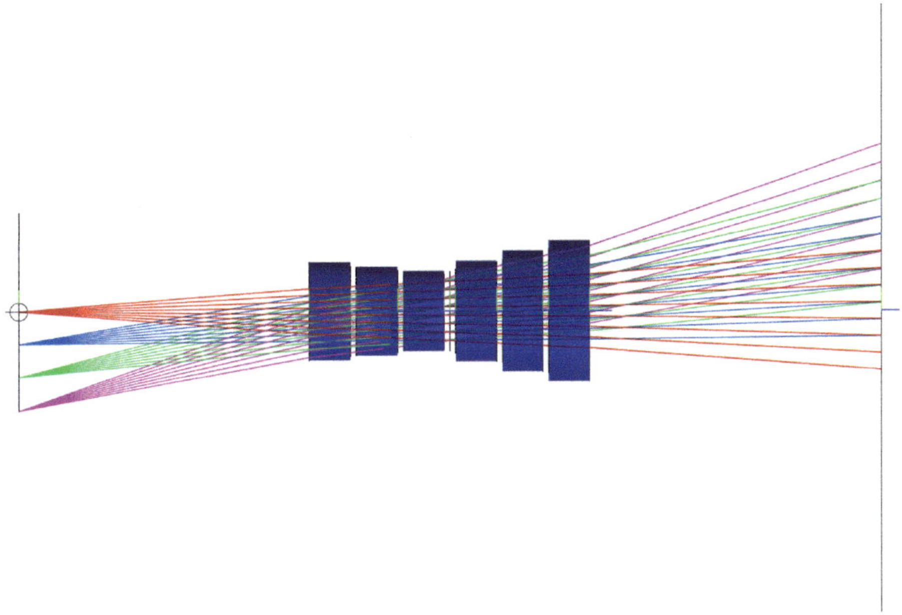

FIGURE 6.2
Ray-tracing of `Optimize_5Let`. The source has three different wavelengths, 500nm, 600nm, and 700nm. The lenses are made by glass N-F2.

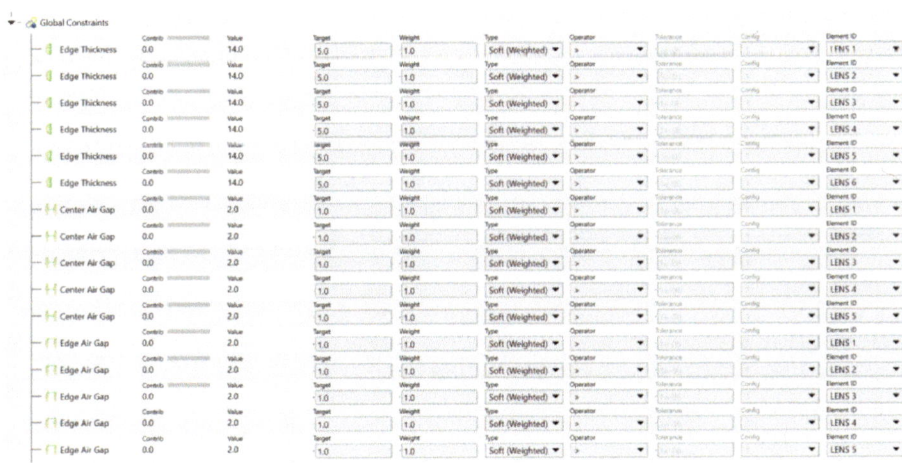

FIGURE 6.3
`Global Constrains` of `Optimize_5Let`.

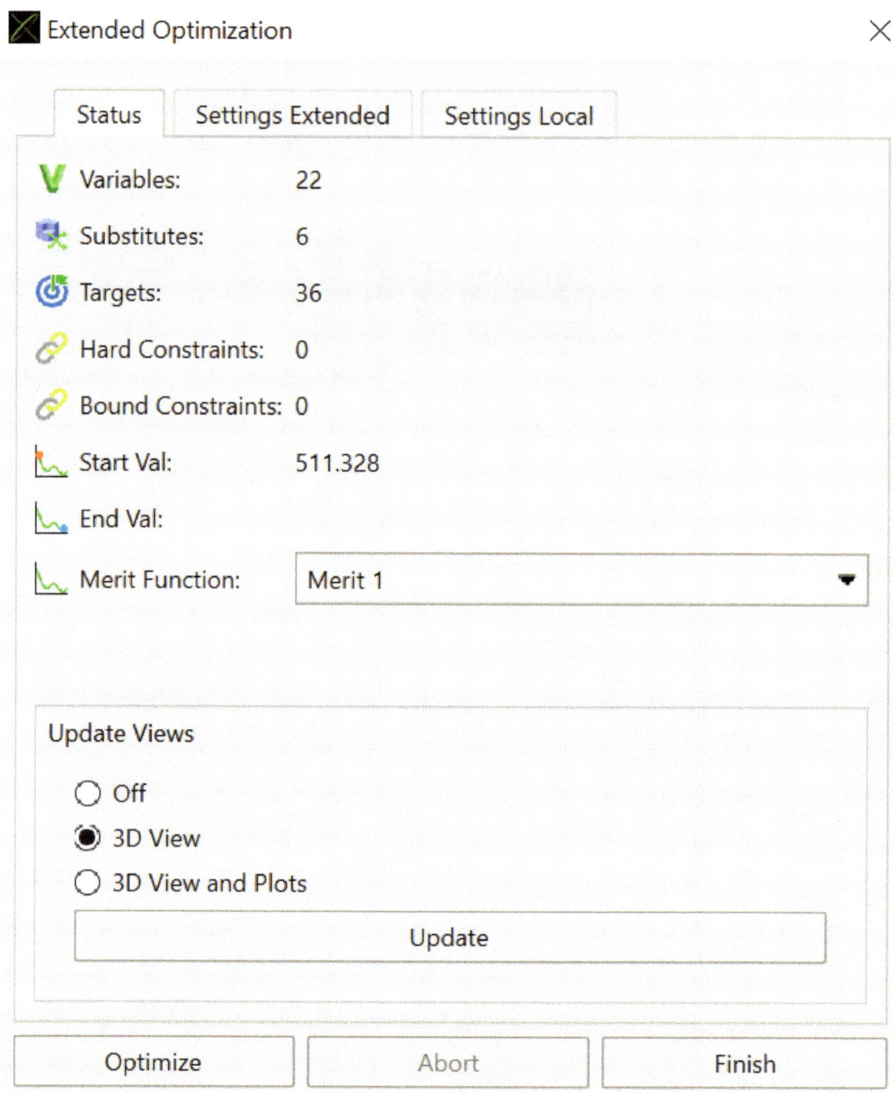

FIGURE 6.4
`Extended Optimization` of `Optimize_5Let`.

given by the user. The edge thickness should not be less than 5mm, while the other two operands should be more than 1mm.

Let's optimize it with `Extended Optimization`. If we press that button appears the following figure, figure 6.4. From it, we note that in total, there are 22 variables and 6 substitute materials. The total targets of the merit function are 36. In it, we press the button `Optimize`.

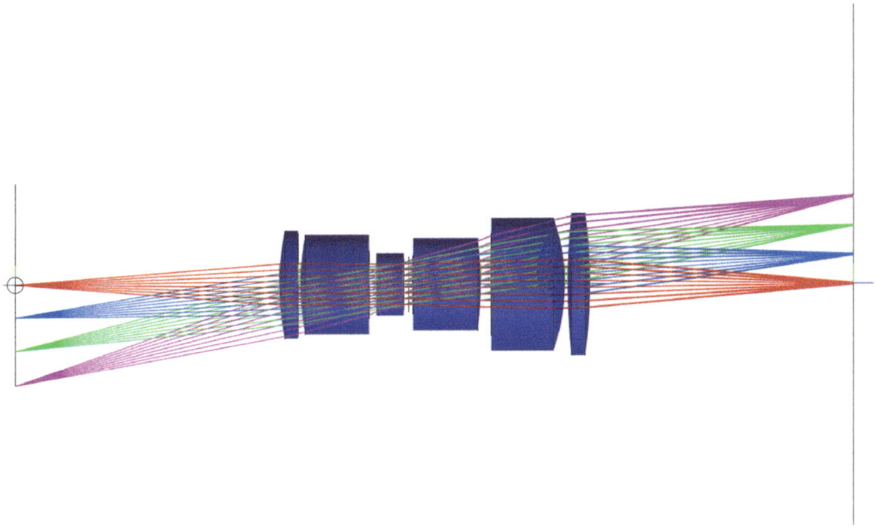

FIGURE 6.5
Ray-tracing of `Optimize_5Let` after a first optimization. The system now focuses on all the fields.

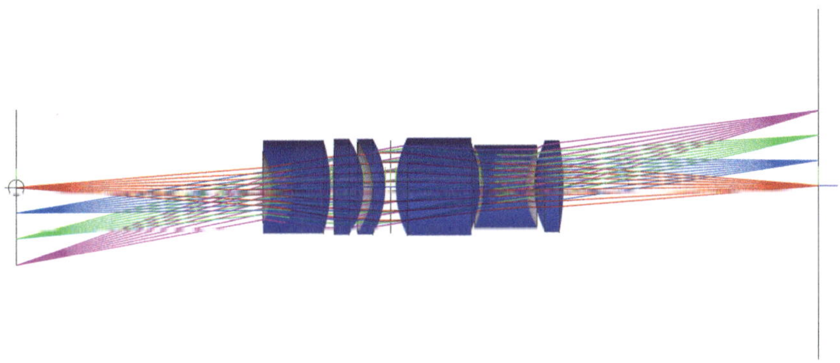

FIGURE 6.6
Ray-tracing of `Optimize_5Let` after a second optimization.

If we stop the optimization, we can see that the system has changed a lot, now the system is focusing, on the fields. This process happened so fast, just a question of second. The ray-tracying is presented in figure 6.5.

Let's continue optimizing the system, so we go back and `Extended Optimization` panel. We press `Optimize` once more. We wait for more seconds and the results are presented in figure 6.6.

After the second optimization, the system has changed significantly. Figure 6.7 shows the change in terms of the data that constitute the system. In the

Surface List

Radius	Thickness	Material	Index (Abs)	Index (Rel)	Aperture
inf	100.0	Air	1.00027	1.0	30.8364
-29.7494	22.7371	N-SF66	1.94928	1.94875	14.0875
-63.1728	0.980662	Air	1.00027	1.0	18.3763
321.472	11.9794	FK5HTi	1.49184	1.49143	18.722
-34.5748	3.57573	Air	1.00027	1.0	18.9304
-29.9889	4.97993	N-LAK34	1.73733	1.73685	17.9063
-38.688	2.0	Air	1.00027	1.0	18.8622
inf	2.0	Air	1.00027	1.0	18.2801
44.1483	32.6377	N-PSK53A	1.62405	1.62361	19.2498
-45.3388	1.37197	Air	1.00027	1.0	16.9157
-40.9508	16.6307	N-LAF33	1.79663	1.79613	16.3307
32.3889	3.00767	Air	1.00027	1.0	15.8301
39.7271	11.4566	N-FK5	1.49186	1.49145	17.3891
-98.8542	100	Air	1.00027	1.0	18.1189
inf					131.532

FIGURE 6.7
Table of `Optimize_5Let` after a second optimization. Radii, thickness, positions, apertures, and materials have been changed.

table, presented in the mentioned figure, appears the new radii, thickness, positions, and materials of the new system. Also, it presents the new aperture sizes of all the surfaces. We can notice that new materials have appeared like `N-SF66`, `FK5HTi`, `N-LAK34`, `N-PSK53A`, `N-LAF33`, and `N-FK5`.

It seems that this optimization has changed so much the design that we can analyze it. Let's start with `Spot Diagram Plot`. The `Spot Diagram Plot` is presented in figure 6.8. In it is presented the spot of the four `Fields`. The first row, from left to right, shows the spot of the on-axis field and the field with 5°. The second row, from left to right, shows the spots for the 10° and 15°. The plots are color-coded in terms of wavelength. Red, blue, and green for 500nm, 600nm, and 700nm.

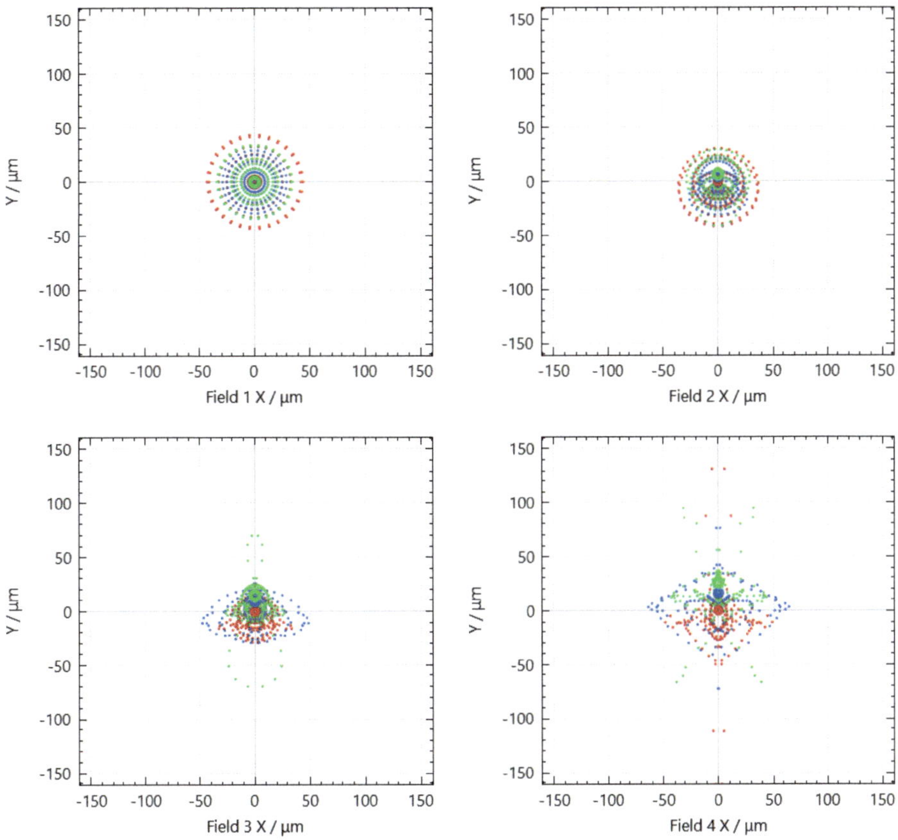

FIGURE 6.8
`Stop Diagram Plot` of `Optimize_5Let`. From left to right, the first row shows the spot of the on-axis field and the field with 5°. From left to right, the second row shows the spots for the fields 10° and 15°. The plots are color-coded in terms of wavelength. Red, blue, and green for 500nm, 600nm, and 700nm. The on-axis field suffers from spherical aberration. The off-axis fields suffer from coma and astigmatism. From the plots we can see the aberration increases with the field of view.

We can notice that the on-axis field suffers from spherical aberration. The sizes of the stops differ considerably too. In the case of the field with 5° angles of incidence, it suffers from coma. The other two off-axis fields suffer from coma and astigmatism. The spot for each wavelength differs from its position to the position of other wavelengths. So we shall study the `Longitudinal Aberration Plot`, which is presented in figure 6.9.

Figure 6.9 shows how the defocus appears which respect to the wavelength. The vertical axis shows the wavelengths from 500nm to 700nm, scanning all

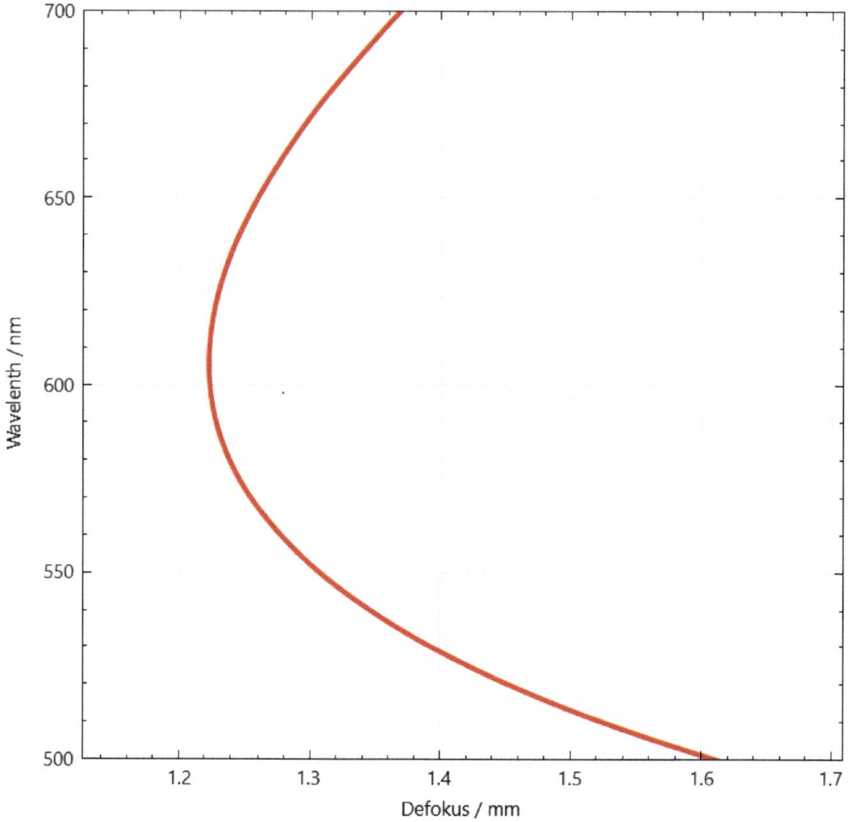

FIGURE 6.9
`Longitudinal Aberration Plot Fields` of `Optimize_5Let` after the second optimization. The vertical axis shows the wavelength. The horizontal axis shows the defocus.

the wavelengths between the fields in the source of `Optimize_5Let`. The horizontal axis shows the total amount of defocus in millimeters. The magnitude of the defocus can be as big as 1.6mm. This information is something we should consider for our next optimization.

The `Lateral Aberration Plot` is displayed in figure 6.10. The primary wavelength (500nm) is taken as the reference, so it does not have any deviation. From the plot, we can observe that the lateral aberration, in this example, increases with the field of view, and its maximum can be approximately 0.025 mm. This is important information, to be considered in our next merit function.

The distortion does not seem to be a problem in this design, this is confirmed with the `Distortion 1D` plot. The plot, in figure 6.11, shows that the distortion is below 3%. This value can be considered acceptable for most

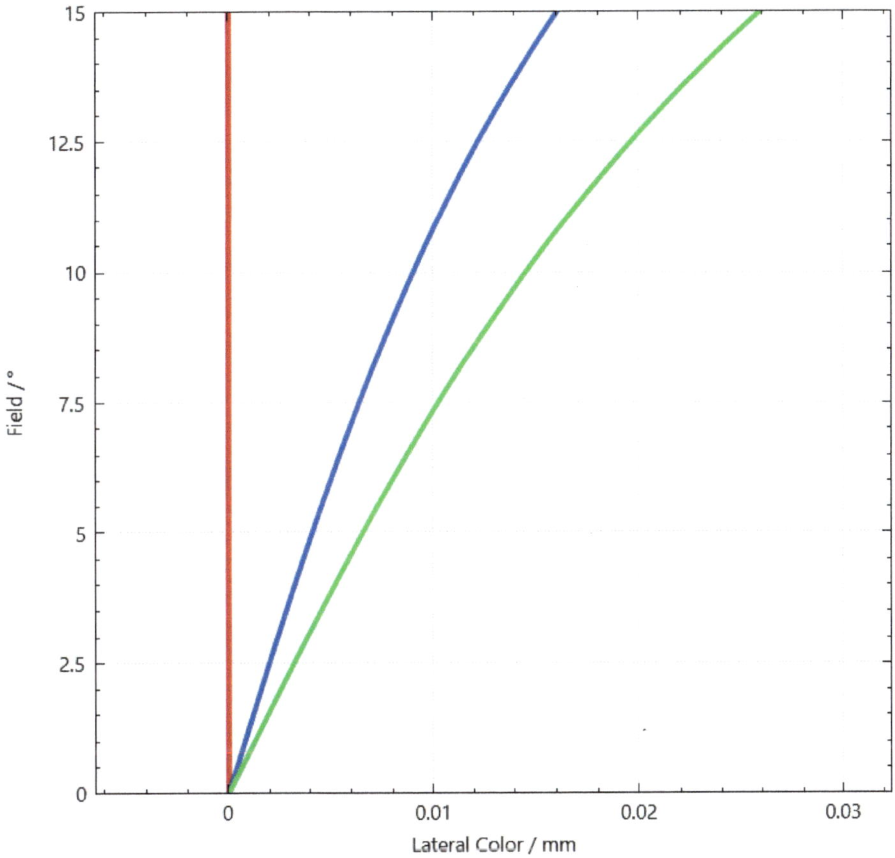

FIGURE 6.10
`Lateral Aberration Plot` of `Optimize_5Let` after the second optimization. The primary wavelength is 500nm. The lateral aberration increases with the field of view and its maximum can be around 0.025 mm.

optical systems. So, we are not going to focus too much on the distortion for our next merit function.

From our previous analysis, we can modify the merit function in order to improve the performance of `Optimize_5Let`. The new permit function's `Trance Goals` is displayed in figure 6.12. The `Global Constrains` are not modified, it was not necessary. Notice that in the `Trance Goals`, it is repeated twice in the operand `Lateral Aberration Plot` . The two operands have the same weight, which is one. We can remove one of them and double the weight of the other to two and we will have the same contribution to the merit function.

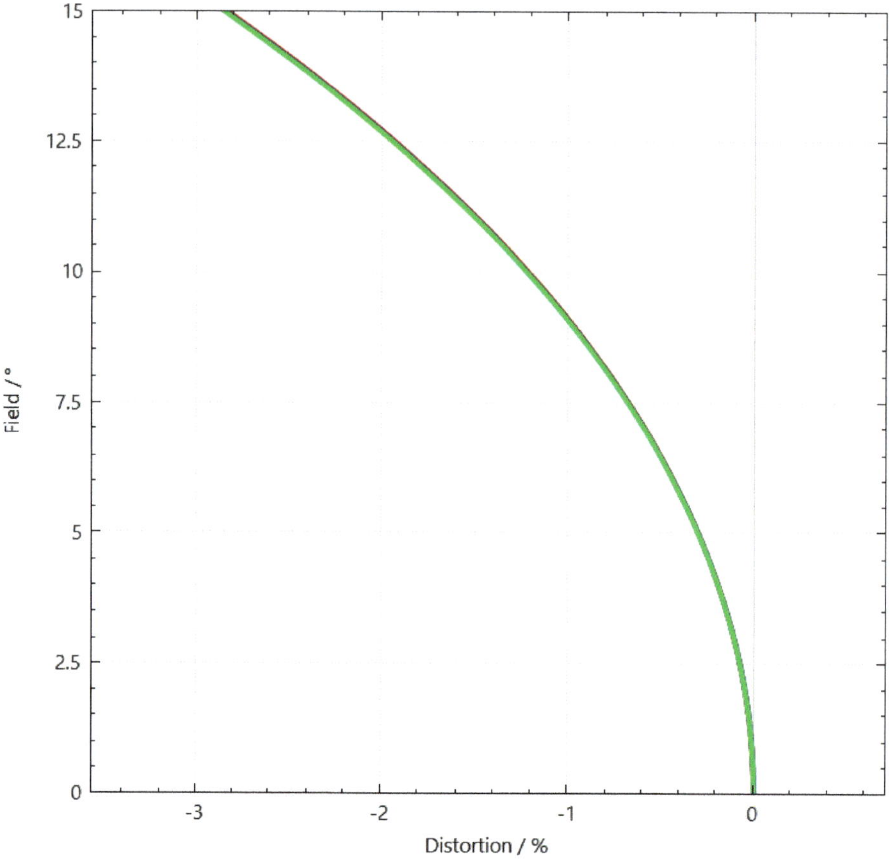

FIGURE 6.11
`Distortion 1D` of `Optimize_5Let` after the second optimization. The distortion is below 3% for all fields.

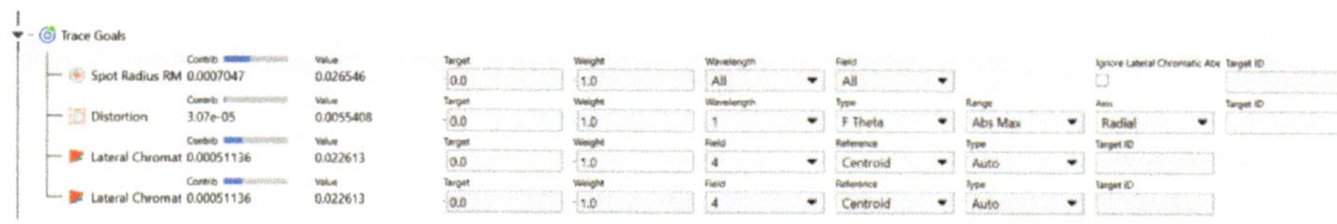

FIGURE 6.12
Modification of the merit function.

Let's optimize `Optimize_5Let` now with the metif function displayed in figure 6.12. The most prominent results are shown in figure 6.13, the `Lateral Aberration Plot` chromatic aberration has diminished its maximum magnitude just to around 0.008mm.

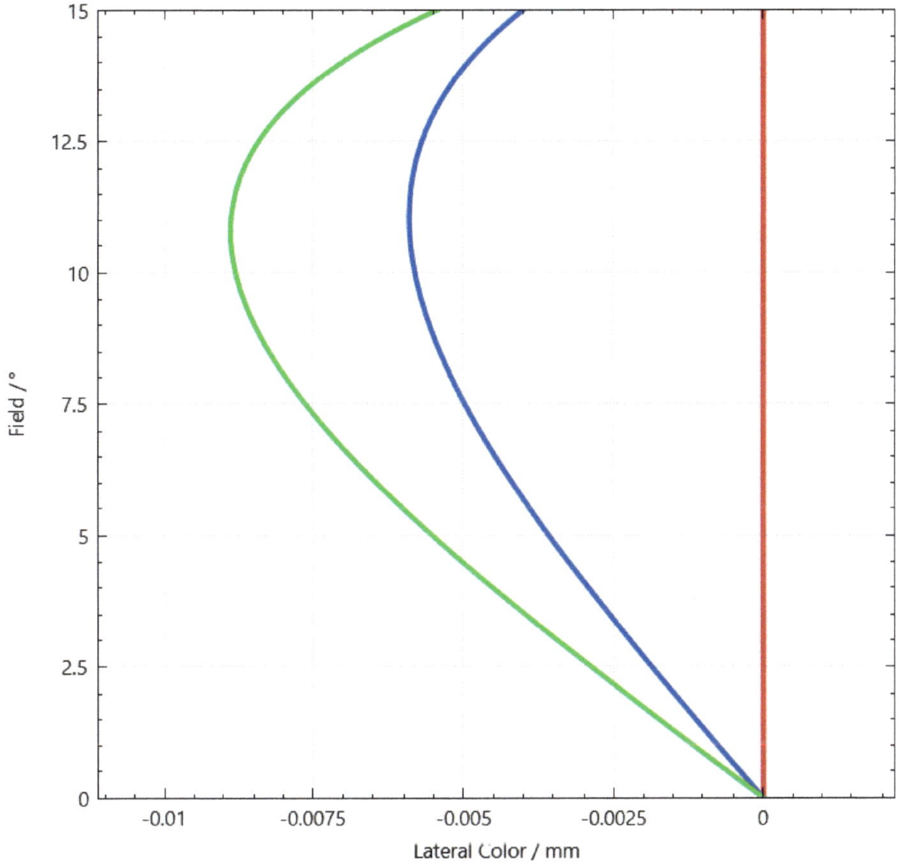

FIGURE 6.13
`Lateral Aberration Plot` of `Optimize_5Let` after the third optimization.

This result means that the modification of the metif function was successful. The reader is invited to add the `Longitudinal Chromatic Aberration` operand on the merit function and test its performance once, `Optimize_5Let` is optimized. In general, this illustrative example shows the process of the design of the optimize-analyze loop until the goals are met.

6.6 Summary

In this chapter, we deeply focus on the concept of sequential raytracing and how `Quadoa Optical CAD` manage it.

By studying the properties of sequential raytracing we also give time to study the source features like wavelength, field of view, distribution, and polarization.

Finally, we study an example of optimization of a sequential raytracing system.

7

Optical Surface Description

> How quick are we to learn: that is, to imitate what others have done or thought before. And how slow to understand: that is, to see the deeper connections.
>
> Frits Zernike

7.1 Saggital Description

Throughout history, lens designers became aware that spherical surfaces were not the best surfaces to design optical systems in general. In some simple cases systems made by spherical surfaces can be enough, but in more general terms a more general description of the profile of a surface is needed [1, 4, 5, 15].

There are many ways to describe the profile of a surface in `Quadoa Optical CAD`. This profile/sag description is made with `Form Objects`. `Form Objects` are utilized to define the sag $z(x, y)$ of a surface within the surface Coordinate System. The values of x, y, and the sag z are consistently expressed in the length unit specified in the settings, typically millimeters. To incorporate a `Form Object` into a `Surface`, first, select the `Surface` in the `Optical Design Editor`. Then, choose the desired `Form Object` to be added. Subsequently, the object will be displayed as a child element of the selected Surface. If no `Surface` is selected, an error message will be prompted. Alternatively, users can utilize the right-click menu, accessible by right-clicking on the `Surface` in the `Optical Design Editor`.

In cases where a `Surface` lacks any `Form Objects`, it will be treated as a default `Spherical Surface`. However, if the `Surface` contains a single `Form Object`, the sag will be computed based on the specific object's sag formula $z(x, y)$.

DOI: 10.1201/9781003506355-7

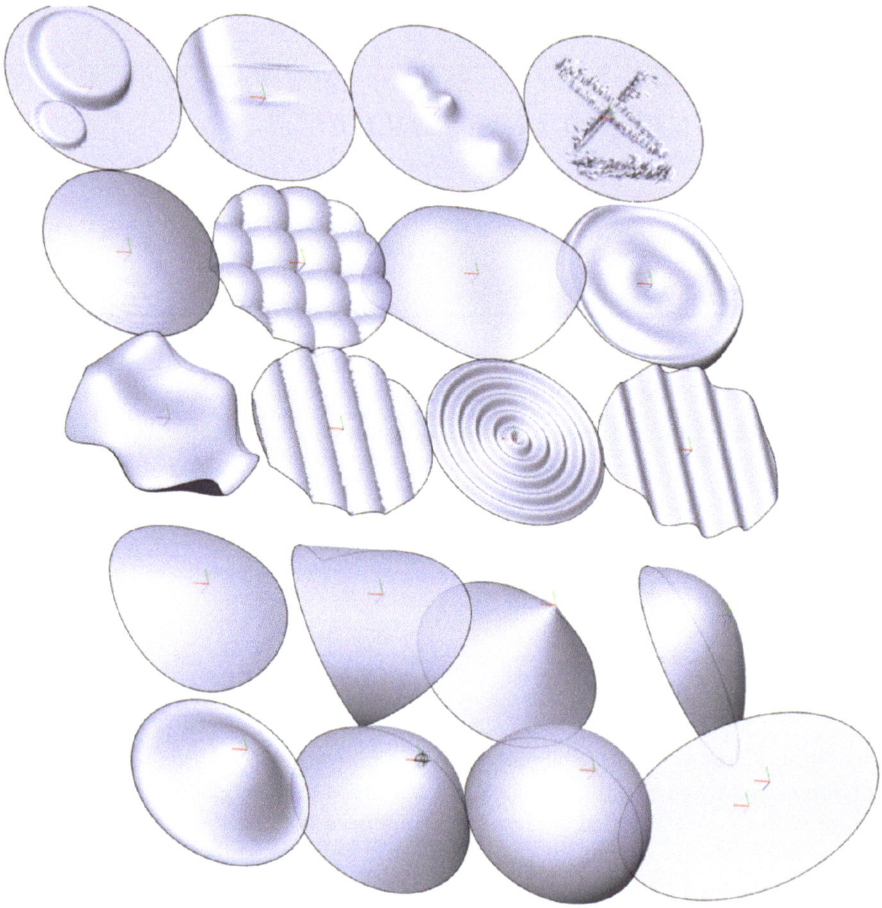

FIGURE 7.1
Gallery of surfaces that can be obtained in `Quadoa Optical CAD`. In `Quadoa Optical CAD` surfaces have described an object called `Surface`. `Surface` has a child object called `Form Object`. The shape of a `Surface` is computed by using the computed based on the specific sag formula $z(x, y)$ provided in `Form Object`.

Figure 7.1 shows a gallery of several surfaces that can be obtained in `Quadoa Optical CAD`. The diversity of surfaces in `Quadoa Optical CAD` is so big that it is needed a whole chapter to study them [18]. That chapter is this one.

In general, in `Quadoa Optical CAD` a surface object, `Surface`, can have more than one `Form Object` assigned. This can lead to several convention of `Form Object` to describe a single `Surface`.

7.1.1 Basic Forms

- **Sphere:** The `Sphere Surface Form Object` delineates a basic spherical surface and comprises a solitary parameter, namely the radius of the sphere. Notably, the sag at the center of the surface coordinate system consistently measures zero.

- **Tilt:** The `Tilt Surface Form Object` characterizes a tilted plane determined by two tilt parameters: Tilt X and Tilt Y. These parameters denote the slope of the surface in the x and y directions, respectively.

- **Paraboloid:** The `Paraboloid Surface Form Object` delineates a straightforward paraboloid surface. The paraboloid equation is,

$$z(x, y) = A^2(x^2 + y^2) + s, \tag{7.1}$$

  where s is an offset and the amplitude A. Both given by the user.

- **Asphere:** The `Asphere Surface Form Object` delineates a standard aspheric `Surface`, possessing a variable number of parameters contingent on the value of the polynomials parameter. Notably, the sag at the center of the surface coordinate system is consistently zero. The standard aspheric surface is described as follows,

$$z(x, y) = \frac{h^2/r}{1 + \sqrt{1 - (1 + c)(h/r)^2}} + \sum_{i=1}^{n} A_{2i} h^{2i}, \tag{7.2}$$

where $h = \sqrt{x^2 + y^2}$, the radius is r and A_{2i} are the coefficients.

- **Off-Axis-Asphere:** The `Off-Axis-Asphere Surface Form Object` shares the characteristics of the asphere surface form description, with the inclusion of three additional parameters. `Decenter` x and `Decenter` y specify the displacement of the vertex in the x and y directions, respectively. The parameter z Reference determines the offset of the surface sag. When set to none, the vertex of the surface is situated at z=0. The `Off-Axis-Asphere` is expressed as,

$$z(x, y) = \frac{h^2/r}{1 + \sqrt{1 - (1 + c)(h/r)^2}} + \sum_{i=1}^{n} A_{2i} h^{2i}, \tag{7.3}$$

where $h = \sqrt{(x - d_x)^2 + (y - d_y)^2}$. d_x and d_y are decenters.

- **Axicon:** The `Axicon Surface Form Object` defines a cone-shaped surface form, commonly employed in laser beam shaping to produce annular intensity distributions or Bessel beams. Notably, the sag at the center of the surface `Coordinate System` consistently measures zero.

$$z(h) = h \tan \alpha \tag{7.4}$$

where $h = \sqrt{x^2 + y^2}$ and α si the sloope.

- **Cylinder:** The `Cylinder Surface Form Object` characterizes a cylindrical surface form. Importantly, the sag at the center, as well as along the axis of the cylinder, within the surface `Coordinate System`, is consistently zero.

- **Biconic:** The `Biconic Surface Form Object` defines a surface based on a conic section separately in the x and y directions. Parameters for each direction are independent of one another.

$$z(x, y) = \frac{\frac{x^2}{r_x} + \frac{y^2}{r_y}}{1 + \sqrt{1 - (1 + c_x)\frac{x^2}{r_x} - (1 + c_y)\frac{y^2}{r_y}}} \tag{7.5}$$

where r_x, r_y, c_x and c_y are adjustable parameters.

- **Toroidal:** The `Toroidal Surface Form Object` defines a surface shaped like a torus. This torus is created by extruding a circle, whose radius is determined by the radius of circle R_S, along another circle, whose radius is specified by the radius of revolution R_L. If the axis of revolution is in y,

$$z(x, y) = R_L - \sqrt{\left(R_L - \frac{y^2}{R_s\sqrt{1 - y^2/R_S^2}}\right)^2 - x^2} \tag{7.6}$$

If the axis of revolution is in x,

$$z(x, y) = R_L - \sqrt{\left(R_L - \frac{x^2}{R_s\sqrt{1 - x^2/R_S^2}}\right)^2 - y^2} \tag{7.7}$$

- **Cosine:** The `Cosine Surface Form Object` characterizes a periodic harmonic surface sag. Primarily, this `Form Object` is utilized to emulate the behavior of optical systems concerning the spatial frequency of surface errors. If the axis in y direction

$$z(x, y) = a \cos\left(\frac{2\pi}{p}(x - \phi)\right), \tag{7.8}$$

and if the axis is in x direction

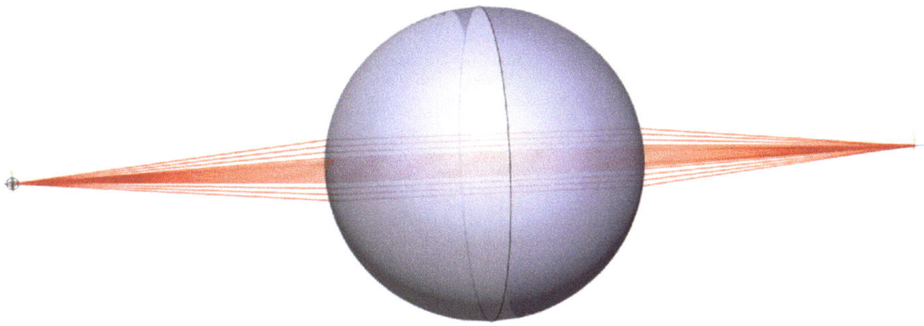

FIGURE 7.2
Ray-tracing of a ball lens.

$$z(x, y) = a \cos\left(\frac{2\pi}{p}(y - \phi)\right). \tag{7.9}$$

where a, p and ϕ are adjustable parameters.

- **Gauss:** The `Gauss Surface Form Object` defines a localized surface sag bump, which is modeled based on a Gaussian function. This functionality serves to depict local surface defects in simulations and is primarily intended for simulating existing systems, rather than design purposes.

- **Grid:** The `Grid Form Object` is employed to define a surface sag using a 2D lookup table, where the sag value is determined. Interpolation options for the table include *Nearest Neighbor*, *Linear*, or *Bicubic*.

7.1.1.1 Examples of Basic Forms

At this moment, we have seen several mathematical expressions of basic forms of `Form Object`. Let now focus now on how these mathematical expressions look like in `Quadoa Optical CAD`. Our first example is a ball lens. A ball lens it's just a lens whose both surfaces are `Sphere Form` with the same radius. The aperture of both surfaces is the same and it has the maximum aperture possible. The ball lens is presented in figure 7.2.

The next example is an `Aspheric Form`, more precisely a parabola. A parabola is a conic surface whose conic constant is equal to one. This optical system consists of a parabolic mirror with a single field on-axis field. The ray-tracing of this design is presented in figure 7.3. The vertex of the parabola is not shown in figure 7.3, just a section of the parabolic surface, since this surface is `Off-Axis-Asphere Surface Form Object`. The file of this example is presented in:
`C:\Program Files\Quadoa\examples\Surface Definitions`
`\OffAxisAsphere`

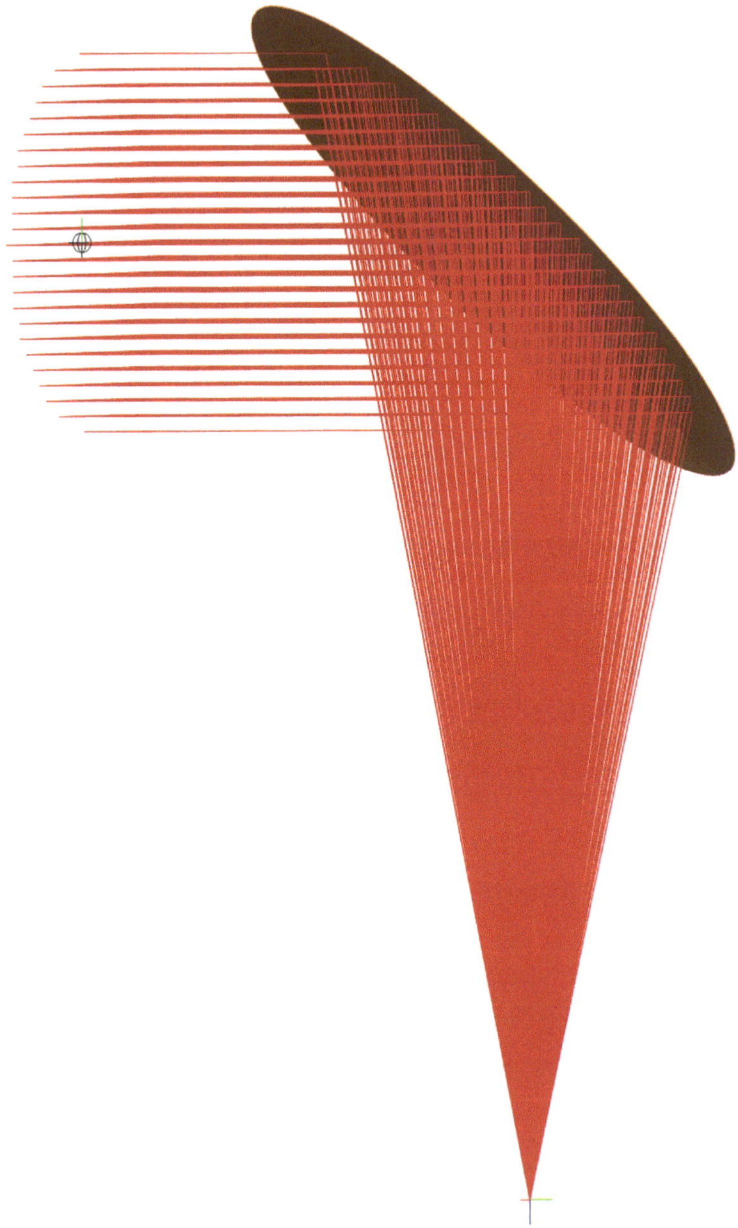

FIGURE 7.3
Ray-tracing of `OffAxisAsphere`.

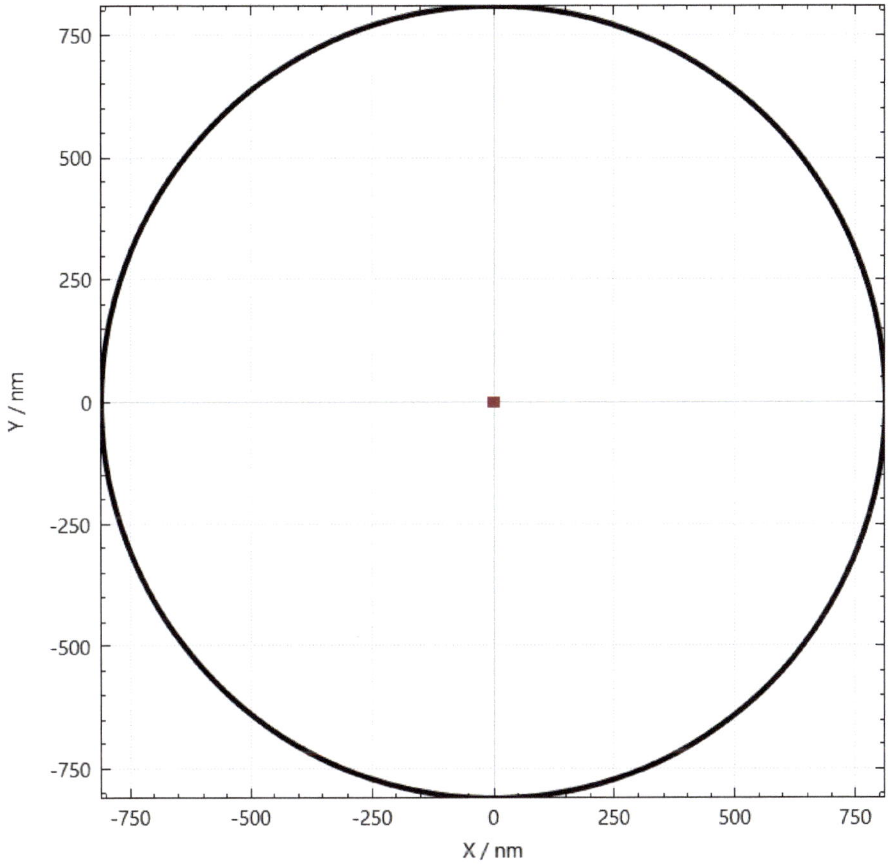

FIGURE 7.4
`Spot diagram Plot` of `OffAxisAsphere`.

During this explanation, we are going to call the design figure 7.3 as `OffAxisAsphere`. The main property, as seen before, of a parabolic mirror is to format the perfect image [7, 24]. In the case of `OffAxisAsphere` it forms the perfect image point. This can be confirmed with the `Spot diagram Plot` in figure 7.4.
The next example is a system composed of two lenses with a `Cosine Form` each. The ray tracing is presented in figure 7.5. The file path is:
`C:\Program Files\Quadoa\examples\Surface Definitions\Cosine`

The cosine patterns of the surfaces are orthogonal to each other, which made a rectangular distribution of energy in the image plane. The `Irradiance Plot Incoherent`, of `Cosine`, confirms this. It is shown in figure 7.6.

The next example is a single lens. The first surface is made with several `Gauss Surface Form Object`. The second surface is flatt. The ray-tracing is presented in figure 7.7. The path of the file is

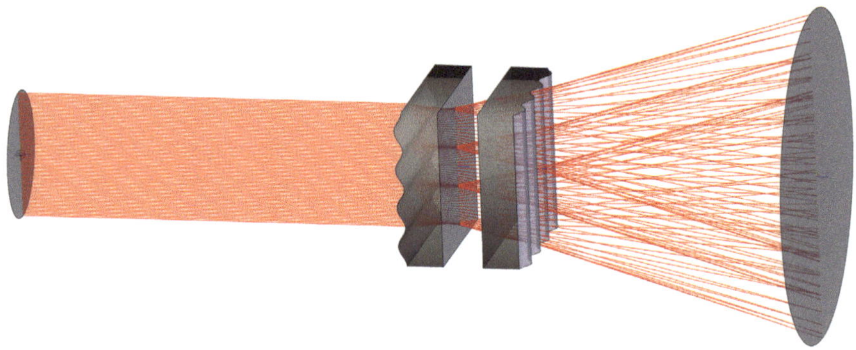

FIGURE 7.5
Ray-tracing of `Cosine`.

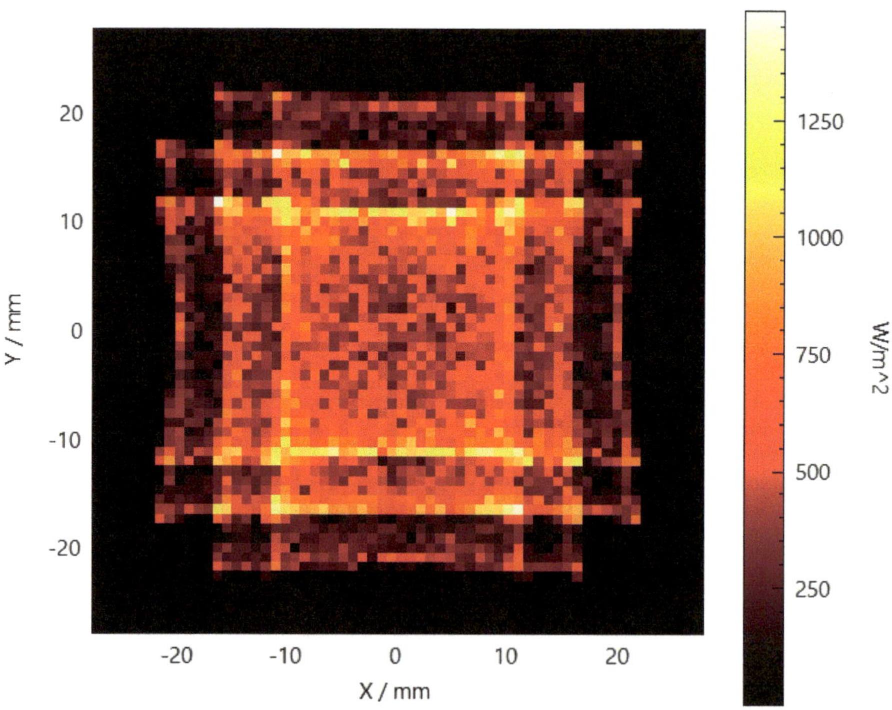

FIGURE 7.6
`Irradiance Plot Incoherent` of `Cosine`.

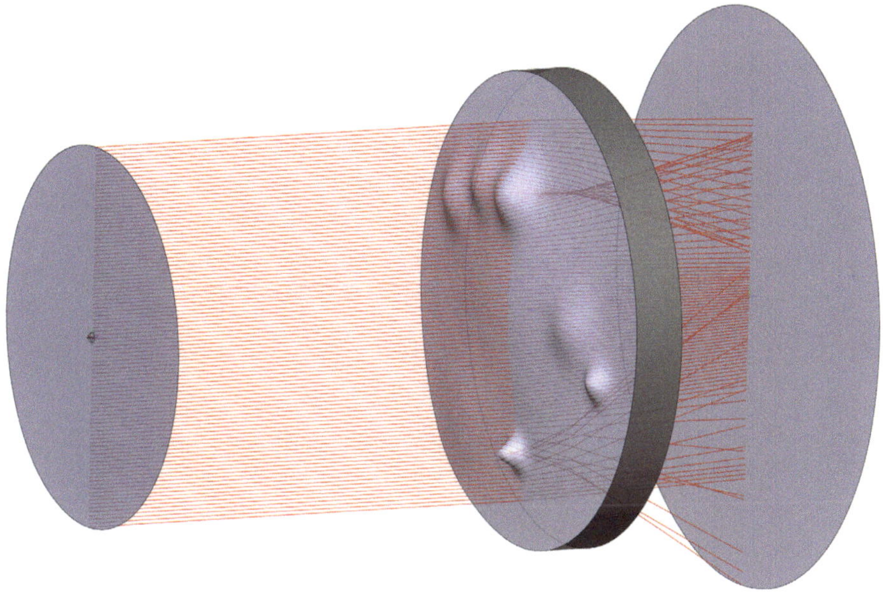

FIGURE 7.7
Ray-tracing of `Gauss`.

`C:\Program Files\Quadoa\examples\Surface Definitions\Gauss`

As it has been our local tradition we are going to call the design presented in figure 7.7 as `Gauss`.

The several Gauss peaks in the first surface are decentred and they also have different amplitudes. This particular distribution makes this lens so unique. The `Surface Form` of this lens is plotted in figure 7.8. This example in particular does not show a useful lens at all but shows the robustness for defining a surface in `Quadoa Optical CAD`.

The next example is a surface used a lot in optical science and engineering an `Axicon Form` [59, 60, 61]. The ray training is presented in figure 7.9. The path file is: `C:\Program Files\Quadoa\examples\Wave Optics\Axicon`

The system consists of two axicons facing each other. The aperture of the second axicon is bigger than the first. This configuration is such that they expand an on-axis field.

7.1.2 Other Basic Shapes

`Quadoa Optical CAD` offer other basic shape which are not displayed here because they are not used so frequently. But the reader is invited to see the `Manual`. Some of these other shapes are different versions of Gaussian functions and cylinders.

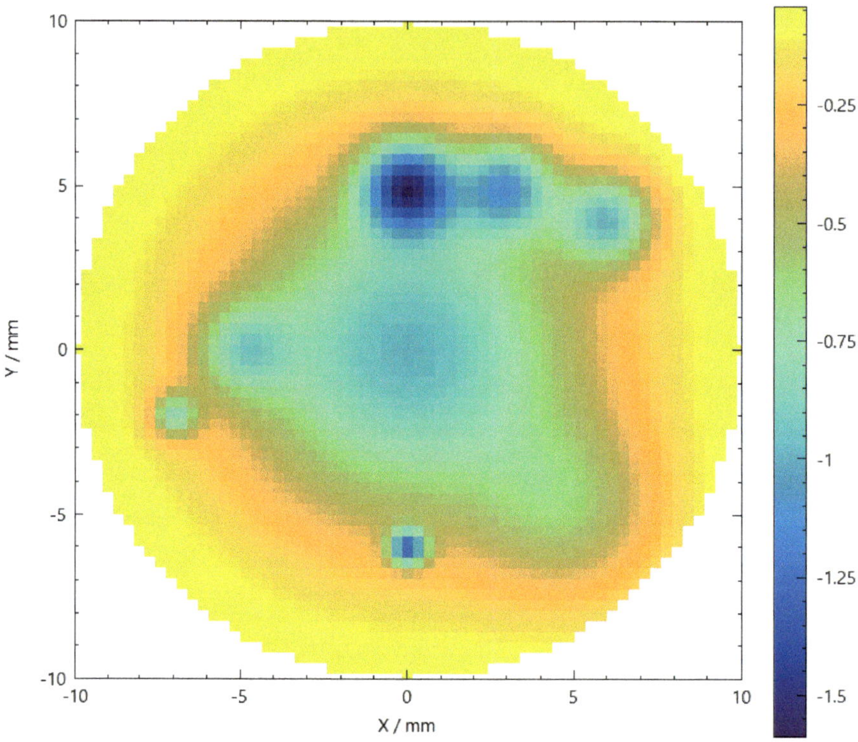

FIGURE 7.8
`Surface Form` of `Gauss`.

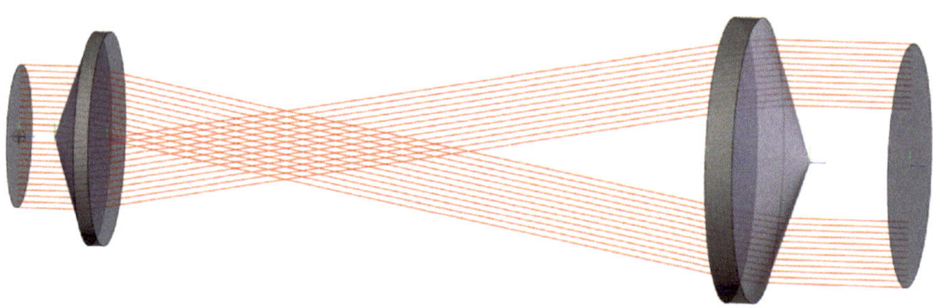

FIGURE 7.9
Ray-tracying of `Axicon`.

7.1.3 Polynomials Form

`Quadoa Optical CAD` also contains polynomials which are used to describe a surface which can be optimized. In this section we are going to explore in detail the Forbes polynomials [62, 63, 64] and then we will focus on polynomials forms which include the Zernike polynomials [47, 48, 49]. We will not spend to much time already with Zernike polynomials because we have talked about them, when we explain them that they are also used to fit the wavefront at the exit pupil of an optical system.

7.1.4 Forbes Polynomials

Forbes aspheres offers several potential advantages in optical design and analysis. The Forbes aspheres are divided into `QCon` and `Qbfs`.

7.1.4.1 Asphere QCon

The `QCon` asphere is defined as,

$$z(h) = \frac{ch^2}{1+\sqrt{1-\epsilon c^2 h^2}} + u^2 \sum_{m=0}^{M} a_m Q_m^{\text{con}}(u^2) \tag{7.10}$$

Here, c is the paraxial curvature of the surface and ϵ the conic parameter, and $h = \sqrt{x^2+y^2}$.

$$u = \frac{h}{h_{\max}} \tag{7.11}$$

Q_m^{con} based on the be a particular case of the Jacobi polynomials,

$$Q_m^{\text{con}}(x) = P_m^{(0,4)}(2x-1), \tag{7.12}$$

where,

$$P_m^{(\alpha,\beta)}(z) = \frac{(-1)^m}{2^m m!}(1-z)^{-\alpha}(1+z)^{-\beta}\frac{d^m}{dz^m}\left\{(1-z)^{\alpha}(1+z)^{\beta}\left(1-z^2\right)^m\right\}. \tag{7.13}$$

Therefore, the first `QCon` asphere are given by,

$$\begin{aligned}
Q_0^{\text{con}}(x) &= 1\\
Q_1^{\text{con}}(x) &= -(5-6x)\\
Q_2^{\text{con}}(x) &= 15-14x(3-2x)\\
Q_3^{\text{con}}(x) &= -\{35-12x[14-x(21-10x)]\}\\
Q_4^{\text{con}}(x) &= 70-3x\{168-5x[84-11x(8-3x)]\}\\
Q_5^{\text{con}}(x) &= -[126-x(1260-11x\{420-x[720-13x(45-14x)]\})]
\end{aligned} \quad . \tag{7.14}$$

- **Asphere Qcon:** The `Asphere QCon Surface Form Object` delineates an aspheric surface characterized by the `QCon` polynomials pioneered by G. Forbes. Its parameter count varies depending on the specified value of the polynomials parameter. Notably, the sag at the center of the surface coordinate system remains consistently zero. Moreover, the `Asphere QCon` introduces additional parameters, ranging from n equals 1 to the designated value of the polynomials parameter. The count of polynomial coefficients can be determined by adjusting the parameter `Polynomials`. For the sag formula definition, please refer to [2]. It's important to note that the conic term is absent in the original polynomial definition and should be omitted, as its inclusion adversely impacts the numerical characteristics of the asphere, particularly during optimization. Although included for compatibility with other optical design software, it's highly advised to always set the conic term to zero.

7.1.4.2 Asphere Qbfs

The `Qbfs` asphere is defined as,

$$z(h) = \frac{c_{\text{bfs}} h^2}{1 + \sqrt{1 - \epsilon c_{\text{bfs}}^2 h^2}} + \frac{u^2(1-u^2)}{\sqrt{1 - c_{\text{bfs}}^2 h_{\max} u^2}} \sum_{m=0}^{M} a_m Q_m^{\text{bfs}}(u^2) \tag{7.15}$$

where, $Q_m^{\text{bfs}}(u)$ is explicitly forced to vanish at the aperture's center and edge, in other means, at $u = 0$ and $u = 1$ can be converted to a deviation measured along the surface normal by multiplying it by the cosine of the angle between the optical axis and the local normal to the best-fit sphere. This cosine factor is precisely the square root that appears in the denominator of Eq. (7.15), and this is the justification for the square root's presence. It is now straightforward to choose $Q_m^{\text{bfs}}(u)$ to be polynomials of order m and to configure them so that the weighted `rms` slope of the departure along the normal is just the sum of the squares of amplitudes. To achieve this, the elements of the normal-departure slope are written as,

$$S_m(u) = \frac{\mathrm{d}}{\mathrm{d}u}\{u^2(1-u^2)Q_m^{\text{bfs}}(u^2)\} \tag{7.16}$$

Therefore, the first `Qbfs` asphere are given by,

$$\begin{aligned}
Q_0^{\text{con}}(x) &= 1 \\
Q_1^{\text{con}}(x) &= \frac{1}{\sqrt{19}}(13 - 16x), \\
Q_2^{\text{con}}(x) &= \sqrt{\frac{2}{95}}[29 - 4x(25 - 19x)] \\
Q_3^{\text{con}}(x) &= \sqrt{\frac{2}{2545}}\{207 - 4x[315 - x(577 - 320x)]\} \\
Q_4^{\text{con}}(x) &= \frac{1}{3\sqrt{131831}}(7737 - 16x4653 - 2x[7381 - 8x(1168 - 509x)])
\end{aligned} \tag{7.17}$$

- **Asphere Qbfs:** The `Asphere QBfs Surface Form Object` defines an aspheric surface using the `QBfs` polynomials pioneered by G. Forbes. The number of parameters it employs varies depending on the specified value of the `Polynomials` parameter. Notably, the sag at the center of the surface coordinate system remains consistently zero. Additionally, the `Asphere QBfs` introduces extra parameters, ranging from n equals 1 to the designated value of the "polynomials" parameter. The coefficients are editable. The count of polynomial coefficients can be adjusted using the parameter `Polynomials`. Please refer to [2] for the sag formula definition. It's important to note that the conic term is not originally part of the polynomial definition and should be avoided due to its detrimental impact on the numerical properties of the asphere, particularly during optimization. While included for compatibility with other optical design software, it's highly advisable to always set the conic term to zero.

7.1.4.3 Other Polynomials Forms

- **Zernike:** The `Zernike Surface Form Object` is a form of polynomial object that utilizes a Zernike polynomial. Just like the one used to fit the wavefront surface at the exit pupil in chapter 4.

- **Zernike Fringe:** The `Zernike Fringe Surface Form Object` is a polynomial form object that employs a Zernike Fringe polynomial, akin to the one utilized for fitting the wavefront surface at the exit pupil as described in chapter 4.

- **XYZ Polynomial:** The `XYZ Polynomial Surface Form Object` is a type of polynomial form object that employs an `XYZ polynomial`. The `XYZ polynomial` constitutes a basic power series over the variables x and y. The nomenclature of coefficients, such as $X5Y2$, implies that the formula for this particular coefficient is $z(x, y) = x^5y^2$. The polynomial's order denotes the highest power present in the series. Notably, the coordinates (x, y) used for evaluating the polynomial are always normalized to the norm radius r_n. In figure 7.12, a `XYZ polynomial` of degree 4 is depicted. The arrangement in the `Optical Design Tree` follows a row-by-row sorting, from top to bottom and left to right.

The plot of the first terms of `Zernike Surface Form Object` is presented in figure 7.10.

The plot of the first terms of `Zernike Fringe Surface Form Object` is shown in figure 7.11.

Finally, the first terms of `Zernike Fringe Surface Form Object` are plotted in figure 7.11.

7.1.5 Array Forms

In `Quadoa Optical CAD` array of lenses can be constructed. In this section we study the most common arrays.

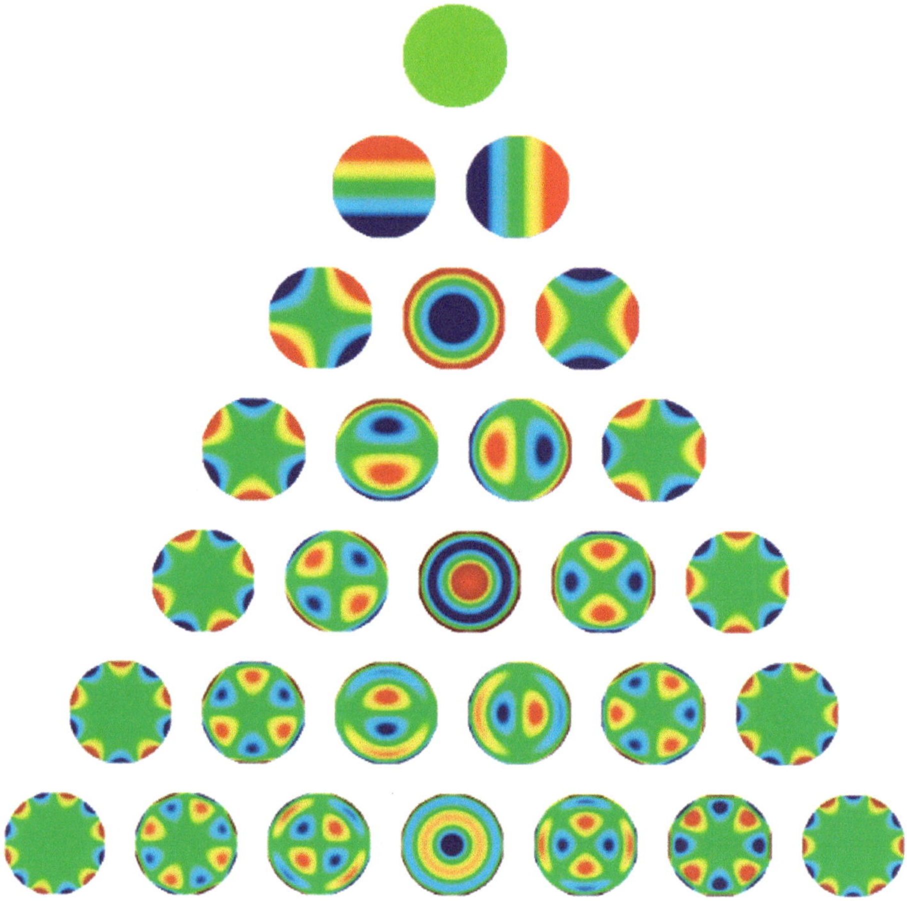

FIGURE 7.10
Plot of `Zernike Surface Form Object`.

- **Array Operator:** The `Form Array Operator` serves to create arrays of surface Form objects. Its common application involves generating microlens arrays, which may include spherical, cylindrical, or aspherical lenses. Nevertheless, it remains adaptable for use with any Form object, including Form stacks. Similar to other Form objects, it is capable of being stacked itself.

- **Hex Array Operator:** The `Form Hex Array Operator` is employed to produce hexagonal arrays of surface Form objects. Primarily, this functionality is utilized to generate arrays of microlens arrays, featuring spherical, cylindrical, or aspherical lenses. However, it's also compatible with any other Form object, as well as a Form stack. Similar to other Form objects, it can be stacked as well.

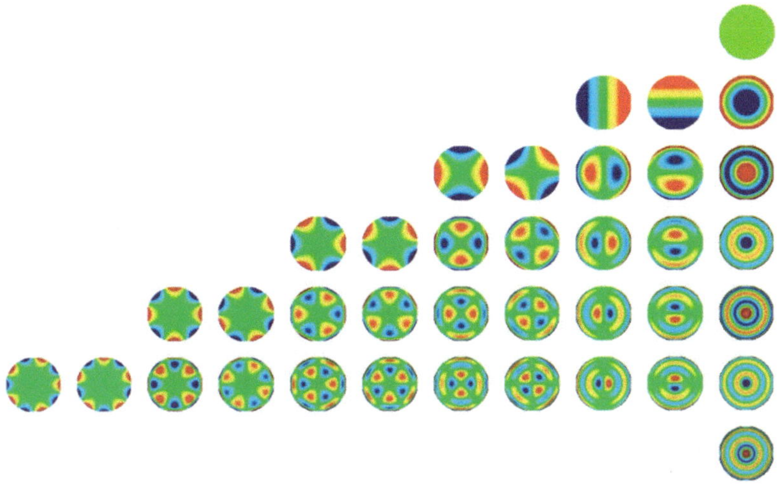

FIGURE 7.11
Plot of `Zernike Fringe Surface Form Object`.

FIGURE 7.12
Plot of `XYZ polynomial`.

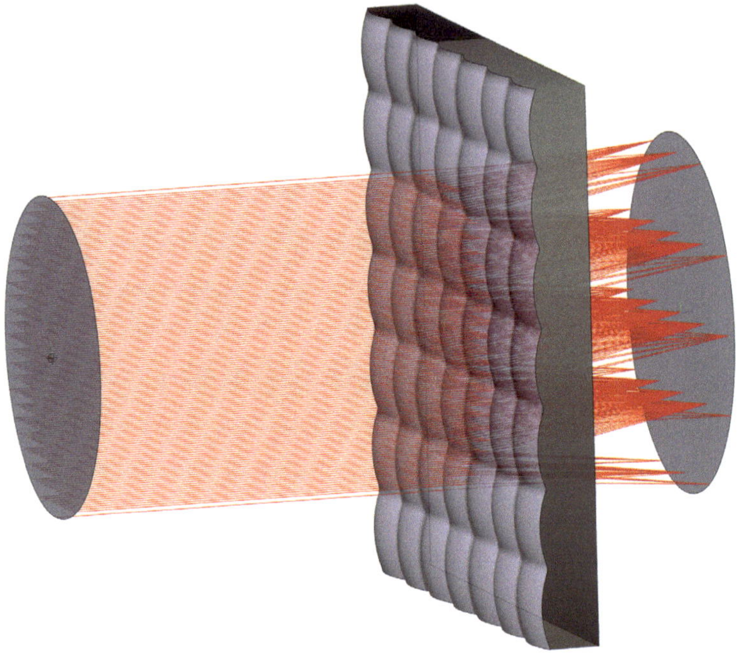

FIGURE 7.13
Ray-tracing of `Mikrolens_Array_Refractive`.

In figure 7.13 is the ray-tracing of a `Form Array Operator`.

The path of the file is: `C:\Program Files\Quadoa\examples\Surface Definitions\ Mikrolens_Array_Refractive`. The design will be called as `Mikrolens_Array_Refractive`.

The system consists of a single lens. The first surface is an array of lenses that focus on an on-axis field. The array of microlens is made `N-BK7`. Each lens of the array has two `Form Objects`. The first `Form Object` is a `Sphere Form` and the surface is a parabolic form. The surface of the lens is flat.

Let us explore the `Surface Form` of `Mikrolens_Array_Refractive`, which is displayed in figure 7.14. The color levels show the deepness of the surface, the units are in millimeters. From the figure it clear the grid shape of the array.

The outcome of `Mikrolens_Array_Refractive` can be seen in the image plane. We can use the `Spot Diagram Plot` to study the shape of the spots of each of the microlens but, usually for microlens applications the `FootPrint Diagram` is more convenient. The `FootPrint Diagram` is presented in figure 7.15. From it, we can grid of focuses generated by the microlens array.

The classical example of a microlens array is the Shack–Hartmann wavefront sensor. The Shack–Hartmann wavefront sensor is a device used in optics

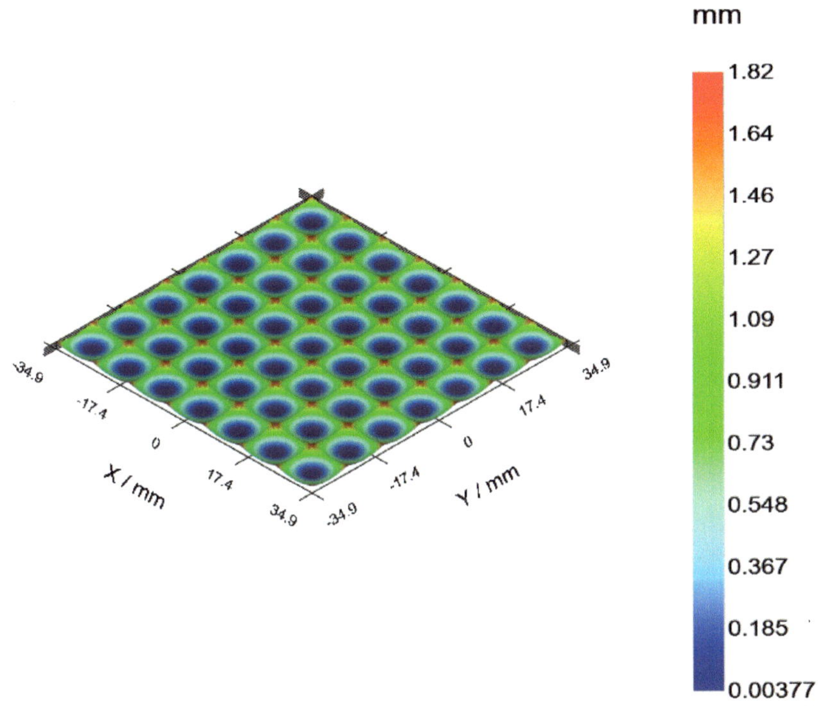

FIGURE 7.14
`Surface Form` of `Mikrolens_Array_Refractive`. The setup comprises a singlet lens. Its initial surface consists of an array of lenses. These microlenses are fashioned from `N-BK7`. Each lens within the array possesses two distinct forms: a spherical form and a parabolic form, designated as the first and second `Form Objects` respectively. The lens surface itself remains flat. The system has only an on-axis field.

to measure the wavefront distortion of light [65]. It's commonly employed in adaptive optics systems, astronomy, and vision science. The sensor consists of an array of small lenses, each focusing a portion of the incoming light onto a detector/image plane. Just like figure 7.15.

When a wavefront of light passes through the lens array, it gets divided into multiple small beams. By measuring the displacement of the centroids of these beams on the detector, the local tilt of the wavefront at each lenslet location can be determined. This information is then used to reconstruct the shape of the wavefront across the entire aperture.

By analyzing the deviations from the ideal wavefront, the Shack–Hartmann sensor can detect aberrations introduced by optical elements or distortions

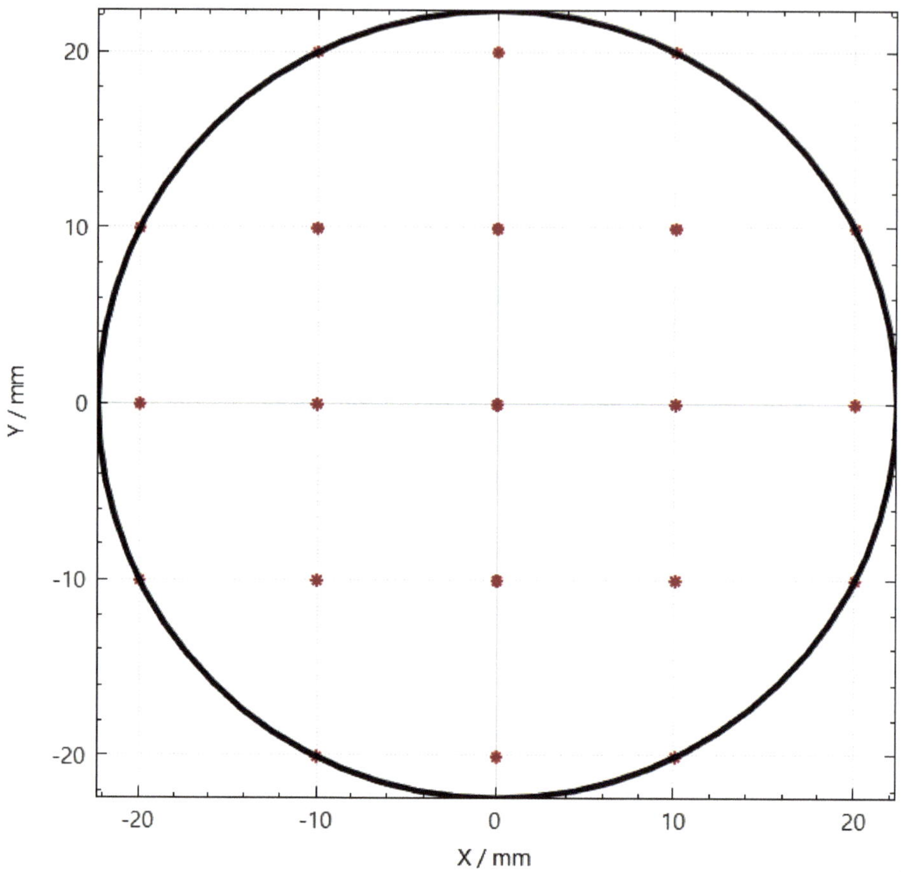

FIGURE 7.15
`FootPrint Diagram` of `Mikrolens_Array_Refractive`.

caused by the atmosphere. This data is crucial for correcting these aberrations in real-time, especially in applications like adaptive optics where precise control of the optical system is necessary to compensate for atmospheric turbulence or other disturbances.

The next example is presented in figure 7.16, it is also a **`Form Array Operator`**. But instead of separating a single field into several fields. This microlens array merges several fields in a single focus. In the figure we have four five fields that go through a microless are the fields are collimated after passing the microlens array. Finally, they are focused in a single focus point after being refracted by an aspherical lens.

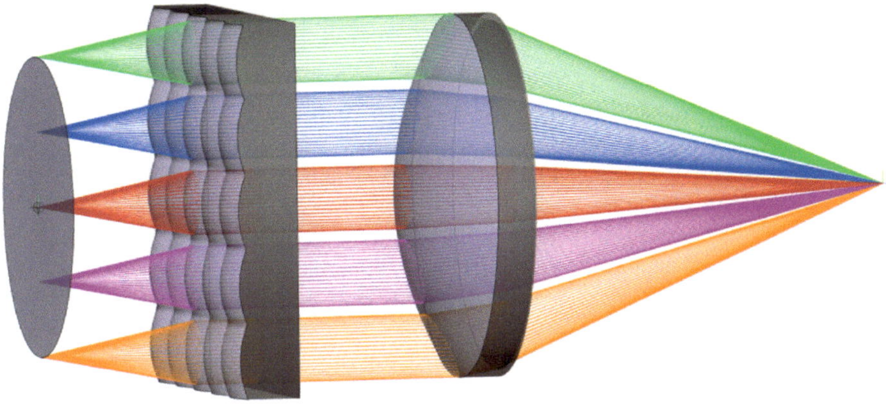

FIGURE 7.16
Ray-tracing of a `Form Array Operator`. The design has four five fields that go through a microless are the fields are collimated after passing the microlens array. Finally, they are focused in a single focus point after being refracted by an aspherical lens.

7.2 Change Between Saggital Forms

`Quadoa Optical CAD` has a power tool to convert from one `Form Object` to another. The tool is called `Convert surface Form`. It is very intuitive and it can be accessed only by pressing right-click on a `Form Object`.

7.3 Phase Description

`Phase Objects` are utilized to define a phase function $\phi(x, y)$ on a surface. Such a phase function can be employed to model diffractive optical elements (DOEs) or computer-generated holograms (CGHs), introduce phase errors in simulations, or address any task involving manipulating the phase of a wavefront [37]. The unit of any amplitude coefficient is, by default, in waves (λ), unless explicitly stated otherwise. When a surface with a `Phase Object` is encountered, the default diffraction order used to calculate rays is typically +1. However, alternative diffraction orders can be explicitly specified using a `User Defined Sequence`.

Similar to `Form Objects`, `Phase Objects` can be stacked to create combinations of various basic or polynomial `Phase Objects`. This allows for the realization of arbitrary combinations of form and `Phase Objects`. Additionally, `Phase Operators` can be employed to modify basic `Phase Objects`, such as generating arrays.

When a `Phase Object` is assigned to a surface that serves as the boundary surface of an element, where the phase function is defined in waves, the phase function is assumed to be on the exterior of the element. Consequently, the wavelength used for diffraction calculations is that of the surrounding environment. Conversely, if the `Phase Object` is assigned to an intermediate surface, such as in a double lens, the wavelength of the material with the lower refractive index is employed for the calculation.

7.3.1 Basic Phase Objects

`Quadoa Optical CAD` has several phase descriptions which can be used in order to improve a lens design, in the following list we show the describe the most common phases.

- **Grating:** The `Grating Phase Object` is employed to characterize a linear grating, with the lines oriented either in the x or y direction. The parameter *lines per mm* is utilized to indicate the number of line pairs of the grating per millimeter, which is equivalent to a phase value in waves per millimeter.

- **Grating 2D:** The `2D Grating Phase Object` serves to depict a `2D` linear grating formed by overlaying two `1D` linear gratings, resulting in a rectangular phase array. The parameter *lines per* mm in x specifies the number of line pairs of the grating per millimeter along the x-axis, while *lines per* mm in y specifies the number of line pairs per millimeter along the y-axis. Notably, unlike other phase functions, the `2D Grating` allows for the selection of two independent diffraction orders.

- **Radial Phase:** The `Radial Phase Object` is employed to depict a rotationally symmetric phase function, typically used to model `Fresnel-Zone-Plates` or rotationally symmetric computer-generated holograms (CGHs), such as those utilized in null test conditions to measure aspheric surfaces. The Radial Phase Object is given by

$$\phi(x, y) = \sum_{i=1}^{n} A_{2i} h^{2i}, \tag{7.18}$$

where A_{2i} is a adjustable coefficient,

$$h = \frac{\sqrt{x^2 + y^2}}{r_n}, \tag{7.19}$$

and r_n is the normalization radius.

- **Axial:** The `Axial Phase Object shares` similarities with the `Radial Phase Object`, but differs in that the phase value is solely dependent on one coordinate, either x or y, rather than the radius. It is commonly employed to create line-focusing diffractive elements or to model a computer-generated hologram (CGH) for the measurement of cylindrical surfaces in null test conditions. If the Axis parameter is set to x-axis, the phase is calculated in terms of the he normalization radius r_n and the adjustable coefficient A_{2i},

$$\phi(x, y) = \sum_{i=1}^{n} A_{2i} \left(\frac{y}{r_n}\right)^2 \tag{7.20}$$

and if it is set to y-axis,

$$\phi(x, y) = \sum_{i=1}^{n} A_{2i} \left(\frac{x}{r_n}\right)^2. \tag{7.21}$$

- **Spiral:** The `Spiral Phase Object` is employed to define a spiral phase function, wherein the phase value is linearly dependent on the angular coordinate Φ when observed in a polar coordinate system. The `Spiral Phase Object` is described by

$$\phi(x, y) = \frac{W\Phi}{2\pi} \tag{7.22}$$

where Φ is

$$\Phi(x, y) = \arctan(x, y). \tag{7.23}$$

- **Grid:** The `Grid Phase Object` is utilized to define a phase function, whereby the phase value is determined by a 2D lookup table. The interpolation options are the same as the interpolation with the Grid surface, `Nearest Neighbor`, `Linear` or `Bicubic`.

7.3.2 Polynomial Phase Objects

`Quadoa Optical CAD` offers various polynomial phase descriptions that can enhance lens designs. The description of these polynomials is similar to the ones presented in section tal. Below, we present the most common ones,

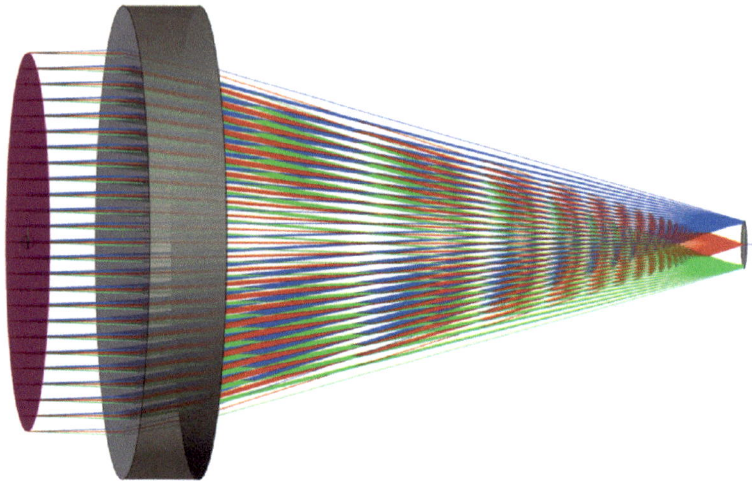

FIGURE 7.17
Ray-tracing of `CGH`.

- **ZZernike:** The `Zernike Surface Phase Object` is a polynomial form object. The same Zernike polynomials used to describe a surface and fit the wavefront at the exit pupil are used here.

- **FZernike Fringe:** The `Zernike Fringe Surface Phase Object` is a form of polynomial object that utilizes a `Zernike Fringe Polynomial` that we have discussed before.

- **XYZ Polynomial:** The `XYZ Polynomial Surface Phase Object` is a type of polynomial form object that employs `XYZ Polynomials`. The `XYZ Polynomial` represents a fundamental power series involving the variables z and y. We have discussed them previously in this chapter.

In figure 7.17 is the ray tracing of a singlet lens with flat surfaces. The first surface has a `Zernike Fringe Surface Phase Object` such that the lens is available to focus three different fields. The second surface of the singlet is just flat with no `Phase Object`. The files have a field half of view of 0°, 2°, and −2°. The source has a singlet wavelength, which is 285nm. The material of the lens is `C79-80`. The path of the file is: `C:\Program Files\Quadoa\examples\Surface Definitions\ CGH`. The design will be called as `CGH`.

The `Surface Phase` shows a surface plot of the phase along the lens aperture. The `Surface Phase` of the design of figure 7.17 is presented in figure 7.18. From the figure, we can see how the phase changes along the

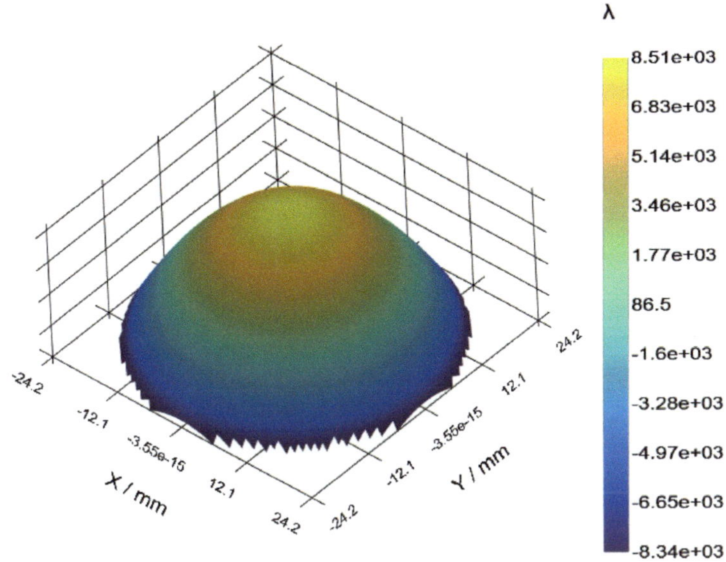

FIGURE 7.18
`Surface Phase` of CGH.

aperture of the lens. The highest value is at the center of the aperture and radially starts decreasing from that point. All Zernike coefficients of the `Zernike Fringe Surface Phase Object` are zero with the expectation of 4th and 12th coefficients. That reason for radial symmetry.

7.3.3 Phase Array Objects

In `Quadoa Optical CAD` array of phases can be constructed. In this section, we study the most common of them.

- **Phase Rect Array Operator:** The `Phase Rect Array Operator` is employed to produce arrays of surface phase functions, commonly utilized for generating diffractive micro-lens arrays. It is compatible with any Phase object as well as a phase stack.

- **Phase Hex Array Operator:** The `Phase Hex Array Operator` is utilized to create hexagonal arrays of surface phase functions, commonly employed for generating diffractive micro lens arrays.

Our next example considers a `Phase Rect Array Operator`. The `Surface Phase` of this example is presented in figure 7.19. The ray tracing of the example is displayed in figure 7.20. The `Phase`

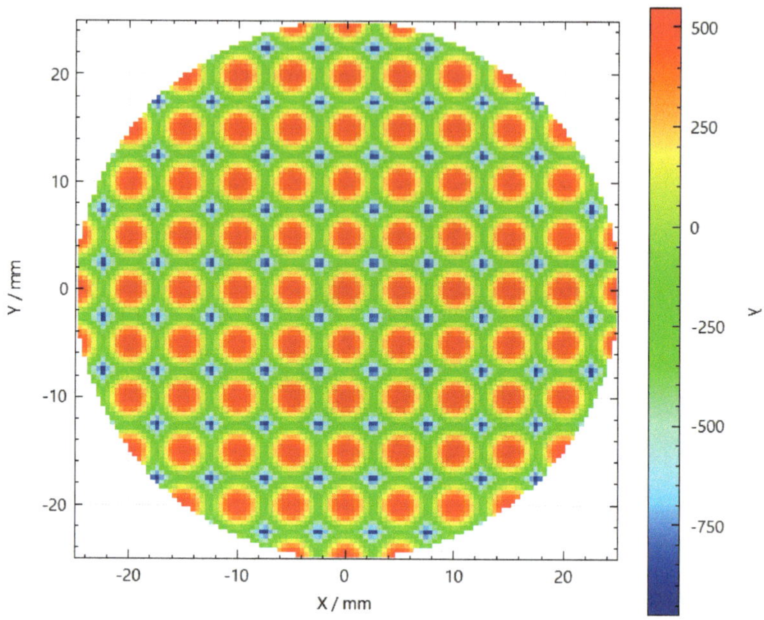

FIGURE 7.19
`Surface Phase` of `Diffractive_Ghost_Orders`.

`Rect Array Operator` spreads the on-axis fields in several focal points along the aperture of the second surface. Each of the elements of the `Phase Rect Array Operator` forms its focus. The file path is `C:\Program Files\Quadoa\examples\Surface Definitions\Diffractive_Ghost_Orders`.

7.4 Aperture and Obscuration Description

`Aperture` and `Obscuration` objects are utilized to designate the aperture on a surface. When a surface lacks an aperture as a child object, the default circular aperture, defined as a parameter in the surface, is employed. The default aperture radius can be either locked or unlocked. When unlocked, it dynamically adjusts its size to encompass the smallest possible circle around all rays.

It's important to note that if the surface is not part of any sequence and the aperture is unlocked, the radius of the floating aperture will be zero, rendering the surface invisible.

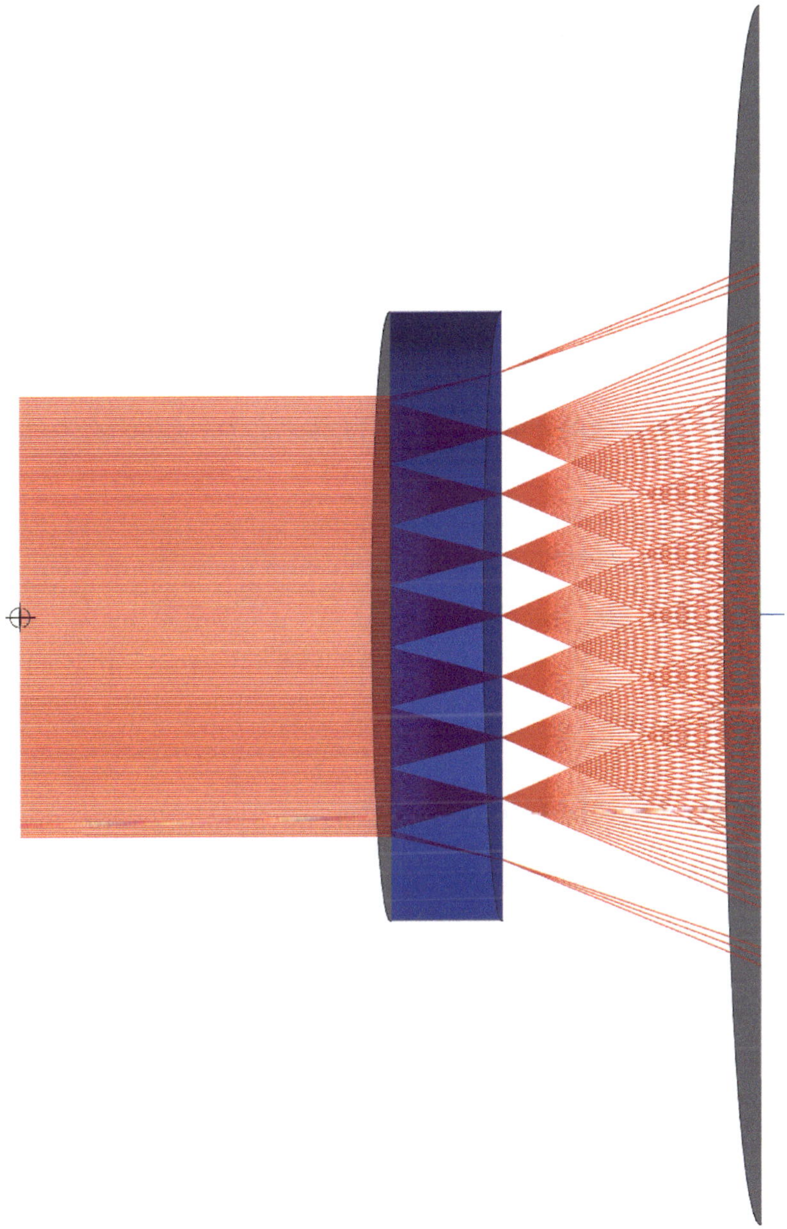

FIGURE 7.20
Ray-tracing of `Diffractive_Ghost_Orders`.

In the absence of additional Aperture objects, the default aperture specifies both the physical and clear aperture of the surface. Consequently, in the `3D-view` or when exporting an element to a `CAD` file, the aperture radius serves as the dimension for mechanical purposes.

When adding the first Aperture or Obscuration to a surface, it replaces the default aperture and delineates the physical size of the surface. If multiple apertures are specified, additional apertures define the clear aperture and further constrain the area through which rays may pass a given surface.

7.4.1 Aperture

An `Aperture` permits the passage of any rays located within its area. `Rays` positioned outside the `Aperture` will be vignetted.

- **Ciruclar:** The circular aperture represents the most common type of aperture. As its name suggests the shape is the shape of a circle.
- **Rectangular:** The rectangular aperture has two parameters the half size x and half size y specify the half size of the sides of the rectangle.
- **Elliptic:** The elliptic aperture has two parameters the half size x and half size y specify the two half axes of the ellipse.
- **Annular:** The annular aperture has two parameters the inner and outer radius.
- **Hexagonal:** The hexagonal aperture has three parameters the size of the side of the hexagon and the centers in x and y of the hexagon in the local surface coordinate system.
- **Array:** The aperture array operator is utilized to create a rectangular linear array of aperture objects. The parameters are count, the pitch and the centers. The count parameters determine the number of rows and columns in the array. The pitch parameter defines the spacing of the array in both the x and y directions. The center parameters enable the decentering of the array within the local surface coordinate system.

7.4.2 Obscuration

An `Obscuration` permits the passage of any rays positioned outside its area. `Rays` located inside the `Obscuration` will be vignetted.

- **Circular:** The circular obscuration has three parameters, the radius of the obscuration, and the x and y position of the center of the obscuration in the local surface coordinate system.

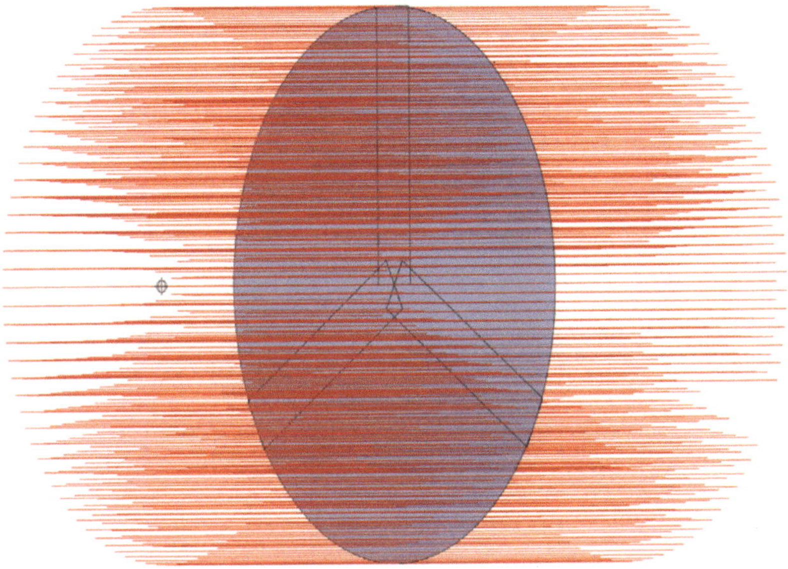

FIGURE 7.21
Example of `Rectangular Oscuarations`

- ■ **Rectangular:** The rectangular obscuration has two parameters the half size x and half size y specify the half size of the sides of the rectangle.

- ● **Elliptic:** The elliptic obscuration has two parameters the half size x and half size y specify the two half axes of the ellipse.

The next example shows an example of an aperture. The ray tracing is placed in figure 7.21.

The aperture is composed of three different `Rectangular Oscuarations` ■ which are decentered and rotated.

The `Spot Diagram Plot` of the design presented in show the effect on the aperture composed by their different `Rectangular Oscuarations` ■. The ray passes through the aperture in the zone where the `Rectangular Oscuarations` ■ are not placed. See figure 7.22.

Let's see another example, an aperture with obscurations. This aperture is composed of two different `Circular Oscuarations` ● and one `Rectangular Oscuaration` ■. The obscurations are decentered and rotated. The `Spot Diagram Plot` is shown in figure 7.23 and ray tracing is presented in figure 7.24. The ray passes through the aperture in the zone where the obscurations are not placed. The obscurations form a face. This example does not have a particular application but highlights the robustness of the `Quadoa Optical CAD`.

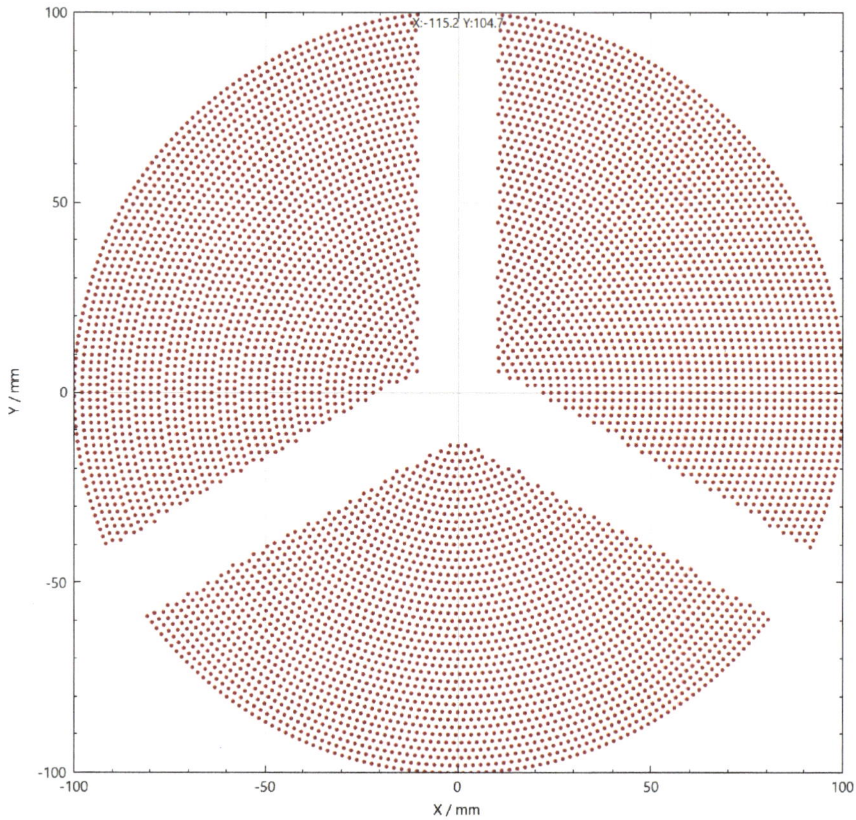

FIGURE 7.22
`Spot Diagram Plot` of the design composed of three different `Rectangular Oscuarations` ■.

The next example is a classic example of classical Fourier optics, see figure 7.25. The system consists of a surface with grid obscurations and a lens made by an `Aperture Array Operator`. The `Aperture Array Operator` is used to generate a rectangular linear array of `Aperture Objects`. The `Aperture Array Operator` considers three parameters. The *count* parameters specify the number of rows and columns of the array. The *pitch* parameter is used to define the pitch of the array in x- and y-direction. The *center* parameters allow to decenter the array inside the local surface coordinate system.

The path of the file is `C:\Program Files\Quadoa\examples\Surface Definitions\Obscuration Grid`. Here we going to call it as `Obscuration Grid`.

The system only considers three fields with an angle of incidence of $0°$, $0.1°$, and $-0.1°$. The source has one wavelength of 285nm. The material of the lens is `C79-80`. The first surface of the setup works as a filter of the

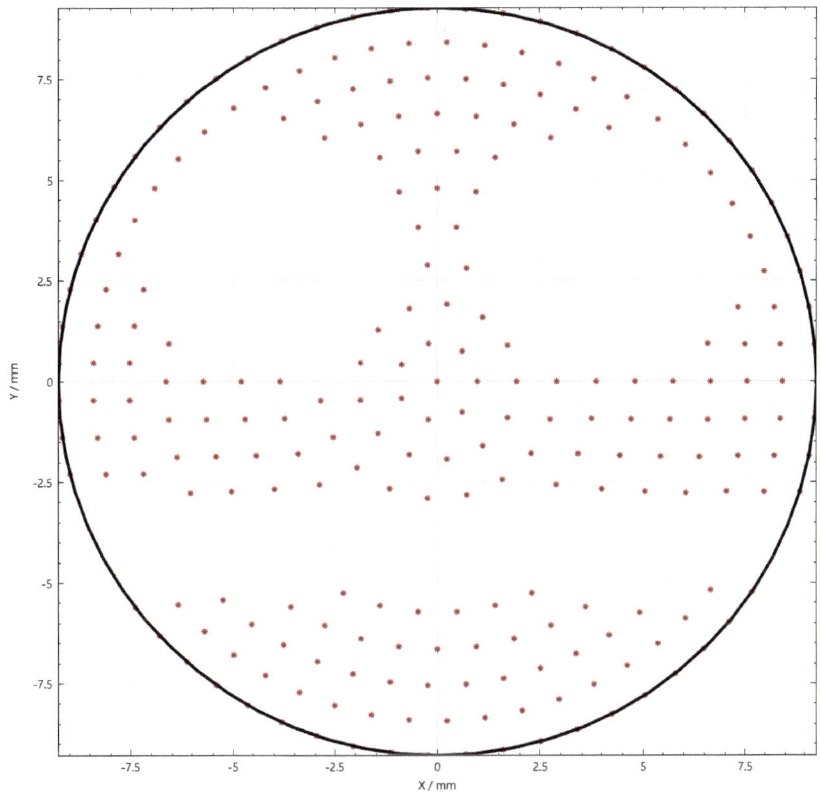

FIGURE 7.23
`Spot Diagram Plot` of a face shape aperture.

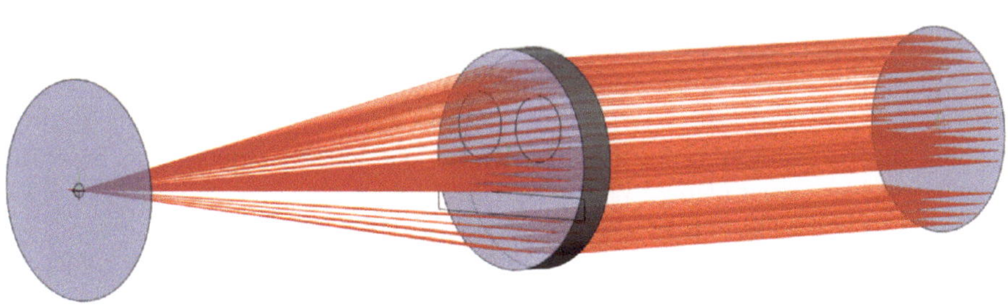

FIGURE 7.24
Ray-tracying of a face shape aperture.

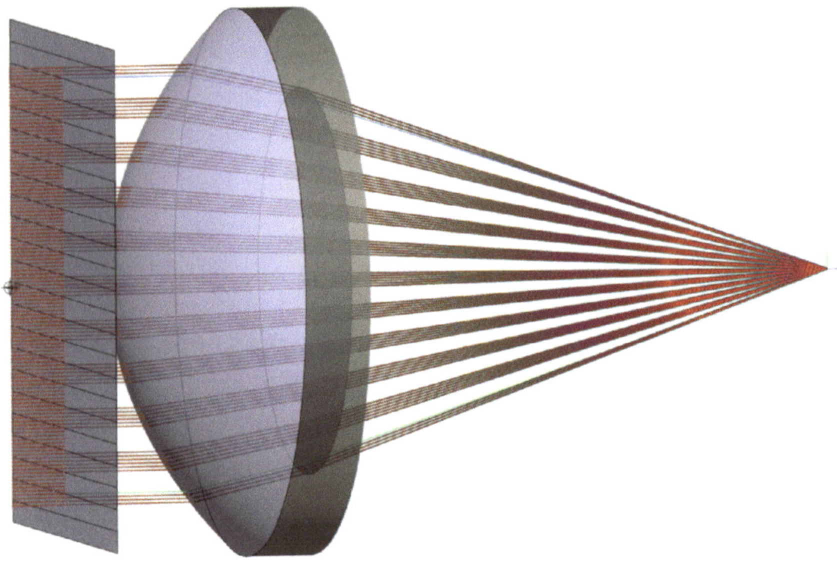

FIGURE 7.25
Ray-tracing of `Obscuration Grid`. The system only considers three fields with the angle of incidence of 0°, 0.1°, and −0.1°. The source has one wavelength of 285nm. The material of the lens is `C79-80`.

fields, it lets just pass the rays in a periodic pattern. The periodic pattern is given by the `Aperture Array Operator`. The thickness of the rectangular obscurations, the number of rectangular obscurations. So at the output of the aperture of the `Aperture Array Operator` the fields are distributed in a periodical pattern. With places on the `Aperture Array Operator` where no rays pass and other places where the rays pass.

This periodic pattern of intensity goes directly to the lens and it is focused. The focus for geometrical optics is just a single spot. But actually, geometrical optics ignore the diffraction effects. If we plot the `FFT Point Spread Function`, we will see that there are three spots. The spot in the middle has the biggest intensity point, while the spots at the side of the less but equal intensity distribution. This result is a classic result of Fourier optics that for sure is considered by `Quadoa Optical CAD` in its definition of `FFT Point Spread Function`. This is an implication of applying the Fraunhofer diffraction integral to the wavefront presented at the exit pupil. The mathematical description of this procedure is not considered a crucial topic for this book but can be found in *Introduction to Fourier Optics* by Joseph W. Goodman [32]. The `FFT Point Spread Function` plot of the on-axis field of the design in question is in figure 7.26.

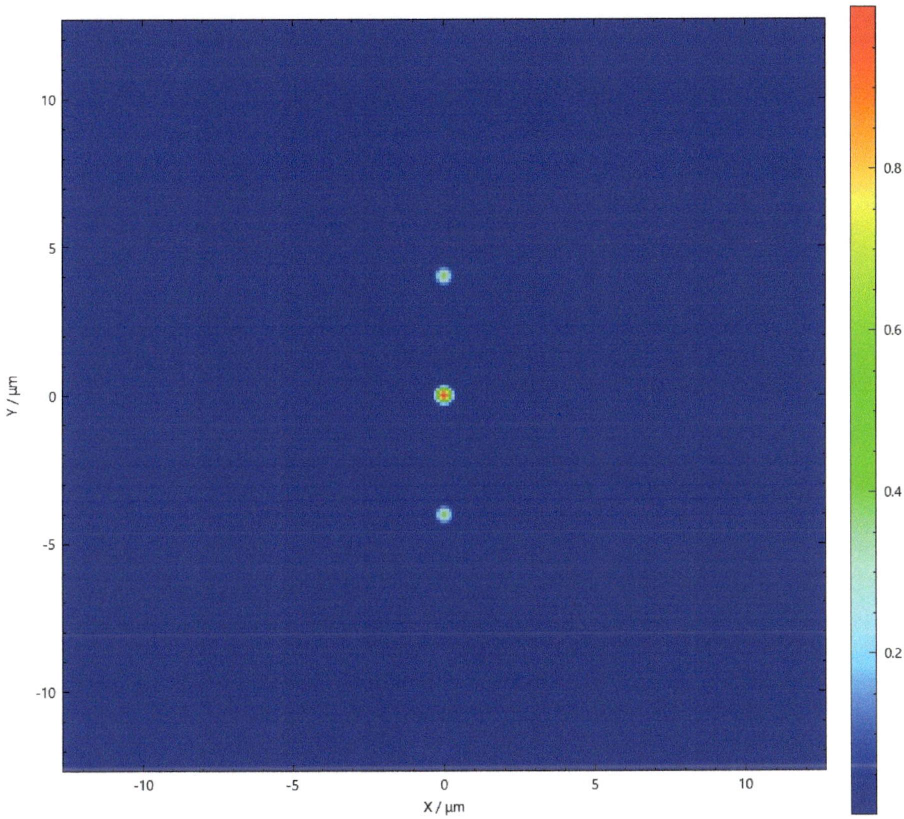

FIGURE 7.26
`FFT Point Spread Function` plot `Obscuration Grid`. The spot in the middle has the biggest intensity point, while the spots at the side of the less but equal intensity distribution. This result is a classic result of Fourier optics.

7.5 Coatings

`Quadoa Optical CAD` offers various `Coating` descriptions that can enhance lens designs. In this section, we are going to discuss the most prominent.

`Coating` objects determine the transmission and reflection properties of a surface. They are assumed to have no thickness and do not affect the refraction or general path of rays at a surface. Depending on the type of coating, reflection and transmission can vary based on the wavelength, incident angle, and polarization of the ray. If a surface lacks a coating, its reflectance and transmittance are calculated using the Fresnel equations (the ones we saw in Chapter 2) for a glass-air interface. In such cases, the reflectance is typically around 4 percent and depends on the incident angle and polarization of the ray. For unpolarized rays, the reflectance is assumed to be the average of horizontally and vertically polarized rays. A surface can have only one coating assigned to it at any given time.

7.5.1 Ideal Coatings

The basic coatings reflect or transmit a ray in an idealized manner, regardless of the incident angle or polarization of the ray.

- **Ideal AR:** The ideal anti-reflective (AR) coating consistently transmits 100% of the rays and reflects 0%.
- **Ideal Mirror:** The Ideal Mirror coating consistently transmits 0% of the rays and reflects 100%
- **Beam Splitter:** In the beam splitter coating can model simple anti-reflection coatings that remain unaffected by the wavelength or incident angle of the ray. The key of the beam splitter is that the transmittance t is in function of the reflectance r as follows,

$$t = 1 - r. \tag{7.24}$$

7.5.2 Interpolated Coatings

`Interpolated Coatings` utilize linear or `2D-Bilinear Interpolation` on the provided coating data. They are suitable for modeling coatings where only transmittance or reflectance data is available from the manufacturer, without disclosing the details of the thin-film coating stack. `Interpolated Coatings` do not include any additional parameters.

- **Wavelength Dependent:** The `Wavelength Dependent` coating reflects a specific portion of light, contingent upon the wavelength of the incident ray. Neither the incident angle nor the polarization influences the reflectance of this coating. Each data point in the wavelength range is specified in nanometers (nm) and must be arranged in ascending order. While the spacing between data points can vary, it is not required to be equally spaced. Reflectance r values range from 0.0 to 1.0. The transmittance t is determined by $t = 1.0 - r$. If a wavelength falls outside the specified data range available in the interpolation table, the intensity of both the reflected and transmitted ray will be set to `NaN` (not a number).

- **Incident Angle Dependent:** The `Incident Angle Dependent` coating reflects a certain proportion of light, contingent upon the incident angle of the incoming ray. Neither the wavelength nor the polarization affects the reflectance of this coating. The incident angle at each data point is specified in degrees and must be arranged in ascending order, within the range of $0.0\circ$ to $90.0\circ$. While the spacing between data points can vary, it is not required to be equally spaced. Reflectance r values range from 0.0 to 1.0. The transmittance is calculated using the formula $t = 1.0 - r$. If a ray has an incident angle outside the specified data range available in the interpolation table, the intensity of both the reflected and transmitted ray will be set to `NaN`.

- **Incident Angle and Wavelength Dependent:** The `Incident Angle and Wavelength Dependent` coating reflects a certain portion of light, determined by both the incident angle and the wavelength of the incoming ray. The polarization does not affect the reflectance of this coating. The incident angle at each data point is specified in degrees and must be arranged in ascending order within the range of $0.0\circ$ to $90.0\circ$. Similarly, the wavelength at each data point is specified in nanometers (nm) and must also be sorted in ascending order. While the spacing for both values can vary, it is not required to be equally spaced. Reflectance r values range from 0.0 to 1.0, and transmittance is computed, as before, using the formula $t = 1.0 - r$. If a ray has an incident angle or a wavelength that falls outside the specified data range available in the interpolation table, the intensity of both the reflected and transmitted ray will be set to `NaN`.

- **Pupil Position Dependant:** The `Pupil Position Dependent` coating reflects a specific portion of light based on the ray's position on the surface. Neither the wavelength, incidence angle, nor polarization affect the reflectance of this coating. The x and y sample points for the interpolation table must be specified in millimeters and in ascending order. Transmittance t is given by $t = 1.0 - r$ where r is the reflectance. To generate periodic patterns, the pattern can either be directly entered into the interpolation table.

The next example is a Mach-Zehnder interferometer. See figure 7.27. The basic setup of a Mach-Zehnder interferometer consists of a light source, typically a laser [66]. This beam is split into two separate paths by a beam splitter. Each path then travels through its own series of optical components, before recombining at a second beam splitter. The recombined beams then interfere with each other, creating an interference pattern that can be observed at the output of the interferometer. Mach-Zehnder interferometers have a wide range of applications in various fields, including metrology, telecommunications, quantum optics, and microscopy. In figure 7.27 is a Mach-Zehnder interferometer, and the beam splitter is a `Costum Elemetns` where the surface that decides the beam splitter has a `Beam Splitter` coating.

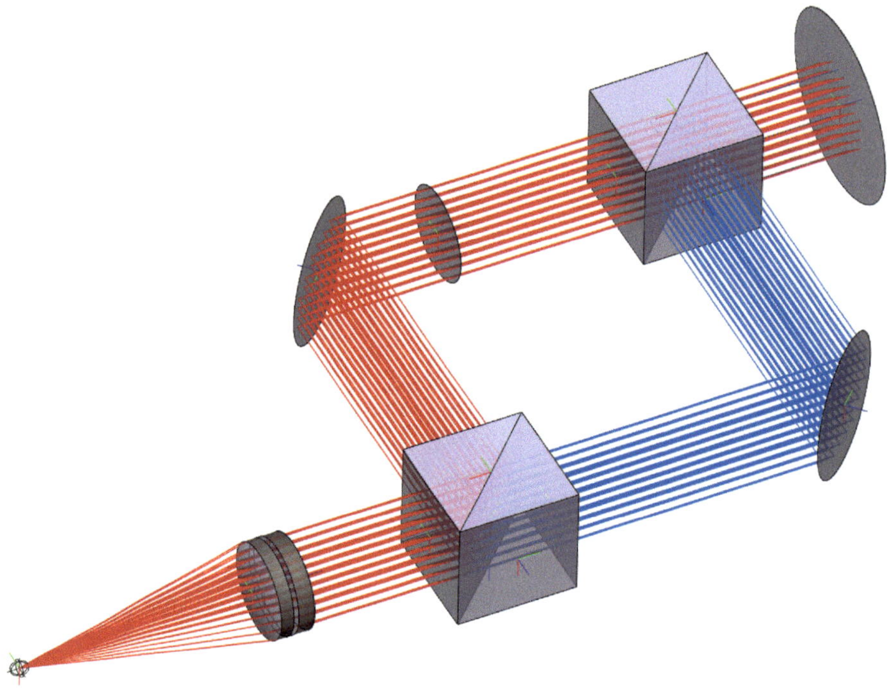

FIGURE 7.27
Mach-Zehnder interferometer. It consists of a light source. This beam is split into two separate paths by a beam splitter. Each path then travels through its path, before recombining at a second beam splitter.

7.5.3 Thin Film

The transfer-matrix method is a technique utilized in optics to examine the propagation of electromagnetic through a layered medium, typically a stack of thin films. This method is particularly relevant for applications such as the design of anti-reflective coatings and dielectric mirrors. `Thin Film Stack Coatings` are employed to simulate coatings calculated using the transfer-matrix method (the one we saw in chapter 2) [8].

The reflection of light at a single interface between two media is governed by the Fresnel equations [8]. However, in scenarios involving multiple interfaces, the reflections are not only partially transmitted but also partially reflected themselves. Depending on the precise path length, these reflections can interfere constructively or destructively. The overall reflection of a layered structure results from the summation of an infinite number of such reflections.

The transfer-matrix method operates on the premise that, as per Maxwell's equations, there exist straightforward continuity conditions for the electric field across boundaries between different media. Given the electric field at the

$$a(x,y) = A_b + (A_p - A_b)\exp^{-\frac{x^2+y^2}{2\sigma^2}}, \tag{7.25}$$

where A_b, A_p, A_b and σ are adjustable paramters.

- **Periodic:** The `Periodic Attenuator` diminishes light based on the position of the surface, with the local attenuation coefficient described by a cosine function. When the axis is set to y, the attenuation coefficient a is calculated as follows,

$$a(x,y) = A_{\text{min}} + (A_{\text{max}} - A_{\text{min}})\cos\left(\frac{2\pi}{p}(y-\phi)\right), \tag{7.26}$$

and if the axis in x,

$$a(x,y) = A_{\text{min}} + (A_{\text{max}} - A_{\text{min}})\cos\left(\frac{2\pi}{p}(y-\phi)\right). \tag{7.27}$$

A_{min}, A_{max}, p and ϕ are adjustable parameters.

7.6 Polarization Elements

`Polarization Elements` are employed to alter the polarization state of a ray. They can be incorporated into any `Surface`.

7.6.1 Static Polarization Elements

- **Linear Polarizer:** A linear polarizer is an optical device that selectively transmits light waves with a specific polarization orientation while blocking or attenuating light waves with other orientations [38, 39]. An ideal polarizer selectively allows the passage of the electric field component aligned with its transmission axis. To simulate an ideal polarizer, one of the values for transmission along the x-axis T_x or the y-axis T_y is set to 1.0, while the other value is set to 0.0.

$$\begin{pmatrix} T_x & 0 \\ 0 & T_y \end{pmatrix}, \qquad (\text{‾}$$

- **Circular Polarizer:** A circular polarizer, also known as a circul izing filter, is an optical device that selectively transmits circularly light while blocking unpolarized or linearly polarized light [38, ? `Optical CAD` supports left-handed circular polarizer, commencement of a layer, the field after the layer can be determined throu a simple matrix operation. Consequently, a stack of layers can be depicted a system matrix, derived from the product of the individual layer matric The concluding phase of this method entails transforming the system mat back into reflection and transmission coefficients.

- **From Catalog:** The `From Catalog` coating is utilized to reference coating that is defined within the `Coating Catalog`.

- **Thin Film:** The Thin `Film Coating` feature enables the direct defi tion of a thin film stack within the model. The benefit of defining the sta in the model is the ability to modify the layers, such as for optimizati or tolerancing of the coating stack. In most cases, it is advisable to defi the coating in the `Coating Catalog`. The coating consists of a series layers, with each layer specifying the thickness and material. For any inte face—whether material to air or air to material—the first layer in the li corresponds to the interface layer interfacing with the lens material, whi the last layer represents the top layer of the coating stack interfacing wi the environment. If the coating is applied within a kitted element, such between the two parts of a doublet lens, the first layer interfaces with th material of the first part of the lens, and the last layer interfaces with th second part of the doublet. In the calculation of the optical path length a system, the coating is assumed to be infinitely thin and therefore do not contribute. Similarly, for the calculation of the size of an element, th coating is disregarded.

7.5.4 Attenuators

`Attenuators` are optical elements that reduce the amount of light that passe throughout them and they just transmit another amount. `Quadoa Optica CAD` has three attenuates, constants, Gaussian and periodic. Next, thei main properties are explained.

- **Constant:** The constant attenuator reduces the intensity of light by a certain amount, with the transmitted intensity t calculated as $t = 1.0 - a$, where a represents the attenuation coefficient. Here a is a constant given by the user.

- **Gaussian:** The `Gaussian Attenuator` diminishes light based on the surface's position, where the local attenuation coefficient follows a Gaussian distribution. This method is commonly employed to produce a flat-top intensity distribution from a system with a non-uniform pupil apodization. The transmitted intensity t calculated as $t = 1.0 - a$, where a represents the attenuation coefficient. Here a is given by

$$\frac{1}{2}\begin{pmatrix}1 & -i \\ i & 1\end{pmatrix}, \tag{7.29}$$

and right-handed,

$$\frac{1}{2}\begin{pmatrix}1 & i \\ -i & 1\end{pmatrix}, \tag{7.30}$$

- **Quarter-wave plate:** A quarter-wave plate is an optical device that introduces a quarter-wavelength phase difference between two orthogonal components of linearly polarized light [38, 39]. When not rotated, the fast axis is aligned horizontally along the x-axis.

$$e^{-\frac{i\pi}{4}}\begin{pmatrix}1 & 0 \\ 0 & i\end{pmatrix}, \tag{7.31}$$

- **Half-wave plate:** A half-wave plate is an optical device that introduces a half-wavelength phase difference between two orthogonal components of linearly polarized light [38, 39]. When not rotated, the fast axis is aligned horizontally along the x-axis.

$$e^{-\frac{i\pi}{2}}\begin{pmatrix}1 & 0 \\ 0 & -1\end{pmatrix}, \tag{7.32}$$

- **General Retarder:** A half-wave plate is an optical device that introduces a reatardation r in waves in the polarized ray.

$$e^{-\frac{i\pi r}{2}}\begin{pmatrix}1 & 0 \\ 0 & e^{-ir}\end{pmatrix}, \tag{7.33}$$

- **General Jones Matrix:** The Jones Matrix is directly constructed from parameters specifying the real and imaginary parts for each matrix element and is defined as:

$$\begin{pmatrix}J_{00r} + iJ_{00i} & J_{01r} + iJ_{01i} \\ J_{10r} + iJ_{10i} & J_{11r} + iJ_{11i}\end{pmatrix}, \tag{7.34}$$

The first example of a polarization element is in figure 7.28. The path for this example is:
`C:\Program Files\Quadoa\examples\Polarization\HalfWavePlate`. Here we are going to call it as `HalfWavePlate`.

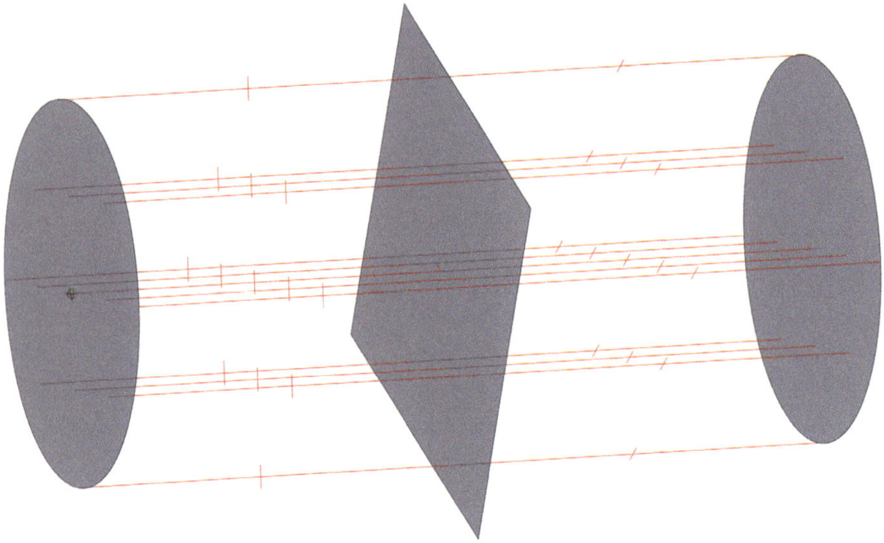

FIGURE 7.28
Ray-tracing of `HalfWavePlate`. It is composed of a single on-axis field, which is collimated. The first surface is the object, then a half-wave plate and the image plane.

It is composed of a single on-axis field, which is collimated. The first surface is the object, then a half-wave plate and the image plane. The half-wave plate is made in `Quadoa Optical CAD` with a surface which has `Half Wave Plate Polarization` element.

The source has an ideal linear polarization with an angle of 90°. The surface with the `Half Wave Plate Polarization` element is rotated 30°. Therefore, the polarization of the rays at the image plane is rotated −30°.

The next example uses `Quarter Wave Plate Polarization` element. See figure 7.29. A quarter-wave plate is an optical tool designed to create a phase difference equivalent to a quarter-wavelength between two perpendicular components of linearly polarized light.

The path for this example is:
`C:\Program Files\Quadoa\examples\Polarization\QuarterWavePlate`.
Here we are going to call it as `QuarterWavePlate`.

The setup consists of an on-axis field which is polarization is ideal linear polarization at 45°. This field passes through a surface with a `Quarter Wave Plate Polarization`. This surface is rotated 16°. Then finally, it is the image plane.

The output of the setup can be observed in the image plane. Since the only optical element is a flat surface, the systems do not form an image. There is no need to analyze any tool for image formation. But we can see how the `Quarter`

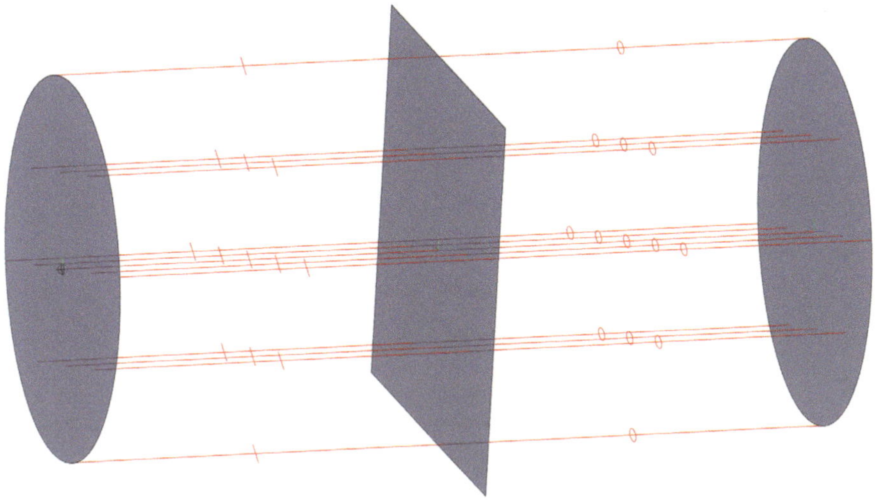

FIGURE 7.29
Ray-tracing of system with a `Quarter Wave Plate Polarization`. The setup consists of an on-axis field which is polarization is ideal linear polarization at 45°. This field passes through a surface with a `Quarter Wave Plate Polarization`. This surface is rotated 16°. Then finally, it is the image plane.

`Wave Plate Polarization` changes the polarization with `Polarization Map` plot.

The `Polarization Map` of `QuarterWavePlate` is displayed in figure 7.30. As expected the polarization across the aperture is the same for all the rays, since the `Quarter Wave Plate Polarization` does not depend on the position of the aperture. The polarization is elliptic, since the surface where the `Quarter Wave Plate Polarization` is rotated by 16°. The polarization can be circular if the `Quarter Wave Plate Polarization` was not rotated.

The next example is the Twyman-Green interferometer. Named after its inventors, Frank Twyman and Arthur Green, the Twyman-Green interferometer consists of several key components: Typically, a monochromatic light source such as a laser is used to provide coherent light for the interference pattern. The light source is divided into two parts by a beam splitter [67]. One part serves as a reference beam, while the other interacts with the test object. The optical component being tested is placed in the path of one of the split beams. This component could be a lens, mirror, or any other optical surface whose quality needs evaluation. After interacting with the test object, the beam is recombined with the reference beam. The two beams overlap and interfere with each other, creating an interference pattern. Finally, a detector array is placed at the interference pattern's focal plane to observe and analyze

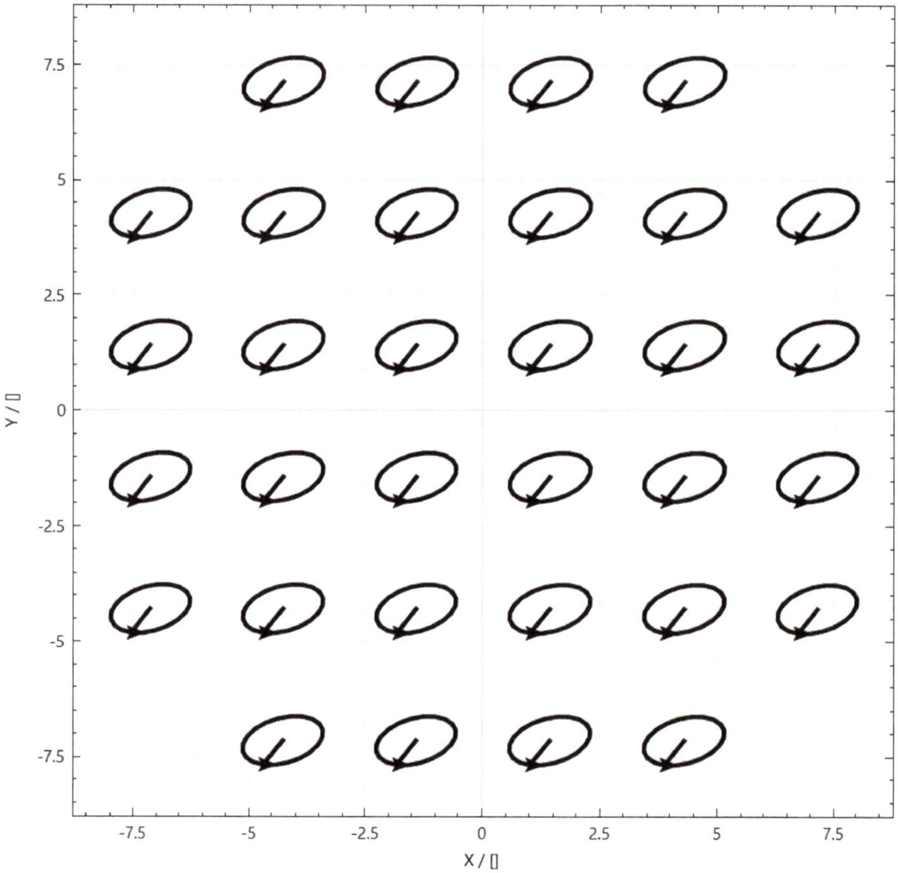

FIGURE 7.30
`Polarization Map` of `QuarterWavePlate`. The polarization is elliptic, since the surface where the `Quarter Wave Plate Polarization` is rotated by 16°. The polarization can be circular if the `Quarter Wave Plate Polarization` was not rotated.

the interference fringes. Any deviations from a perfectly flat surface will result in alterations in the interference pattern.

The path for this example is:
`C:\Program Files\Quadoa\examples\Polarization\Polarization_Beam_Splitter`. Here we are going to call it as `Polarization_Beam_Splitter`.

In figure 7.31 is the ray-tracing of a Twyman-Green interferometer. The particularity of this Twyman-Green interferometer is that it has a beam splitter with two linear polarizers one at 90° and the other at 0°, for transmision and reflection, respectively. The transmitted beam has polarization at 90°. The reflected beam has polarization at 0°.

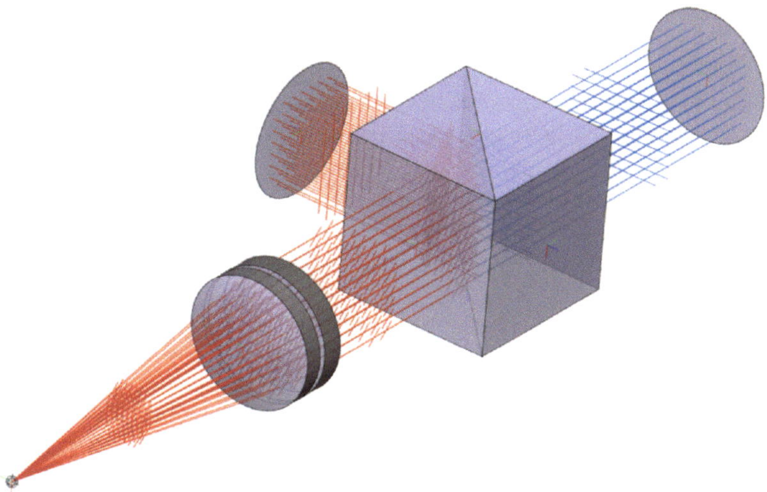

FIGURE 7.31
Ray-tracing of `Polarization_Beam_Splitter`. The particularity of this setup is that it has a beam splitter with two linear polarizers one at 90° and the other at 0°, for transmission and reflection, respectively.

7.6.2 Interpolated Polarization Elements

`Interpolated Polarization Elements` utilize `Linear` or `2D-bilinear` interpolation on the provided polarization data. These elements do not require any additional parameters.

- **Wavelength Dependent Retarder:** The `Wavelength Dependent Retarder` is employed to simulate non-achromatic retarders, where the retardance typically varies with the wavelength of the ray. The wavelength at each data point is specified in nanometers (nm) and must be arranged in ascending order. The retardance is specified in units of waves. For each incident angle, an interpolated value is computed and utilized as input for calculating the Jones matrix, similar to how it's done for the `General Retarder`.

- **Incident Angle Dependent Retarder:** The `Incident Angle Dependent Retarder` specifies a retarder whose retardance varies with the angle of incidence of the ray. This characteristic is typical for any retarder because the optical path length through the birefringent plan-parallel material, and thus the retardance, is contingent upon the angle of incidence. The data of the incident angle-dependent retarder is treated as the general `General Retarder`.

- **Pupil Position Dependent Retarder:** The `Pupil Position Dependent Retarder` is characterized by a retarder whose retardance varies based on the position of the ray within the `Surface` coordinate system. The positional data for both the x and y coordinates is expressed in millimeters and must be organized in ascending order. While the spacing between positions may or may not be uniform, it is essential for the retardance to be specified in terms of waves. For each position, an interpolated value is computed and utilized as input for the calculation of the Jones matrix, which remains consistent with that of the `General Retarder`.

- **Wavelength and Incident Angle Dependent Retarder:** The `Wavelength and Incident Angle Dependent Retarder` specifies a retarder whose retardance is contingent upon both the angle of incidence and the wavelength of the ray.

- **Pupil Position and Wavelength Dependent Retarder:** The `Pupil Position and Wavelength Dependent Retarder` describes a retarder whose retardance is contingent upon both the position of the ray within the `Surface` coordinate system and the wavelength. The position data for the x and y coordinates is specified in millimeters and must be arranged in ascending order. Similarly, the wavelength at each data point is specified in nanometers and must also be arranged in ascending order.

The next example uses `Pupil Position and Wavelength Dependent Retarder`. See figure 7.32. The `Pupil Position and Wavelength Dependent Retarder` describes a retarder whose retardance is contingent upon both the position of the ray in the local coordinate system of the surface.

The path is: `C:\Program Files\Quadoa\examples\Polarization\PupilAndWavelenghtDependantRetarder`. Here we are going to call it as `PupilAndWavelenghtDependantRetarder`.

The setup consists of an on-axis field which is polarization is ideal linear polarization at 45°. The source has three wavelengths which are 500nm, 600n, and 700nm, respectively. The only field in the system is on-axis. This field passes through a surface with a `Pupil Position and Wavelength Dependent Retarder`. After this surface is the image plane.

The `Polarization Map` of `PupilAndWavelenghtDependantRetarder`. is presented in figure 7.33. The polarization changes across the aperture, from elliptic, circular, and mostly linear. The plot presented is for 600nm. The respective plots for 500nm and 700nm vary a little from each other.

7.6.3 Mueller Matrix Element

Müller matrices are a mathematical framework used in optics to describe the polarization transformations undergone by light as it propagates through optical systems [38, 39]. Müller matrices describe how the components of the

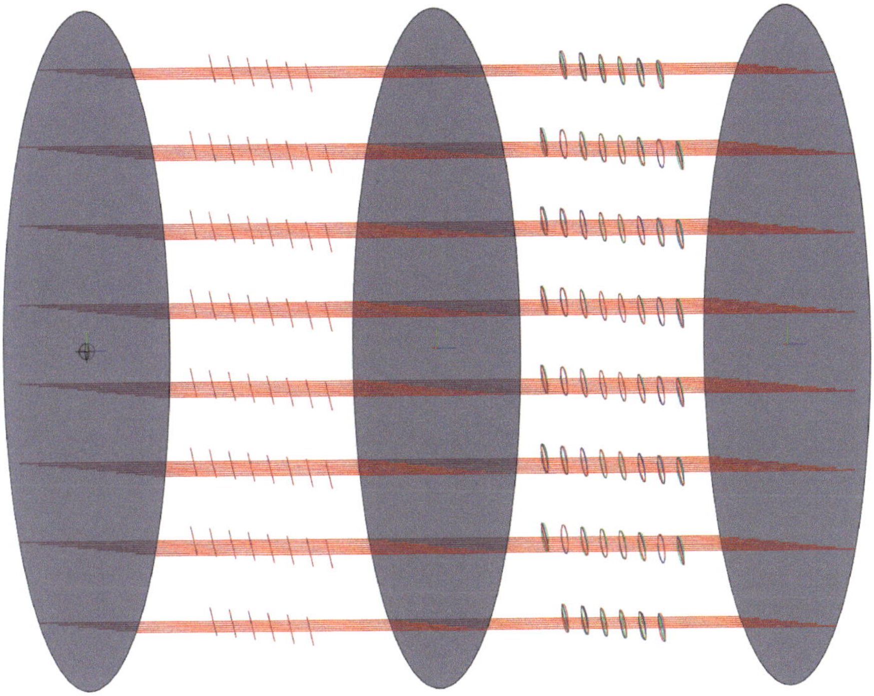

FIGURE 7.32
Ray-tracing of `PupilAndWavelenghtDependantRetarder`. The setup consists of an on-axis field which is polarization is ideal linear polarization at 45°. The source has three wavelengths which are 500nm, 600n, and 700nm, respectively. The field is on-axis. This field passes through a surface with a `Pupil Position and Wavelength Dependent Retarder`. After this surface is the image plane.

Stokes vector representing the polarization state of incident light change as it passes through various optical elements such as wave plates, retarders, polarizers, and birefringent materials. Each Müller matrix represents a specific optical element or combination of elements and quantifies their effect on the polarization state of light.

- The elements of a Müller matrix represent the optical properties of the system, such as the transmission and phase retardance of optical elements, the orientation of polarizers, and the birefringence of materials. By examining the elements of the Müller matrix, one can gain insight into how these properties affect the polarization state of light.

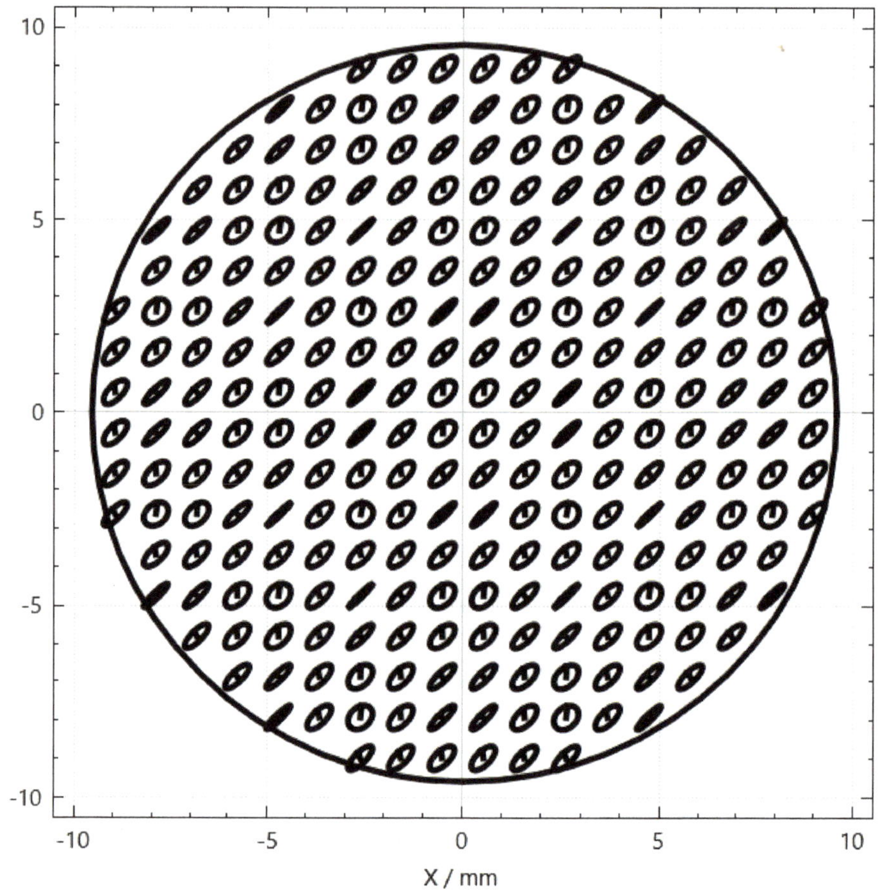

FIGURE 7.33
`Polarization Map` of `PupilAndWavelenghtDependantRetarder`. The polarization change across the aperture, from elliptic, circular and mostly linear.

The Mueller Matrix Element enables the direct input of a Mueller matrix.

$$\begin{pmatrix} M_{00} & M_{01} & M_{02} & M_{03} \\ M_{10} & M_{11} & M_{32} & M_{43} \\ M_{20} & M_{21} & M_{32} & M_{43} \\ M_{30} & M_{31} & M_{32} & M_{43} \end{pmatrix}. \tag{7.35}$$

Now, let's focus on a classical example of optics. Show in figure 7.34, it is the ray-tracying of the Brewster angle.

Brewster angle corresponds to the angle of incidence of light on a surface that cancels out the component with polarization parallel to the plane of incidence. The result when applying an unpolarized light ray on a surface

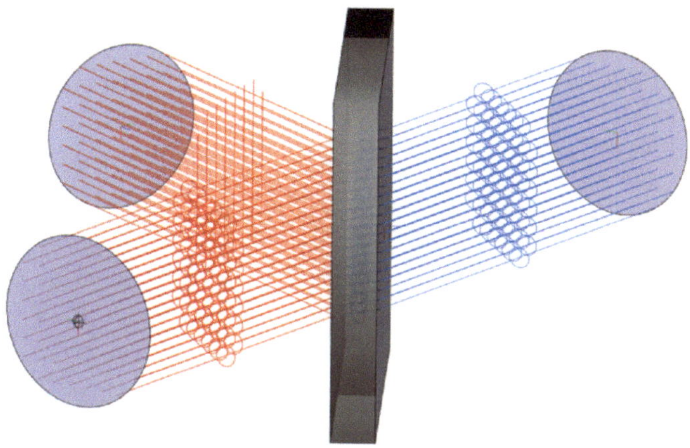

FIGURE 7.34
Ray-tracing of `Brewster`.

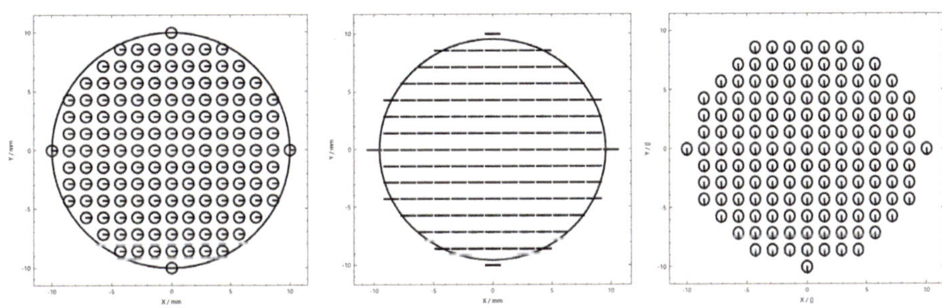

FIGURE 7.35
The `Polarization Map` of the input rays (*left*), the reflected rays (*right*), and the transmitted rays (*right*) from `Brewster`. The polarization of the reflected rays is polarization parallel to the plane of incidence. As respected by the Brewster angle. For this example, the source is a collimated on-axis field with a single wavelength, which is 587nm. The source is circularly left polarized.

at the Brewster angle is obtaining a reflected ray of polarized light in one direction [8]. For this example, the source is a collimated on-axis field with a single wavelength, which is 587nm. The source is circularly left polarized.

The path is: `C:\Program Files\Quadoa\examples\Polarization\Brewster`. Here we are going to call it as `Brewster`.

The `Polarization Map` of the input rays, the reflected rays, and the transmitted rays is shown in figure 7.35. The polarization does not change

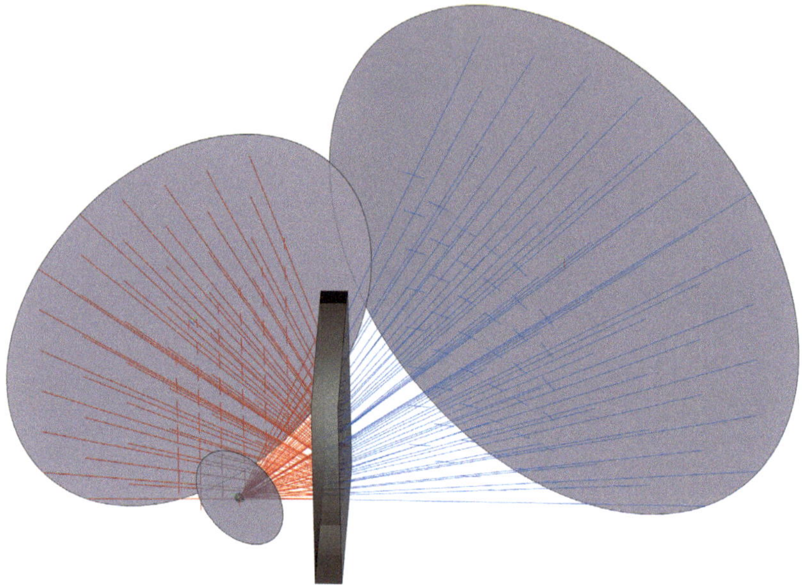

FIGURE 7.36
Ray-tracing of `Brewster_Stokes2`.

since the angles are the same for all the rays. The polarization of the reflected rays is polarization parallel to the plane of incidence. As respected by the Brewster angle.

For the next example, the angles change for all the rays since the field comes from a single-point object. The ray-tracing is presented in figure 7.36. For this example, the source is an on-axis field with a single wavelength, which is 587nm. The source has the following Stokes polarization parameters $S_0 = 0$, $S_1 = 0$, $S_2 = 0$, and $S_3 = 0$.

The path is: `C:\Program Files\Quadoa\examples\Polarization\Brewster_Stokes2`. We shall call this example as `Brewster_Stokes2`.

The `Poincare Sphere` is a graphical tool for visualizing different types of polarized light [39]. The `Poincare Sphere` for the reflecting and refracting fields is shown in the left and right figure 7.37, respectively. In this example, the incident is no longer the Brewster angle, and the incident angle changes for each ray. Therefore the output on the `Poincare Sphere` we can appreciate regions rather than single points in the outcome fields.

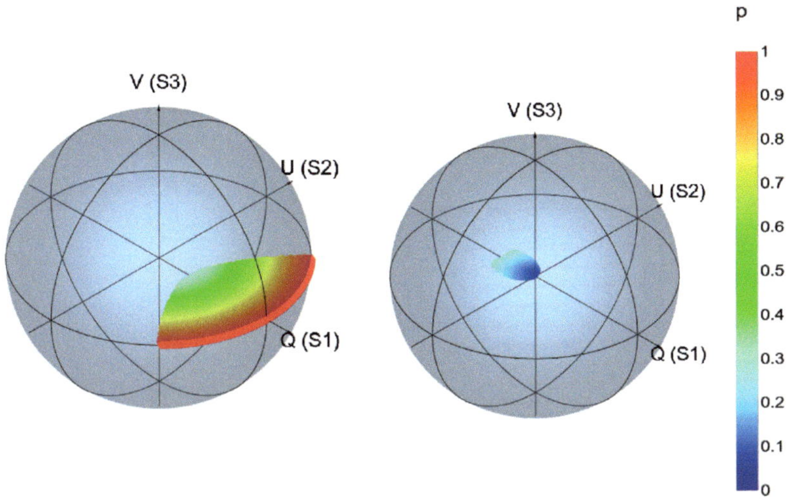

FIGURE 7.37
`Poincare Sphere` of `Brewster_Stokes2`.

7.7 Combination Between Shapes

In addition to freeform surfaces for the upcoming generation of optical devices, `Quadoa Optical CAD` offers a versatile approach to defining arbitrary surface shapes. This capability is enabled by integrated surface stackability, allowing for the creation of a vast array of surface types by combining different surface shapes with various properties such as deviations from measurement data, polarizers, apertures, coatings, and phase-functions.

The next example shows a wired and mostly useless lens in practice, but it is illustrative to show the robustness of the forms of a surface can be achieved by `Quadoa Optical CAD`. The ray-tracing of this optical system is presented in figure 7.38. The first surface is made by combining different surface shapes with various properties. The second surface is flat.

The path is: `C:\Program Files\Quadoa\examples\Surface Definitions\Brewster_Stokes2`. We shall call this design as `Form_Stak`.

The first surface is composed of a `Spherical Form`, `Zernike Form` z, `Cosie Form`, `Guass Peak Form`, and a `Zernike Phase` z, the wired shape can be seen in the `Surface Form` presented in figure 7.39. From the figure, we see the complicated shape of the first surface. But It is easy to get for `Quadoa Optical CAD` due to its surface combination capabilities.

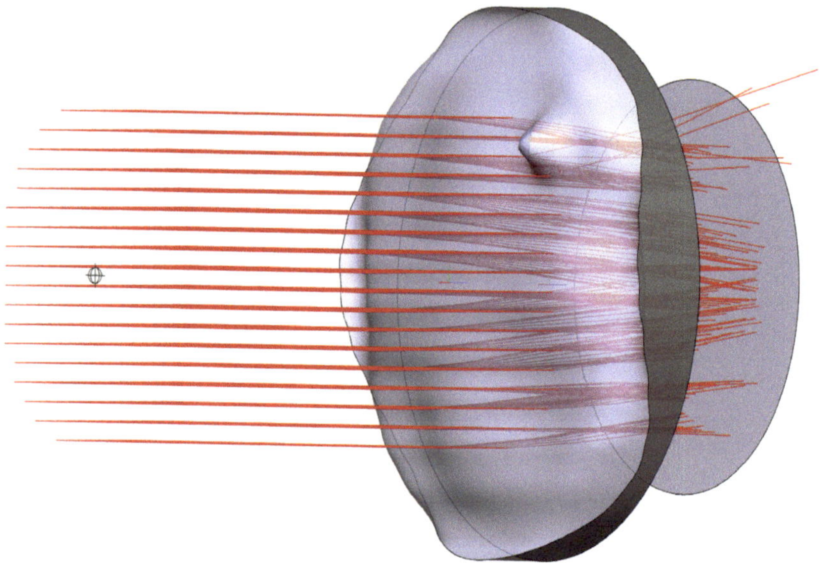

FIGURE 7.38
Ray tracing of `Form_Stak`. The first surface is composed by a `Spherical Form`, `Zernike Form`, `Cosie Form`, `Guass Peak Form`, and a `Zernike Phase`. The second surface is flat.

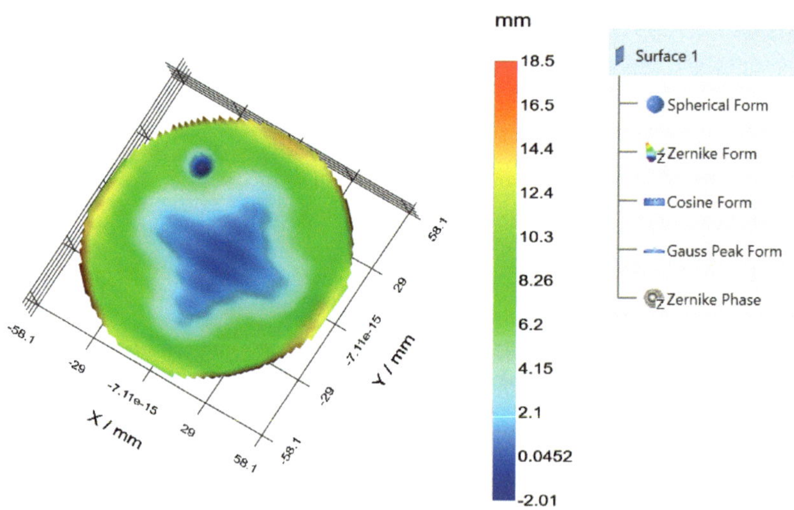

FIGURE 7.39
`Surface Form` of `Form_Stak`.

7.8 Summary

In this chapter, we review several types of forms, phases, coatings and polarization properties in surfaces.

Most of these surfaces and properties, that we seen in this chapter, can be found in optical software. Since they are widely used in optical design. But, to mix/merge them is something that `Quadoa Optical CAD` can make it so far.

8

Lens Design Example 2: Stigmatic Singlet Lens

> Although to penetrate into the intimate mysteries of nature and thence to learn the true causes of phenomena is not allowed to us, nevertheless it can happen that a certain fictive hypothesis may suffice for explaining many phenomena.
>
> Leonhard Euler

8.1 Introduction

In the previous chapter, we study the optimization of an optical system starting from scratch. That procedure is the most common in the industry [5, 58]. Les designers crush numbers with their optical software until they reach a good design [12, 13]. In general, the problem of lens design is just too hard to have an overall theory that explains the problem from its deepest roots to a mass-produced design [1, 3, 40].

Many theories out there try to explain how must be the designing phase of an imaging forming system. The most common is the third aberration theory which usually fails for designs with a big field of view [23, 25, 41]. Nodal aberrations theory which is mathematical to dense and it's just well-known for a few designers [68, 69, 70]. The *first time right* approach, based on the Fermat principle is another example of a dense mathematical theory to designs systems from scratch [71]. The simultaneous multiple surfaces (SMS) method involves the simultaneous calculation of N optical surfaces (refractive or reflective) using N one-parameter bundles of rays for which specific conditions are imposed [72, 73].

Here we are going to study the method provided by *stigmatic optics* [24] and *optical path theory* [74], in order to *analytically lens design* [7] a system from scratch. This theory is also mathemtically dense, but beautiful in its

DOI: 10.1201/9781003506355-8

nature since it does not accept any paraxial approximation nor numerical approach, like the other theories mentioned earlier. *Stigmatic optics* study the imaging forming system which is astigmatic with no paraxial, numerical approximations, just based on the principle of Fermat and Snell's law. `Optical path theory` surpasses `stigmatic optics` in its breadth, as it delves into the trajectory of light governed by the fundamental Fermat principle. This framework extends to examining light paths within optical systems afflicted by aberrations. Aberrations arise when an optical system fails to achieve stigmatic focus for all points of an object. Consequently, it becomes impossible to faithfully resolve a point object at a singular location in the image space, signaling the presence of aberrations. The severity and nature of these aberrations depend on various factors.

After studying the main problem of the mentioned approach we are going yo export the result to `Quadoa Optical CAD`, to show that `Quadoa Optical CAD` is available to work directly with the mentioned dense mathematical theories [34, 35, 36, 45, 75].

8.2 Finite Object Finite/Image

The theory based on *stigmatic optics* and *optical path theory* starts only unsay the Fermat principle and Snell's law, with no approximation. The fundamental problem of this theory is the general equation of the stigmatic lens [45]. In this section, we are going to get such equation, because later we are going to adapt the results to `Quadoa Optical CAD`.

First, we'll examine the scenario where the image object distance is finite. Following that, we'll investigate situations where the object resides at a more distant at infinity.

The chapter presents a problem statement: How should the second surface of the lens be configured, given the specifications of the first surface, along with the positions of the object and image, refractive indexes within the object and image spaces and the glass, and the central thickness of the singlet, to ensure the elimination of spherical aberration?

The answer we are expecting is now the shape of the second surface (z_b, r_b), given the first surface (z_a, r_a), such that the singlet is free of spherical aberration. Where r_a is the only independent variable. Thus, z_b, r_b and z_a are functions of r_a. z_b is the sagita of the second surface and its radius is r_b. Th sagita of the first surface is z_a and the radius is r_a. As usual, the coordinate system is set at the vertex of z_a. See figure 8.1.

The desired outcome is the determination of the second surface's shape, denoted as (z_b, r_b), given the configuration of the first surface, represented as (z_a, r_a), to achieve spherical aberration-free performance, a stigmatic lens. Here, r_a serves as the sole independent variable, with z_b, r_b, and z_a becoming

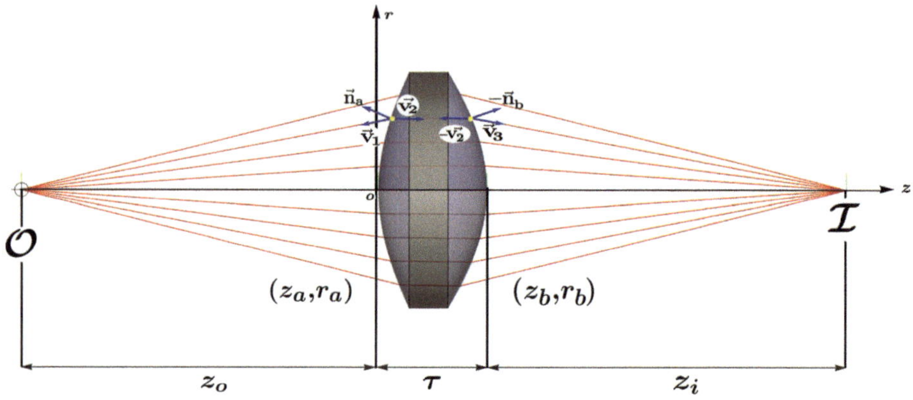

FIGURE 8.1
Illustration of an on-axis stigmatic singlet. The first and second surfaces are denoted by (z_a, r_a) and (z_b, r_b), respectively. The axial separation between the first surface and the object is z_o, while the axial diameter of the lens is τ, and the axial distance between the second surface and the image is z_i. The refractive index in the object space is n_o, within the lens it is n, and in the image space, it is n_i.

functions of r_a. The term z_b denotes the sagitta of the second surface, with a corresponding radius of r_b. Similarly, the sagitta of the first surface is z_a, with a radius of r_a. The coordinate system remains centered at the vertex of z_a, as illustrated in figure 8.1.

8.2.1 Fermat Principle

Previously, in Chapter 2, we observed that for a stigmatic design, the optical path length must be consistent for all rays originating from a point object on the axis and converging onto a point image on the axis. Consequently, the optical path length remains constant for all rays.

Additionally, according to the Fermat principle, light travels along paths where it takes the least amount of time. Thus, by merging both assertions, we can contrast the optical path length of the axial ray with that of the non-axial ray.

Let's begin with the optical path length of the axial ray. Originating from $-z_o$, it traverses through the lens, covering a distance of τ, and finally, from the second surface, it travels a distance of z_i to converge at the point image. Refer to figure 8.1. The optical path of the axial ray is expressed as,

$$-n_o z_o + n\tau + n_i z_i = \text{constant}. \tag{8.1}$$

The optical path length of a non-axial ray is more complex, as it varies depending on whether the object/image is real or virtual.

Let's consider the more straightforward scenario first, where both the object and the image are real, denoted by $z_o < 0$ and $z_i > 0$.

$$
\begin{aligned}
-n_o z_o + n\tau + n_i z_i &= n_o\sqrt{r_a^2 + (z_a - z_o)^2} \\
&+ n\sqrt{(r_b - r_a)^2 + (z_b - z_a)^2} \\
&+ n_i\sqrt{r_b^2 + (z_b - \tau - z_i)^2}, \qquad (8.2)
\end{aligned}
$$

when $z_o > 0$ and $z_i > 0$, it signifies that the object is virtual and the image is real.

$$
\begin{aligned}
-n_o z_o + n\tau + n_i z_i &= -n_o\sqrt{r_a^2 + (z_a - z_o)^2} \\
&+ n\sqrt{(r_b - r_a)^2 + (z_b - z_a)^2} \\
&+ n_i\sqrt{r_b^2 + (z_b - \tau - z_i)^2}. \qquad (8.3)
\end{aligned}
$$

For a real object and virtual image, we have $z_o < 0$ and $z_i < 0$.

$$
\begin{aligned}
-n_o z_o + n\tau + n_i z_i &= n_o\sqrt{r_a^2 + (z_a - z_o)^2} \\
&+ n\sqrt{(r_b - r_a)^2 + (z_b - z_a)^2} \\
&- n_i\sqrt{r_b^2 + (z_b - \tau - z_i)^2}. \qquad (8.4)
\end{aligned}
$$

Finally, if both the object and the image are virtual, then $z_o > 0$ and $z_i < 0$

$$
\begin{aligned}
-n_o z_o + n\tau + n_i z_i &= -n_o\sqrt{r_a^2 + (z_a - z_o)^2} \\
&+ n\sqrt{(r_b - r_a)^2 + (z_b - z_a)^2} \\
&- n_i\sqrt{r_b^2 + (z_b - \tau - z_i)^2}, \qquad (8.5)
\end{aligned}
$$

The four previous equations can be unified into a single expression, represented by,

$$
\boxed{
\begin{aligned}
-n_o z_o + n\tau + n_i z_i = \; & -n_o \mathrm{sgn}(z_o)\sqrt{r_a^2 + (z_a - z_o)^2} \\
& +n\sqrt{(r_b - r_a)^2 + (z_b - z_a)^2} \\
& +n_i \mathrm{sgn}(z_i)\sqrt{r_b^2 + (z_b - \tau - z_i)^2}.
\end{aligned}
} \qquad (8.6)
$$

Please observe that $\mathrm{sgn}(\cdot)$ denotes the sign of its argument, where the function is undefined if the argument is zero. It's important to observe that the two unknowns are (z_b, r_b), and we currently have only one equation. Therefore, we require another equation, which we will derive from Snell's law.

8.2.2 Snell's Law

Let's shift our attention to Snell's law, which describes how light behaves when transitioning from one medium to another. The vector form of Snell's law at the first surface is given by

$$\boxed{\vec{\mathbf{v_2}} = \frac{n_o}{-\mathrm{sgn}(z_o)n}[\vec{\mathbf{v_1}} - (\vec{\mathbf{n_a}} \cdot \vec{\mathbf{v_1}})\vec{\mathbf{n_a}}] - \vec{\mathbf{n_a}}\sqrt{1 - \frac{n_o^2}{n^2}(\vec{\mathbf{n_a}} \times \vec{\mathbf{v_1}})^2}} \tag{8.7}$$

here, $\vec{\mathbf{v_1}}$ represents the unit vector of the incident ray, $\vec{\mathbf{v_2}}$ denotes the unit vector of the refracted ray, and $\vec{\mathbf{n_a}}$ stands for the normal vector of the first surface. The term $-\mathrm{sgn}(z_o)$ accounts for whether the object is real or virtual. Please refer to figure 8.1.

The unit vector of the incident ray $\vec{\mathbf{v_1}}$ is,

$$\vec{\mathbf{v_1}} = \frac{r_a\vec{\mathbf{e_1}} + (z_a - z_o)\vec{\mathbf{e_2}}}{\sqrt{r_a^2 + (z_a - z_o)^2}}. \tag{8.8}$$

The unit vector of the refracted ray $\vec{\mathbf{n_a}}$ is expressed by,

$$\vec{\mathbf{v_2}} = \frac{(r_b - r_a)\vec{\mathbf{e_1}} + (z_b - z_a)\vec{\mathbf{e_2}}}{\sqrt{(r_b - r_a)^2 + (z_b - z_a)^2}}, \tag{8.9}$$

Finally, we have the normal vector of the first surface $\vec{\mathbf{n_a}}$ as,

$$\vec{\mathbf{n_a}} = \frac{z_a'\vec{\mathbf{e_1}} - \vec{\mathbf{e_2}}}{\sqrt{1 + z_a'^2}}, \tag{8.10}$$

certainly, $\vec{\mathbf{e_1}}$ represents the unit vector in the r direction, while $\vec{\mathbf{e_2}}$ denotes the unit vector in the z-direction.

Let's concentrate initially on evaluating $(\vec{\mathbf{n_a}} \times \vec{\mathbf{v_1}})^2$ as part of the replacement process in equation (8.7).

$$(\vec{\mathbf{n_a}} \times \vec{\mathbf{v_1}}) = \frac{[r_a + (z_a - z_o)z_a']}{\sqrt{r_a^2 + (z_a - z_o)^2}\sqrt{1 + z_a'^2}}(\vec{\mathbf{e_1}} \times \vec{\mathbf{e_2}}), \tag{8.11}$$

where $(\vec{\mathbf{e_1}} \times \vec{\mathbf{e_2}})^2 = \vec{\mathbf{e_3}}^2 = 1$,

$$(\vec{\mathbf{n_a}} \times \vec{\mathbf{v_1}})^2 = \frac{[r_a + (z_a - z_o)z_a']^2}{[r_a^2 + (z_a - z_o)^2](1 + z_a'^2)}. \tag{8.12}$$

Now let's concentrate squared root in (8.7),

$$-\vec{\mathbf{n_a}}\sqrt{1+\frac{1}{n^2}(\vec{\mathbf{n_a}}\times\vec{\mathbf{v_1}})^2} = -\frac{(z_a'\vec{\mathbf{e_1}} - \vec{\mathbf{e_2}})}{\sqrt{1 + z_a'^2}}\sqrt{1-\frac{n_o^2[r_a + (z_a-z_o)z_a']^2}{n^2[r_a^2 + (z_a-z_o)^2](1 + z_a'^2)}}. \tag{8.13}$$

In equation (8.7), we have $\vec{\mathbf{v_1}} - (\vec{\mathbf{n_a}} \cdot \vec{\mathbf{v_1}})\vec{\mathbf{n_a}}$,

$$\vec{\mathbf{v_1}} - (\vec{\mathbf{n_a}} \cdot \vec{\mathbf{v_1}})\vec{\mathbf{n_a}} = \frac{r_a\vec{\mathbf{e_1}} + (z_a - z_o)\vec{\mathbf{e_2}}}{\sqrt{r_a^2 + (z_a - z_o)^2}} - \left[\frac{r_a z_a' - (z_a - z_o)}{\sqrt{r_a^2 + (z_a - z_o)^2}\sqrt{1 + z_a'^2}}\right]\frac{(z_a'\vec{\mathbf{e_1}} - \vec{\mathbf{e_2}})}{\sqrt{1 + z_a'^2}}. \quad (8.14)$$

After simplification and multiplication by $\frac{n_o}{(-\text{sgn}(z_o))n}$ above's expression,

$$\frac{n_o}{-\text{sgn}(z_o)n}[\vec{\mathbf{v_1}} - (\vec{\mathbf{n_a}} \cdot \vec{\mathbf{v_1}})\vec{\mathbf{n_a}}] = \frac{n_o[(z_a - z_o)z_a' + r_a]}{-\text{sgn}(z_o)n\sqrt{r_a^2 + (z_o - z_a)^2}\,(1 + z_a'^2)}\vec{\mathbf{e_1}} + \frac{n_o\,[r_a + (z_a - z_o)z_a']\,z_a'}{-\text{sgn}(z_o)n\sqrt{r_a^2 + (z_o - z_a)^2}\,(1 + z_a'^2)}\vec{\mathbf{e_2}}. \quad (8.15)$$

Combining equations (8.13) and (8.15), we obtain $\vec{\mathbf{v_2}}$ as the result of the equation (8.7).

$$\vec{\mathbf{v_2}} = \frac{n_o[(z_a - z_o)z_a' + r_a]}{-\text{sgn}(z_o)n\sqrt{r_a^2 + (z_o - z_a)^2}\,(1 + z_a'^2)}\vec{\mathbf{e_1}} + \frac{n_o\,[r_a + (z_a - z_o)z_a']\,z_a'}{-\text{sgn}(z_o)n\sqrt{r_a^2 + (z_o - z_a)^2}\,(1 + z_a'^2)}\vec{\mathbf{e_2}} - \frac{(z_a'\vec{\mathbf{e_1}} \quad \vec{\mathbf{e_2}})}{\sqrt{1 + z_a'^2}}\sqrt{1 - \frac{n_o^2[r_a + (z_a - z_o)z_a']^2}{n^2[r_a^2 + (z_a - z_o)^2](1 + z_a'^2)}}. \quad (8.16)$$

separating terms multiplied by, $\vec{\mathbf{e_1}}$ and $\vec{\mathbf{e_2}}$,

$$\frac{r_b - r_a}{\sqrt{(z_b - z_a)^2 + (r_b - r_a)^2}} = \frac{n_o[(z_a - z_o)z_a' + r_a]}{-\text{sgn}(z_o)n\sqrt{r_a^2 + (z_o - z_a)^2}\,(1 + z_a'^2)} - z_a'\frac{\sqrt{1 - \frac{n_o^2\,[r_a + (z_a - z_o)z_a']^2}{n^2\,[r_a^2 + (z_o - z_a)^2]\,(1 + z_a'^2)}}}{\sqrt{1 + z_a'^2}}, \quad (8.17)$$

and,

$$\frac{z_b - z_a}{\sqrt{(z_b - z_a)^2 + (r_b - r_a)^2}} = \frac{n_o\left[r_a + (z_a - z_o)z_a'\right] z_a'}{-\mathrm{sgn}(z_o) n \sqrt{r_a^2 + (z_o - z_a)^2}\,(1 + z_a'^2)} + \frac{\sqrt{1 - \dfrac{n_o^2\left[r_a + (z_a - z_o)z_a'\right]^2}{n^2\left[r_a^2 + (z_o - z_a)^2\right](1 + z_a'^2)}}}{\sqrt{1 + z_a'^2}}, \tag{8.18}$$

The unknowns z_b and r_b appear only on the left side of equations (8.17) and (8.18). It's worth noting that the right side of equations (8.17) and (8.18) correspond to the cosine directors within the lens, which implies $\wp_r^2 + \wp_z^2 = 1$. $\wp_r$ represents the director in the direction of $\vec{\mathbf{e_1}}$, and $\wp_z$ denotes the director along $\vec{\mathbf{e_2}}$. Consequently, we derive the following expression by substituting $\wp_r$ and $\wp_z$ into the right side of equations (8.17) and (8.18):

$$\wp_r = \frac{n_o[(z_a - z_o)z_a' + r_a]}{-\mathrm{sgn}(z_o) n \sqrt{r_a^2 + (z_o - z_a)^2}\,(1 + z_a'^2)} - z_a' \frac{\sqrt{1 - \dfrac{n_o^2\left[r_a + (z_a - z_o)z_a'\right]^2}{n^2\left[r_a^2 + (z_o - z_a)^2\right](1 + z_a'^2)}}}{\sqrt{1 + z_a'^2}}, \tag{8.19}$$

and,

$$\wp_z = \frac{n_o\left[r_a + (z_a - z_o)z_a'\right] z_a'}{-\mathrm{sgn}(z_o) n \sqrt{r_a^2 + (z_o - z_a)^2}\,(1 + z_a'^2)} + \frac{\sqrt{1 - \dfrac{n_o^2\left[r_a + (z_a - z_o)z_a'\right]^2}{n^2\left[r_a^2 + (z_o - z_a)^2\right](1 + z_a'^2)}}}{\sqrt{1 + z_a'^2}}. \tag{8.20}$$

The distance between the first and second surface is implicit in (8.17) and (8.18) with the following parameter,

$$\vartheta \equiv \sqrt{(z_b - z_a)^2 + (r_b - r_a)^2}. \tag{8.21}$$

Substituting ϑ into equations (8.17) and (8.18), we have:

$$\frac{z_b - z_a}{\vartheta} = \wp_z, \tag{8.22}$$

and,

$$\frac{r_b - r_a}{\vartheta} = \wp_r. \tag{8.23}$$

From the preceding equation, we can straightforwardly solve for z_b and r_b, giving,

$$z_b = z_a + \vartheta \wp_z, \tag{8.24}$$

and,

$$r_b = r_a + \vartheta \wp_r. \tag{8.25}$$

The expression above provides the solution we seek. It delineates, point by point, the configuration required for the second surface of the lens to achieve stigmatic performance. The only uncertainty lies in the length of ϑ. However, we can resolve this by combining this result with the Fermat principle. In the subsequent section, we'll proceed to do just that!

8.2.3 Solution

In the preceding section, we determined the expressions for z_b and r_b, albeit in terms of ϑ. Therefore, we require an additional equation, which is provided by the Fermat principle. It's crucial to note that we have only two unknowns, z_b and r_b, but we have a system with three equations. Two equations are derived from Snell's law at the first surface, while another is furnished by the Fermat principle. This discrepancy arises because the equations derived from Snell's law are not independent. Let's recall equation (8.6):

$$\begin{aligned} -n_o z_o + n\tau + n_i z_i = \; & -n_o \operatorname{sgn}(z_o)\sqrt{r_a^2 + (z_a - z_o)^2} \\ & + n\sqrt{(r_b - r_a)^2 + (z_b - z_a)^2} \\ & + n_i \operatorname{sgn}(z_i)\sqrt{r_b^2 + (z_b - \tau - z_i)^2}. \end{aligned} \tag{8.6}$$

To simplify the last expression, let's introduce the following parameter,

$$z_f = -n_o z_o + n\tau + n_i z_i + \operatorname{sgn}(z_o) n_0 \sqrt{r_a^2 + (z_o - z_a)^2} \tag{8.26}$$

combining equation (8.26) and equation (8.6),

$$z_f = n\sqrt{(r_b - r_a)^2 + (z_b - z_a)^2} + n_i \sqrt{r_b^2 + (-\tau + z_b - z_i)^2} \tag{8.27}$$

substituting equations (8.24) and (8.25) in (8.27),

$$z_f = n\vartheta + n_i \sqrt{(r_a + \vartheta \wp_r^2 + (-\tau + z_a - z_i + \vartheta \wp_z)^2}. \tag{8.28}$$

Furthermore, let's introduce another parameter to simplify equation (8.28),

$$z_\tau = -\tau + z_a - z_i, \tag{8.29}$$

next, substituting equation (8.29) in equation (8.28),

$$z_f = n\vartheta + n_i \sqrt{(r_a + \vartheta \wp_r)^2 + (z_\tau + \vartheta \wp_z)^2}. \tag{8.30}$$

By squaring both sides of the last equation and performing some manipulation, we have,

$$(z_f - n\vartheta)^2 = n_i^2 \left[(r_a + \vartheta\wp_r)^2 + (z_\tau + \vartheta\wp_z)^2\right] \tag{8.31}$$

expanding,

$$z_f^2 - 2z_f n\vartheta + n^2\vartheta^2 = n_i^2 \left[r_a^2 + 2r_a\vartheta\wp_r + \vartheta^2\wp_r^2 + z_\tau^2 + 2z_\tau\vartheta\wp_z + \vartheta^2\wp_z^2\right] \tag{8.32}$$

we can simplify equation (8.32) since $\vartheta^2\wp_r^2 + \vartheta^2\wp_z^2 = \vartheta^2$,

$$z_f^2 - 2z_f n\vartheta + n^2\vartheta^2 = n_i^2 \left[r_a^2 + 2r_a\vartheta\wp_r + \vartheta^2 + z_\tau^2 + 2z_\tau\vartheta\wp_z\right] \tag{8.33}$$

collecting terms multiplied by ϑ,

$$\vartheta^2\left(n_i^2 - n^2\right) + \vartheta\left(2z_f n + 2n_i^2 r_a\wp_r + 2n_i^2 z_\tau\wp_z\right) + n_i^2(r_a^2 + z_\tau^2) - z_f^2 = 0 \tag{8.34}$$

multiplying equation (8.34) by $4(n_i^2 - n^2)$,

$$\begin{aligned} 4(n_i^2 - n^2)^2\vartheta^2 + 4(n_i^2 - n^2)(2z_f n + 2n_i^2 r_a\wp_r + 2n_i^2 z_\tau\wp_z)\vartheta \\ = -4(n_i^2 - n^2)[n_i^2(r_a^2 + z_\tau^2) - z_f^2] \end{aligned} \tag{8.35}$$

summing $(2z_f n + 2n_i^2 r_a\wp_r + 2n_i^2 z_\tau\wp_z)^2$ in both side of equation (8.35),

$$\begin{aligned} 4(n_i^2 - n^2)^2\vartheta^2 + 4(n_i^2 - n^2)(2z_f n + 2n_i^2 r_a\wp_r + 2n_i^2 z_\tau\wp_z)\vartheta \\ + (2z_f n + 2n_i^2 r_a\wp_r + 2n_i^2 z_\tau\wp_z)^2 \\ = (2z_f n + 2n_i^2 r_a\wp_r + 2n_i^2 z_\tau\wp_z)^2 - 4(n_i^2 - n^2)[n_i^2(r_a^2 + z_\tau^2) - z_f^2] \end{aligned} \tag{8.36}$$

factoring the left side of the previous expression,

$$\begin{aligned} &[2(n_i^2 - n^2)\vartheta + (2z_f n + 2n_i^2 r_a\wp_r + 2n_i^2 z_\tau\wp_z)]^2 \\ &= (2z_f n + 2n_i^2 r_a\wp_r + 2n_i^2 z_\tau\wp_z)^2 - 4(n_i^2 - n^2)[n_i^2(r_a^2 + z_\tau^2) - z_f^2] \end{aligned} \tag{8.37}$$

applying the square root on both sides,

$$\begin{aligned} &2(n_i^2 - n^2)\vartheta + (2z_f n + 2n_i^2 r_a\wp_r + 2n_i^2 z_\tau\wp_z) \\ &= \pm\sqrt{(2z_f n + 2n_i^2 r_a\wp_r + 2n_i^2 z_\tau\wp_z)^2 - 4(n_i^2 - n^2)[n_i^2(r_a^2 + z_\tau^2) - z_f^2]}. \end{aligned} \tag{8.38}$$

Finally we can solve for ϑ,

$$\vartheta = \frac{-[z_f n + n_i^2(r_a \wp_r + z_T \wp_z)]}{n_i^2 - n^2} - \frac{\sqrt{4\left(n_i^2 - n^2\right)\left[n_i^2\left(r_a^2 + z_T^2\right) - z_f^2\right] + 4\left[z_f n + n_i^2(r_a \wp_r + z_T \wp_z)\right]^2}}{2n_i^2 - 2n^2} \tag{8.39}$$

we can simplify a little bit more and remove the number 2,

$$\vartheta = \frac{-[z_f n + n_i^2(r_a \wp_r + z_T \wp_z)]}{n_i^2 - n^2} - \frac{\sqrt{\left(n_i^2 - n^2\right)\left[n_i^2\left(r_a^2 + z_T^2\right) - z_f^2\right] + \left[z_f n + n_i^2(r_a \wp_r + z_T \wp_z)\right]^2}}{n_i^2 - n^2}. \tag{8.40}$$

Using equations (8.24), (8.25), and (8.40), we can express the second surface as, for z_b

$$z_b = z_a + \vartheta \wp_z, \tag{8.41}$$

and for r_b as,

$$r_b = r_a + \vartheta \wp_r. \tag{8.42}$$

Equations (8.41) and (8.42) stand as the most pivotal equations in the chapter. It delineates, point by point, the configuration required for the second surface to achieve a stigmatic lens. It's crucial to note that equations (8.41) and (8.42) apply only to singlet lenses, where rays inside them do not intersect. Put differently, equations (8.41) and (8.42) affirm that for a given point on the first surface, there exists a unique corresponding point on the second surface to attain stigmatism.

It's also important to note that in the expression for ϑ (equation (8.40)), there's no inclusion of $\text{sgn}(t_b)$. This implies that the second surface adjusts its shape based on the image's location, irrespective of whether the image is real or virtual.

Additionally, equation (8.40) features a plus-minus sign $\pm$ preceding the square root, resulting in two potential solutions. However, determining which solution is applicable depends on the sign of the refractive index n and whether the object/image is real or virtual. In the subsequent sections, we'll demonstrate several illustrative examples. But before that, let's present an expanded version of equations (8.41) and (8.42),

$$
\begin{aligned}
z_b(r_a) &= z_a(r_a) \\
&+ \frac{1}{n^2-n_i^2}\left(\frac{n_o\left[z_a'(r_a)\left(r_a+(z_a(r_a)-z_o)z_a'(r_a)\right)\right]}{n\sqrt{r_a+(z_o-z_a(r_a))^2}\left(r_a^2+z_a'(r_a)^2\right)} + \frac{\sqrt{1-\frac{n_o^2\left[r_a+(z_a(r_a)-z_o)z_a'(r_a)\right]^2}{n^2\left[r_a+(z_o-z_a(r_a))^2\right]\left(r_a^2+z_a'(r_a)^2\right)}}}{\sqrt{r_a^2+z_a'(r_a)^2}}\right) \times \\
&\left(
\begin{aligned}
& n\left(n_i z_i - n_o z_o + n\tau + n_o\,\mathrm{Sign}\left[z_o\right]\sqrt{r_a+(z_o-z_a(r_a))^2}\right) \\
& + n_i^2 \left\{
\begin{aligned}
&\left[\begin{aligned} & r_a \times \\ &\left(\frac{n_o\left[z_a'(r_a)\left(r_a+(z_a(r_a)-z_o)z_a'(r_a)\right)\right]}{n\sqrt{r_a+(z_o-z_a(r_a))^2}\left(r_a^2+z_a'(r_a)^2\right)} - \frac{z_a'(r_a)\sqrt{1-\frac{n_o^2\left[r_a+(z_a(r_a)-z_o)z_a'(r_a)\right]^2}{n^2\left[r_a+(z_o-z_a(r_a))^2\right]\left(r_a^2+z_a'(r_a)^2\right)}}}{\sqrt{r_a^2+z_a'(r_a)^2}}\right)\end{aligned}\right] \\
&+ \\
&\left[\begin{aligned} & \left(z_a(r_a)-z_i-\tau\right) \times \\ &\left(\frac{n_o\left[z_a'(r_a)\left(r_a+(z_a(r_a)-z_o)z_a'(r_a)\right)\right]}{n\sqrt{r_a+(z_o-z_a(r_a))^2}\left(r_a^2+z_a'(r_a)^2\right)} + \frac{\sqrt{1-\frac{n_o^2\left[r_a+(z_a(r_a)-z_o)z_a'(r_a)\right]^2}{n^2\left[r_a+(z_o-z_a(r_a))^2\right]\left(r_a^2+z_a'(r_a)^2\right)}}}{\sqrt{r_a^2+z_a'(r_a)^2}}\right)\end{aligned}\right]
\end{aligned}
\right\} \\
& - n_i\,\mathrm{Sign}\left[n_1\right] \times \\
& \sqrt{
\begin{aligned}
& \left(n_i z_i - n_o z_o + n\tau + n_o\,\mathrm{Sign}\left[z_o\right]\sqrt{r_a+(z_o-z_a(r_a))^2}\right)^2 + n^2\left(r_a+(z_a(r_a)-z_i-\tau)^2\right) \\
& - n_i^2 \left\{
\begin{aligned}
&\left[\begin{aligned} & \left(z_a(r_a)-z_i-\tau\right) \times \\ &\left(\frac{n_o\left[z_a'(r_a)\left(r_a+(z_a(r_a)-z_o)z_a'(r_a)\right)\right]}{n\sqrt{r_a+(z_o-z_a(r_a))^2}\left(r_a^2+z_a'(r_a)^2\right)} - \frac{z_a'(r_a)\sqrt{1-\frac{n_o^2\left[r_a+(z_a(r_a)-z_o)z_a'(r_a)\right]^2}{n^2\left[r_a+(z_o-z_a(r_a))^2\right]\left(r_a^2+z_a'(r_a)^2\right)}}}{\sqrt{r_a^2+z_a'(r_a)^2}}\right)\end{aligned}\right] \\
&+ \\
&\left[\begin{aligned} & -r_a \times \\ &\left(\frac{n_o\left[z_a'(r_a)\left(r_a+(z_a(r_a)-z_o)z_a'(r_a)\right)\right]}{n\sqrt{r_a+(z_o-z_a(r_a))^2}\left(r_a^2+z_a'(r_a)^2\right)} + \frac{\sqrt{1-\frac{n_o^2\left[r_a+(z_a(r_a)-z_o)z_a'(r_a)\right]^2}{n^2\left[r_a+(z_o-z_a(r_a))^2\right]\left(r_a^2+z_a'(r_a)^2\right)}}}{\sqrt{r_a^2+z_a'(r_a)^2}}\right)\end{aligned}\right]
\end{aligned}
\right\}^2 \\
& + 2n\left(n_i z_i - n_o z_o + n\tau + n_o\,\mathrm{Sign}\left[z_o\right]\sqrt{r_a+(z_o-z_a(r_a))^2}\right) \\
& \times \left\{
\begin{aligned}
&\left[\begin{aligned} & r_a \times \\ &\left(\frac{n_o\left[z_a'(r_a)\left(r_a+(z_a(r_a)-z_o)z_a'(r_a)\right)\right]}{n\sqrt{r_a+(z_o-z_a(r_a))^2}\left(r_a^2+z_a'(r_a)^2\right)} - \frac{z_a'(r_a)\sqrt{1-\frac{n_o^2\left[r_a+(z_a(r_a)-z_o)z_a'(r_a)\right]^2}{n^2\left[r_a+(z_o-z_a(r_a))^2\right]\left(r_a^2+z_a'(r_a)^2\right)}}}{\sqrt{r_a^2+z_a'(r_a)^2}}\right)\end{aligned}\right] \\
&+ \\
&\left[\begin{aligned} & \left(z_a(r_a)-z_i-\tau\right) \times \\ &\left(\frac{n_o\left[z_a'(r_a)\left(r_a+(z_a(r_a)-z_o)z_a'(r_a)\right)\right]}{n\sqrt{r_a+(z_o-z_a(r_a))^2}\left(r_a^2+z_a'(r_a)^2\right)} + \frac{\sqrt{1-\frac{n_o^2\left[r_a+(z_a(r_a)-z_o)z_a'(r_a)\right]^2}{n^2\left[r_a+(z_o-z_a(r_a))^2\right]\left(r_a^2+z_a'(r_a)^2\right)}}}{\sqrt{r_a^2+z_a'(r_a)^2}}\right)\end{aligned}\right]
\end{aligned}
\right\}
\end{aligned}
}
\end{aligned}
\right)
\end{aligned}
$$

$$
\begin{aligned}
r_b(r_a) &= r_a \\
&+ \frac{1}{n^2-n_i^2}\left(\frac{n_o\left[z_a'(r_a)\left(r_a+(z_a(r_a)-z_o)z_a'(r_a)\right)\right]}{n\sqrt{r_a+(z_o-z_a(r_a))^2}\left(r_a^2+z_a'(r_a)^2\right)} - \frac{z_a'(r_a)\sqrt{1-\frac{n_o^2\left[r_a+(z_a(r_a)-z_o)z_a'(r_a)\right]^2}{n^2\left[r_a+(z_o-z_a(r_a))^2\right]\left(r_a^2+z_a'(r_a)^2\right)}}}{\sqrt{r_a^2+z_a'(r_a)^2}}\right) \times \\
&\Bigg(n\left(n_i z_i - n_o z_o + n\tau + n_o \,\mathrm{Sign}\left[z_o\right]\sqrt{r_a+(z_o-z_a(r_a))^2}\right) \\
&+ n_i^2 \left\{ \begin{aligned} &\left[r_a \times \left(\frac{n_o\left[z_a'(r_a)\left(r_a+(z_a(r_a)-z_o)z_a'(r_a)\right)\right]}{n\sqrt{r_a+(z_o-z_a(r_a))^2}\left(r_a^2+z_a'(r_a)^2\right)} - \frac{z_a'(r_a)\sqrt{1-\frac{n_o^2\left[r_a+(z_a(r_a)-z_o)z_a'(r_a)\right]^2}{n^2\left[r_a+(z_o-z_a(r_a))^2\right]\left(r_a^2+z_a'(r_a)^2\right)}}}{\sqrt{r_a^2+z_a'(r_a)^2}}\right)\right] \\ &+ \\ &\left[\left(z_a(r_a)-z_i-\tau\right) \times \left(\frac{n_o\left[z_a'(r_a)\left(r_a+(z_a(r_a)-z_o)z_a'(r_a)\right)\right]}{n\sqrt{r_a+(z_o-z_a(r_a))^2}\left(r_a^2+z_a'(r_a)^2\right)} + \frac{\sqrt{1-\frac{n_o^2\left[r_a+(z_a(r_a)-z_o)z_a'(r_a)\right]^2}{n^2\left[r_a+(z_o-z_a(r_a))^2\right]\left(r_a^2+z_a'(r_a)^2\right)}}}{\sqrt{r_a^2+z_a'(r_a)^2}}\right)\right] \end{aligned} \right\} \\
&- n_i \,\mathrm{Sign}\left[n_1\right] \times \\
&\sqrt{\begin{aligned} &\left(n_i z_i - n_o z_o + n\tau + n_o \,\mathrm{Sign}\left[z_o\right]\sqrt{r_a+(z_o-z_a(r_a))^2}\right)^2 + n^2\left(r_a+(z_a(r_a)-z_i-\tau)^2\right) \\
&\left\{ \begin{aligned} &\left[\left(z_a(r_a)-z_i-\tau\right) \times \left(\frac{n_o\left[z_a'(r_a)\left(r_a+(z_a(r_a)-z_o)z_a'(r_a)\right)\right]}{n\sqrt{r_a+(z_o-z_a(r_a))^2}\left(r_a^2+z_a'(r_a)^2\right)} - \frac{z_a'(r_a)\sqrt{1-\frac{n_o^2\left[r_a+(z_a(r_a)-z_o)z_a'(r_a)\right]^2}{n^2\left[r_a+(z_o-z_a(r_a))^2\right]\left(r_a^2+z_a'(r_a)^2\right)}}}{\sqrt{r_a^2+z_a'(r_a)^2}}\right)\right] \\ &+ \\ &\left[-r_a \times \left(\frac{n_o\left[z_a'(r_a)\left(r_a+(z_a(r_a)-z_o)z_a'(r_a)\right)\right]}{n\sqrt{r_a+(z_o-z_a(r_a))^2}\left(r_a^2+z_a'(r_a)^2\right)} + \frac{\sqrt{1-\frac{n_o^2\left[r_a+(z_a(r_a)-z_o)z_a'(r_a)\right]^2}{n^2\left[r_a+(z_o-z_a(r_a))^2\right]\left(r_a^2+z_a'(r_a)^2\right)}}}{\sqrt{r_a^2+z_a'(r_a)^2}}\right)\right] \end{aligned} \right\}^2 \\
&-n_i^2 \\
&+ 2n\left(n_i z_i - n_o z_o + n\tau + n_o \,\mathrm{Sign}\left[z_o\right]\sqrt{r_a+(z_o-z_a(r_a))^2}\right) \\
&\times \left\{ \begin{aligned} &\left[r_a \times \left(\frac{n_o\left[z_a'(r_a)\left(r_a+(z_a(r_a)-z_o)z_a'(r_a)\right)\right]}{n\sqrt{r_a+(z_o-z_a(r_a))^2}\left(r_a^2+z_a'(r_a)^2\right)} - \frac{z_a'(r_a)\sqrt{1-\frac{n_o^2\left[r_a+(z_a(r_a)-z_o)z_a'(r_a)\right]^2}{n^2\left[r_a+(z_o-z_a(r_a))^2\right]\left(r_a^2+z_a'(r_a)^2\right)}}}{\sqrt{r_a^2+z_a'(r_a)^2}}\right)\right] \\ &+ \\ &\left[\left(z_a(r_a)-z_i-\tau\right) \times \left(\frac{n_o\left[z_a'(r_a)\left(r_a+(z_a(r_a)-z_o)z_a'(r_a)\right)\right]}{n\sqrt{r_a+(z_o-z_a(r_a))^2}\left(r_a^2+z_a'(r_a)^2\right)} + \frac{\sqrt{1-\frac{n_o^2\left[r_a+(z_a(r_a)-z_o)z_a'(r_a)\right]^2}{n^2\left[r_a+(z_o-z_a(r_a))^2\right]\left(r_a^2+z_a'(r_a)^2\right)}}}{\sqrt{r_a^2+z_a'(r_a)^2}}\right)\right] \end{aligned} \right\} \end{aligned}} \;\Bigg)
\end{aligned}
$$

The journey to derive equation (8.41) may seem straightforward when reading this chapter from start to finish. However, the equation itself is far from simple—it's expansive, spanning the last two pages when fully expanded.

Over centuries, many have attempted to obtain a general analytical closed-form solution, but all efforts failed. The key lies in the fact that throughout the process, no angles are used. Traditional forms of Snell's law complicate matters, as there's no direct correlation between angles and optical paths. However, with Snell's law expressed in its vector form, the association between optical paths and director cosines becomes clear and facilitates the solution process. It's important to note that not only did we fail to derive a general analytical solution in a closed form for the problem, but we also discovered that the solution is unique.

8.3 Stigmatic Aspheric Collector

In the next example, we'll explore the stigmatic aspheric collector. This scenario arises when $z_o \to -\infty$, indicating that the lens collects rays from a distant source at z_o. To configure the collector lens, we must apply the limit as $z_o \to -\infty$ to certain parameters within equation (8.41). These parameters include the cosine directors $\wp_r$, $\wp_z$, and z_f, as described by equations (8.19), (8.20), and (8.26). Let's recall them.

The cosine directors are, equation (8.19),

$$\wp_r = \frac{n_o[(z_a - z_o)z_a' + r_a]}{-\text{sgn}(z_o)n\sqrt{r_a^2 + (z_o - z_a)^2}\,(1 + z_a'^2)} - z_a'\frac{\sqrt{1 - \frac{n_o^2[r_a + (z_a - z_o)z_a']^2}{n^2[r_a^2 + (z_o - z_a)^2](1 + z_a'^2)}}}{\sqrt{1 + z_a'^2}}, \tag{8.19}$$

and equation (8.20),

$$\wp_z = \frac{n_o\,[r_a + (z_a - z_o)z_a']\,z_a'}{-\text{sgn}(z_o)n\sqrt{r_a^2 + (z_o - z_a)^2}\,(1 + z_a'^2)} + \frac{\sqrt{1 - \frac{n_o^2[r_a + (z_a - z_o)z_a']^2}{n^2[r_a^2 + (z_o - z_a)^2](1 + z_a'^2)}}}{\sqrt{1 + z_a'^2}}. \tag{8.20}$$

and z_f is equation (8.26),

$$z_f \equiv n_o\text{sgn}\,(z_o)\,\sqrt{(z_a - z_o)^{\,2} + r_a^2} + n_i z_i - n_o z_o + n\tau. \tag{8.26}$$

Then, we apply the $z_o \to -\infty$ over them. Applying the $z_o \to -\infty$ in equation (8.19),

$$\lim_{z_o \to -\infty} \wp_z = \frac{z_a'\left(n_o - n\sqrt{(z_a')^{\,2} + 1}\sqrt{\frac{(n^2 - n_o^2)(z_a')^2 + n^2}{n^2[(z_a')^2 + 1]}}\right)}{n\,[(z_a')^{\,2} + 1]}. \tag{8.43}$$

Furthermore, by computing the aforementioned limit in equation (8.20), we obtain,

$$\lim_{z_o \to -\infty} \wp_z = \frac{\sqrt{\frac{(n^2-n_o^2)(z_a')^2+n^2}{n^2((z_a')^2+1)}}}{\sqrt{(z_a')^2+1}} + \frac{n_o\,(z_a')^2}{n\,(z_a')^2+n}. \tag{8.44}$$

Certainly, we compute the mentioned limit on the parameter z_f using equation (8.26),

$$\lim_{z_o \to -\infty} z_f = -z_a n_o + n_i z_i + n\tau. \tag{8.45}$$

When $z_o \to -\infty$, ϑ in Equation (8.40) simplifies to,

$$\lim_{z_o \to -\infty} (\vartheta) = \frac{-\beta \pm \sqrt{\beta^2 + (n^2 - n_i^2)\left\{n_i^2 r_a^2 + n_i^2 z_\tau^2 - [\lim_{z_o \to -\infty}(z_f)]^2\right\}}}{(n_i^2 - n^2)}, \tag{8.46}$$

where,

$$\beta \equiv \left\{\left[\lim_{z_o \to -\infty}(z_f)\right] n + n_i^2 r_a \left[\lim_{z_o \to -\infty}(\wp_r)\right] + n_i^2 z_\tau \left[\lim_{z_o \to -\infty}(\wp_z)\right]\right\}. \tag{8.47}$$

Therefore, the second surface is given by,

$$\begin{cases} \lim_{z_o \to -\infty}(z_b) = z_a + [\lim_{z_o \to -\infty}(\vartheta)]\,[\lim_{z_o \to -\infty}(\wp_z)], \\ \lim_{z_o \to -\infty}(r_b) = r_a + [\lim_{z_o \to -\infty}(\vartheta)]\,[\lim_{z_o \to -\infty}(\wp_r)]. \end{cases} \tag{8.48}$$

To properly use equations (8.48) we need to use (8.43), (8.44), (8.45), (8.46), and (8.47). It's crucial to recognize the significance of equation (8.48). This equation represents a novel result, obtained by applying a limit to equation (8.41), without altering the derivation procedure. Remarkably, this limit perfectly supports the solution.

In this scenario, equation (8.48) is valid for real objects and virtual or real images. It's worth noting that evaluating the limit when $z_o \to -\infty$ is feasible with the vector notation proposed. In contrast, using angles with the conventional form of Snell's law would lead to numerous complications.

8.4 Back to Quadoa

Now that we have the expression to design a singlet free of spherical aberration, we can compute the numbers for example in `Mathematica` and

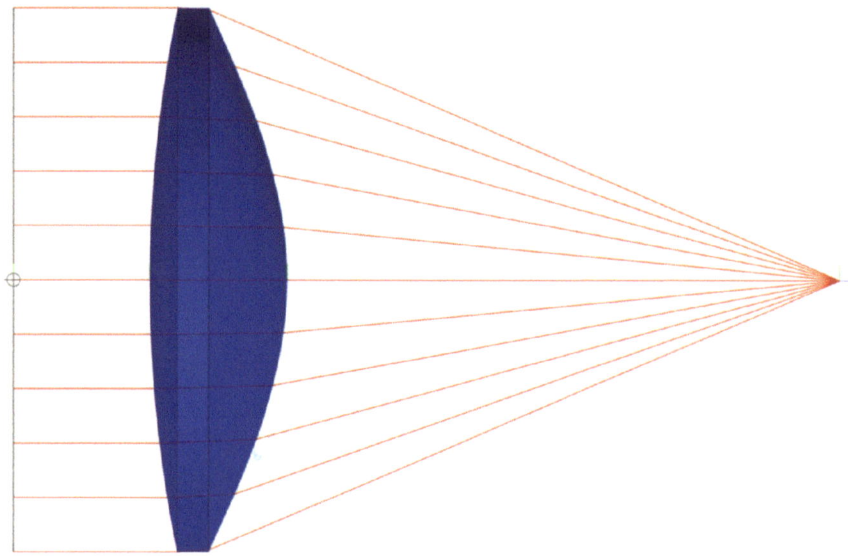

FIGURE 8.2
Ray-tracing of the stigmatic aspherical singlet lens.

then export them into `Quadoa Optical CAD`. To export them to `Quadoa Optical CAD` we use the option `Import CSV Point Cloud` in the `Add Form` section in the second surface [18].

For this example, we are going to design an on-axis astigmatic singlet for an object at infinity so we are going to use the equation of section tal. The material of the lens will be `B-K7` and the wavelength will be 550nm. The first surface of the lens will be `Spherical Form` with a radius of 50mm and an aperture of 10mm. The distance from the second surface to the image is $z_i = 10$mm. The central thickness of the lens is $\tau = 10$mm. The refraction index at object space n_0 and image space n_i will be considered as one.

After exporting the results from `Mathematica` into `Quadoa Optical CAD`, the ray-tracing is displayed in figure 8.2.

The singlet lens of figure 8.2 gives an astigmatic image point. This can be confirmed in figure 8.3 where it is displayed the `Spot Diagram Plot` of the stigmtic aspherical singlet lens. The black circle is the Airy disk and the red tiny point is the astigmatic image point. The results are as predicted by equations of the section 8.3.

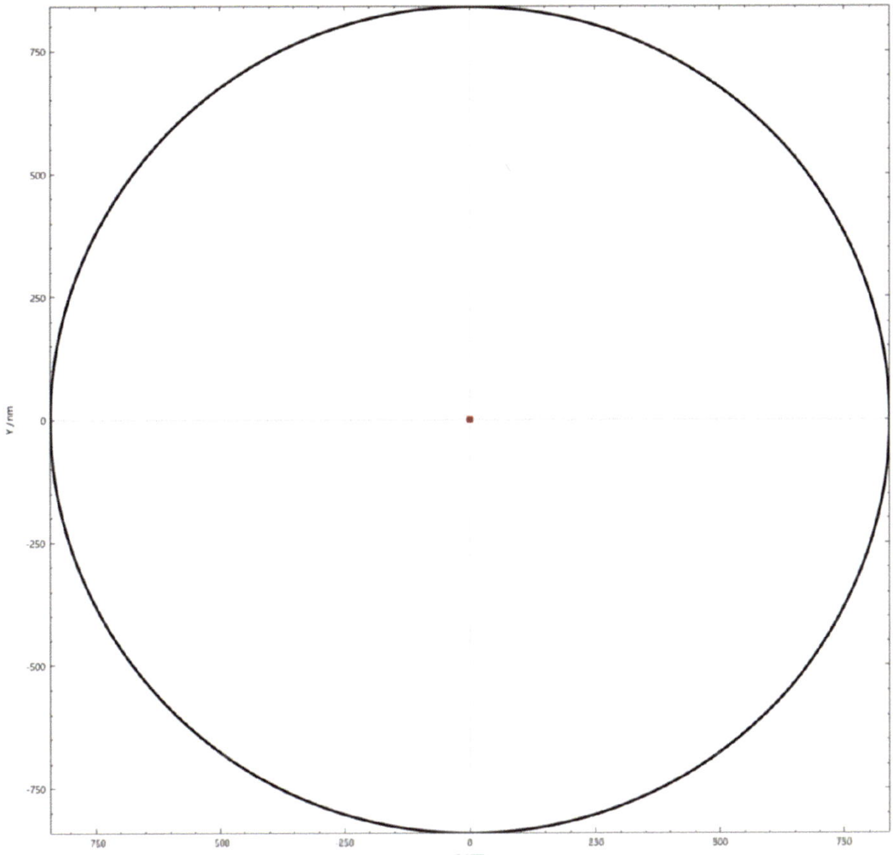

FIGURE 8.3
`Spot Diagram Plot` of the stigmatic aspherical singlet lens.

8.5 Summary

In this chapter, we studied a mathematical dense theory, *stigmatic optics* and *optical path theory*, in order to *analytically lens design* a singlet lens. The theory was exposed in the first three sections of the chapter. Then, from the theory exposed, we compute the data point cloud and export it to `Quadoa Optical CAD` with the tool `Import CSV Point Cloud`. The results were as expected by the theory.

9

Lens Design Example 3: Off-Axis Freeform Camera

In things to be seen at once, much variety makes confusion, another vice of beauty. In things that are not seen at once, and have no respect one to another, great variety is commendable, provided this variety transgress not the rules of optics and geometry

Christopher Wren

9.1 Introduction

This is the third chapter, where we show how to design an optical system step by step. The first chapter, of this kind, is Chapter 6, where we focus on a radially symmetric system from scratch [5, 58]. The illustrative example was an on-axis system. There we illustrate how to make a simple merit function. We correct several aberrations including chromatic aberrations.

The second example, chapter 8, was studied as a very sophisticated theory of stigmatic lenses and was successfully imported to `Quadoa Optical CAD`. The idea of that chapter was to show that the mathematical theories [7, 24, 34, 35, 36, 45, 74, 75] out there can be exported to `Quadoa Optical CAD` without major problems. Then the designs can be optimized with the robust tools provided by `Quadoa Optical CAD`. Now in this chapter, we want to show the design step by step of a more complex system and off-axis freeform camera made only by three mirrors [68, 69, 70, 71]. We show this example because this particular system can be a huge problem with other commercial software that due to no work with global coordinates [13, 15], like `Quadoa Optical CAD` does [18]. They use cumbersome concepts called *coordinate breaks*. Here we are going to show that it is very smooth to design such systems.

DOI: 10.1201/9781003506355-9

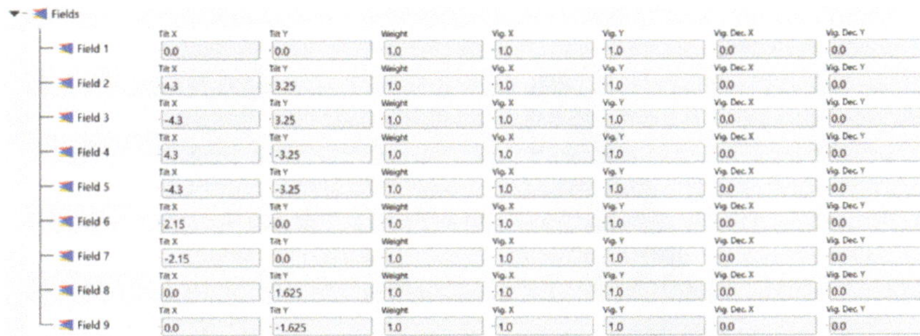

Fields

	Tilt X	Tilt Y	Weight	Vig. X	Vig. Y	Vig. Dec. X	Vig. Dec. Y
Field 1	0.0	0.0	1.0	1.0	1.0	0.0	0.0
Field 2	4.3	3.25	1.0	1.0	1.0	0.0	0.0
Field 3	-4.3	3.25	1.0	1.0	1.0	0.0	0.0
Field 4	4.3	-3.25	1.0	1.0	1.0	0.0	0.0
Field 5	-4.3	-3.25	1.0	1.0	1.0	0.0	0.0
Field 6	2.15	0.0	1.0	1.0	1.0	0.0	0.0
Field 7	-2.15	0.0	1.0	1.0	1.0	0.0	0.0
Field 8	0.0	1.625	1.0	1.0	1.0	0.0	0.0
Field 9	0.0	-1.625	1.0	1.0	1.0	0.0	0.0

FIGURE 9.1
Fields according to the specification of the illustrative example.

9.2 Illustrative Example

Let us consider that we want to design a freeform camera, with the following specifications:

- The camera has an independent surface stop. The aperture of the system should be 5mm.
- Made of three freeform mirrors with `XYZ Polynomial Form`.
- The sensor diagonal of the camera is 5mm. So, the x dimension is 3mm and the y length is 4mm. The mirrors are placed in the coordinates presented in `10xe2_flat`.
- The coordinate of the sensor is in `10xe2_flat`.
- The field of view is $8.6° \times 6.5°$.

The first step is to put this information in `Quadoa Optical CAD`. For that, we will use the file called `10xe2_flat` from the supplementary material.

In this file, the step, the mirrors, and the sensor are already placed in the given place. Please take a look at the tilt and rotation of all the elements. Since the only non-tilded element is the aperture stop.

The next step is to go for the source, since the system is made of mirrors it is not so important to choose the wavelength you can put 550nm. Then, the fields must be added. See figure 9.1, in it, we put the `Fields` according to the description. The first field is the on-axis field. The next fields are the corner fields in the image plane. The last fields are some fields in between the image plane.

I highly encourage the reader to really study the parameters presented in `10xe2_flat`. For example the aperture of each mirror. Since we do not want

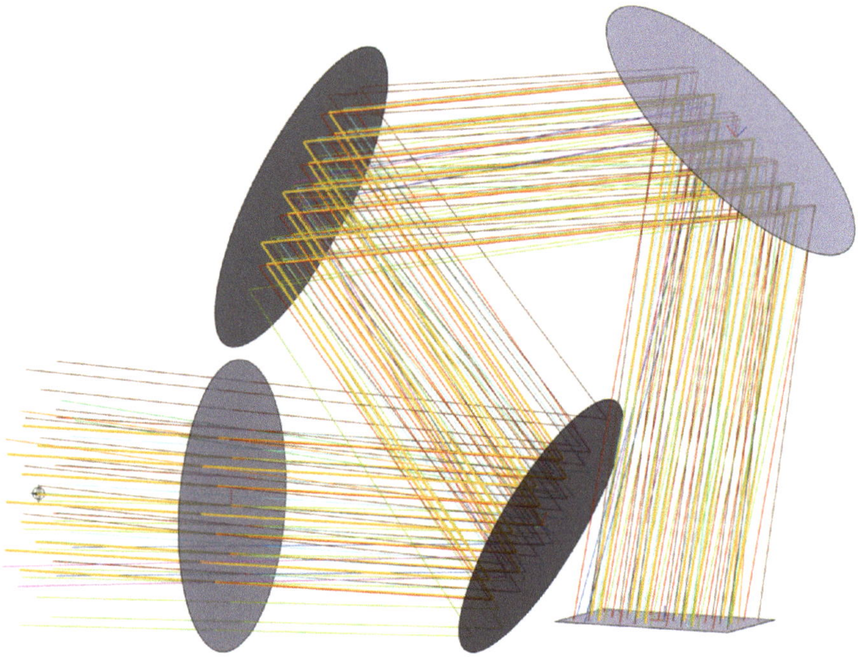

FIGURE 9.2
Illustrative example before the optimization.

an endless text to describe everything that is on `10xe2_flat`, please see this document carefully.

Let's keep going, notice that from the specifications the effective focal length is not given. This can be obtained with the sensor size and the field of view. So the effective focal length is implicit in the speciation of the design. This happens with the f-number.

In industry, it is widely common to work more in terms of the sensor size, the field of view, and the aperture size since there are parameters that can be easily measured. There are just some discrete sizes of sensors, for example.

With the information presented in Chapter 2, it is enough to obtain the effective focal length and the f-number It is a good exercise for the reader to get them from their own. Also, this information is always available in the analysis section of `Quadoa Optical CAD`.

At this point, the design should look like the one displayed in figure 9.2. This is before any optimization. The mirrors are flat and placed in the right position.

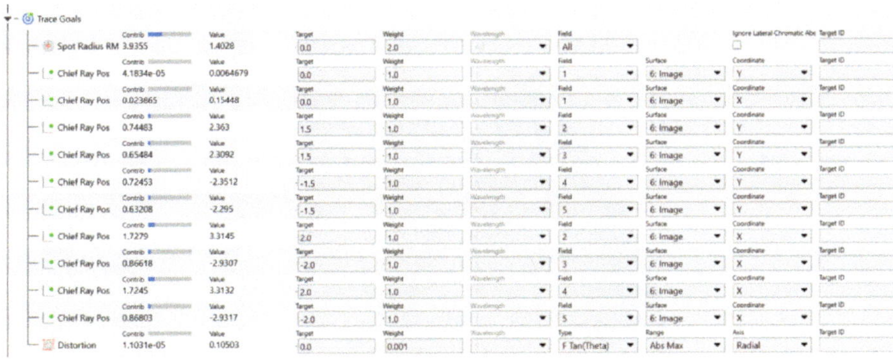

FIGURE 9.3
Merit function before optimization.

FIGURE 9.4
Merit function after optimization.

9.3 Optimization and Analysis

The next step is to set all the mirrors to be xy polynomials and some coefficients and mark them as variables. Then, we shall write the following merit function, presented in figure 9.3. The first-line goal is to reduce the spot size RMS.

The next lines are to put the corner fields in the right place, the respective corners of the sensor. Read the lines in detail, over and over again, if it is needed. The target is the first column, where the coordinates of the corners of the sensor are placed. The is the weight. The field is chosen and the surface, in this case, the image plane. Finally the coordinate x or y. For example, the on-axis field should arrive at the center of the sensor, so see lines 2 and 3 of figure 9.3.

The last line is to control distortion. The blue bars, in figure 9.3, graphically show the contribution to the merit function of each of the lines.

Then, we run the extended optimization. After some seconds, we get the merit function after optimization, which is presented in figure 9.4.

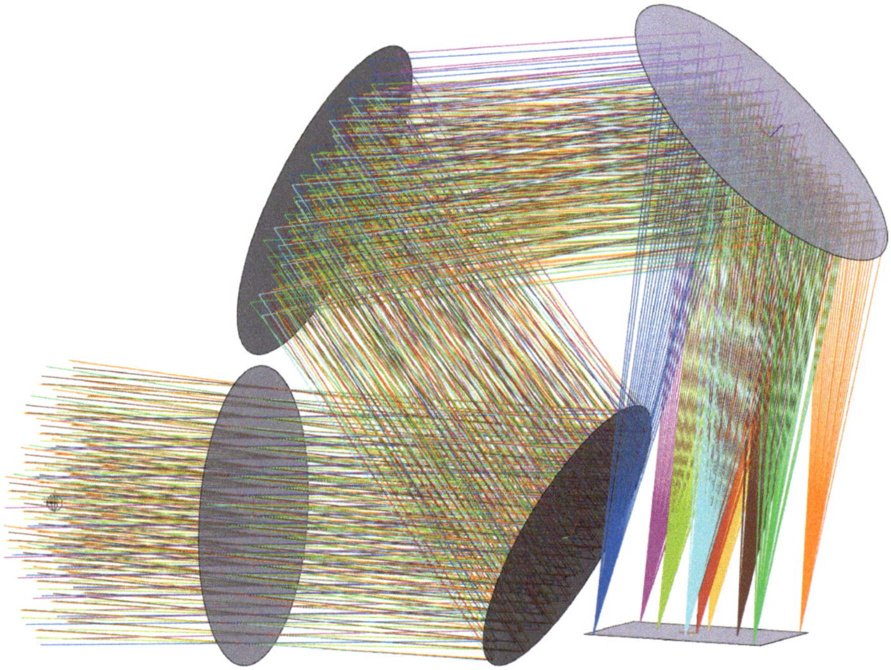

FIGURE 9.5
The ray-tracing of `10xe2`.

The first thing relevant from figure 9.4 is that the value of the commands `Chief Ray Pos` is mostly equal to the target. This means that our fields are landing correctly into the image plane. The Spot Radius RMS has decreased significantly. Also, the distortion has decreased.

Notice that the number of the merit function may change with the number of coefficients used as variables in the three freeform mirrors. By experience, I recommend using orders up-to 7 or 8 and then adding higher terms after the initial optimization. In the file `10xe2_flat`, the order of terms used is 8. The file after optimization is also added, in the supplementary material of this book, under the name `10xe2`. The ray-tracing of `10xe2` is presented in figure 9.5.

In the merit function, we did not put `Global Constrains` because the mirrors are separated enough that it is not necessary. But we can add them if we want to reduce the f-number of the system, by enlarging the apertures of the stop and mirrors. The reader can play with `10xe2` to get a f-number of 3.3, for example.

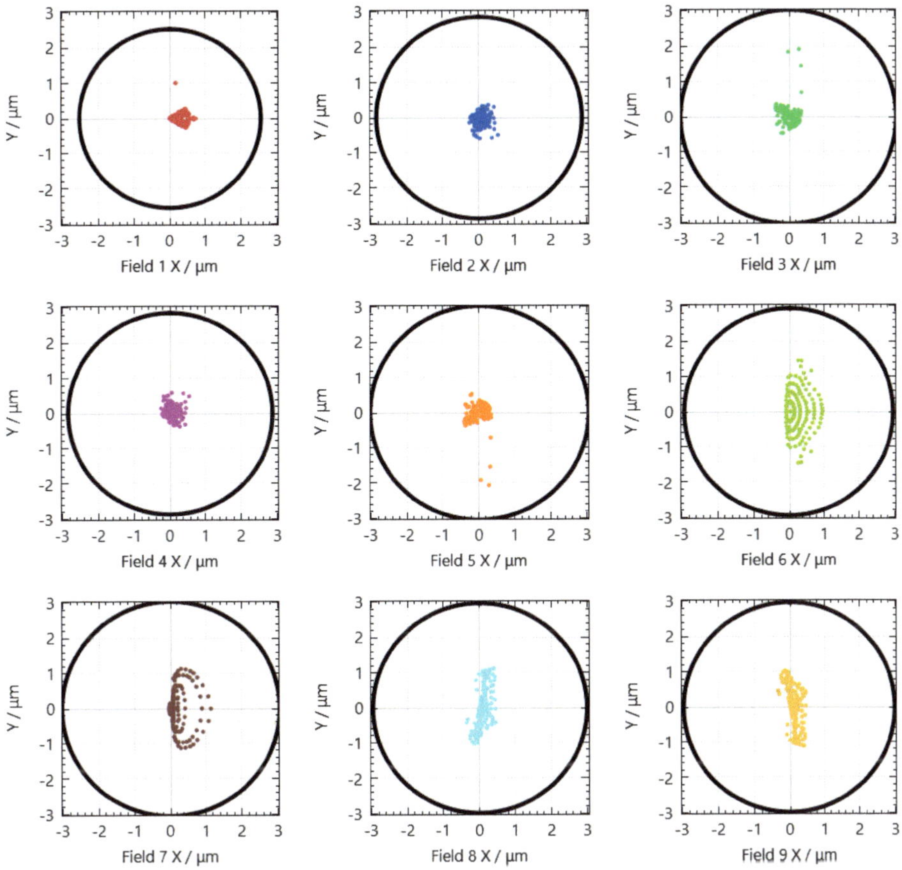

FIGURE 9.6
of `10xe2`.

Also, another tip, for better performance the position and tilts of the mirrors can be set as variables and re-optimized. However, the position and tilts of the mirrors should not consider in the first optimization. This may lead to unexpected results.

But for the moment let's focus on the spot diagram. The `Spot Diagram Plot` is shown in figure 9.6. The results are remarkable since all the fields are diffracted-limited. All the spots are inside the Air disk, which is represented by the black circles. The shape of the spots is not symmetric due to the freeform nature of the system.

In figure 9.7 is the `Footprint Diagram` of the rays. The first row is the stop and first mirror. The second row is the second and third mirror, and

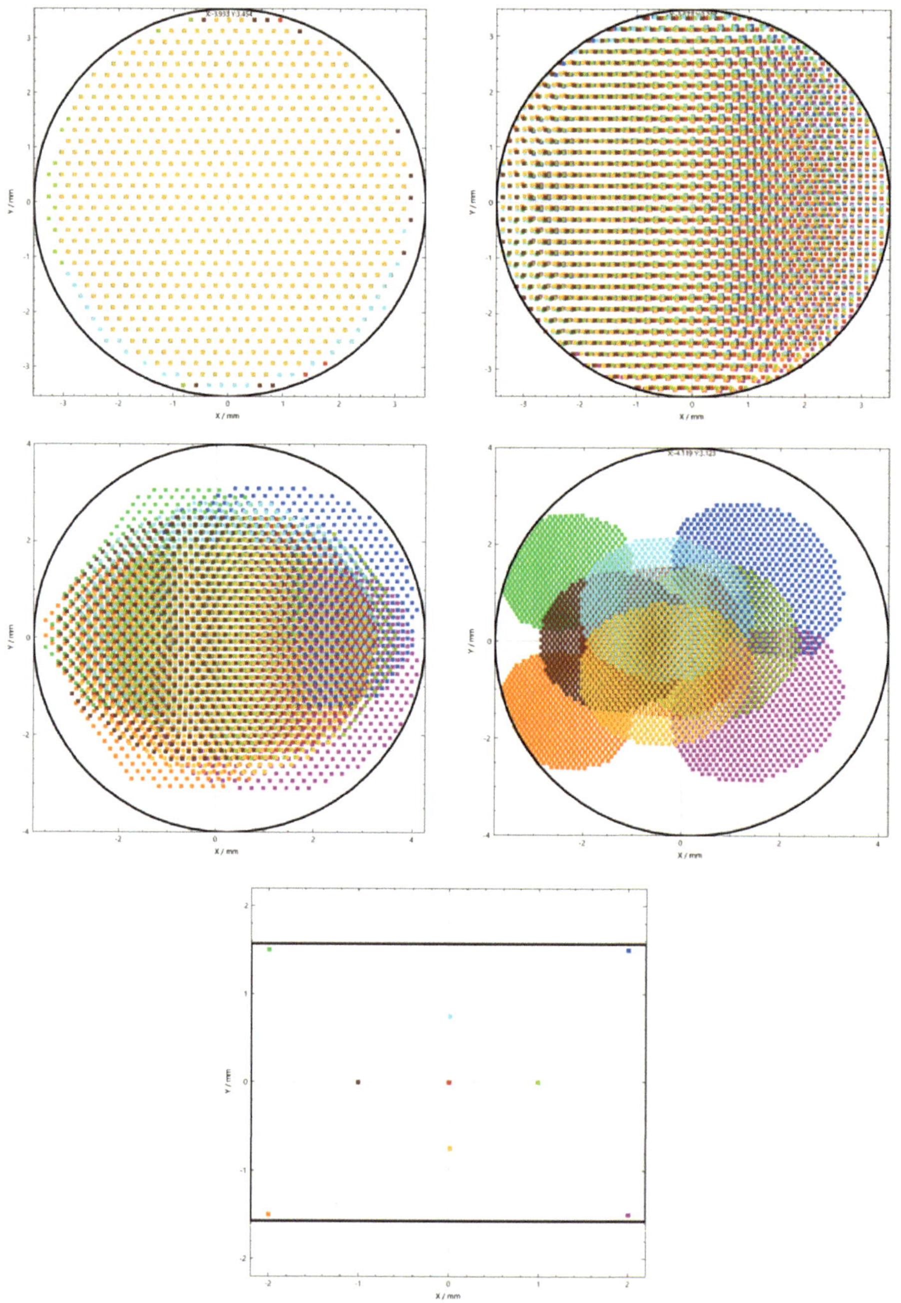

FIGURE 9.7
`Footprint Diagram` of `10xe2`, from *left to right stop:* mirrors 1th, 2th, 3th, and sensor.

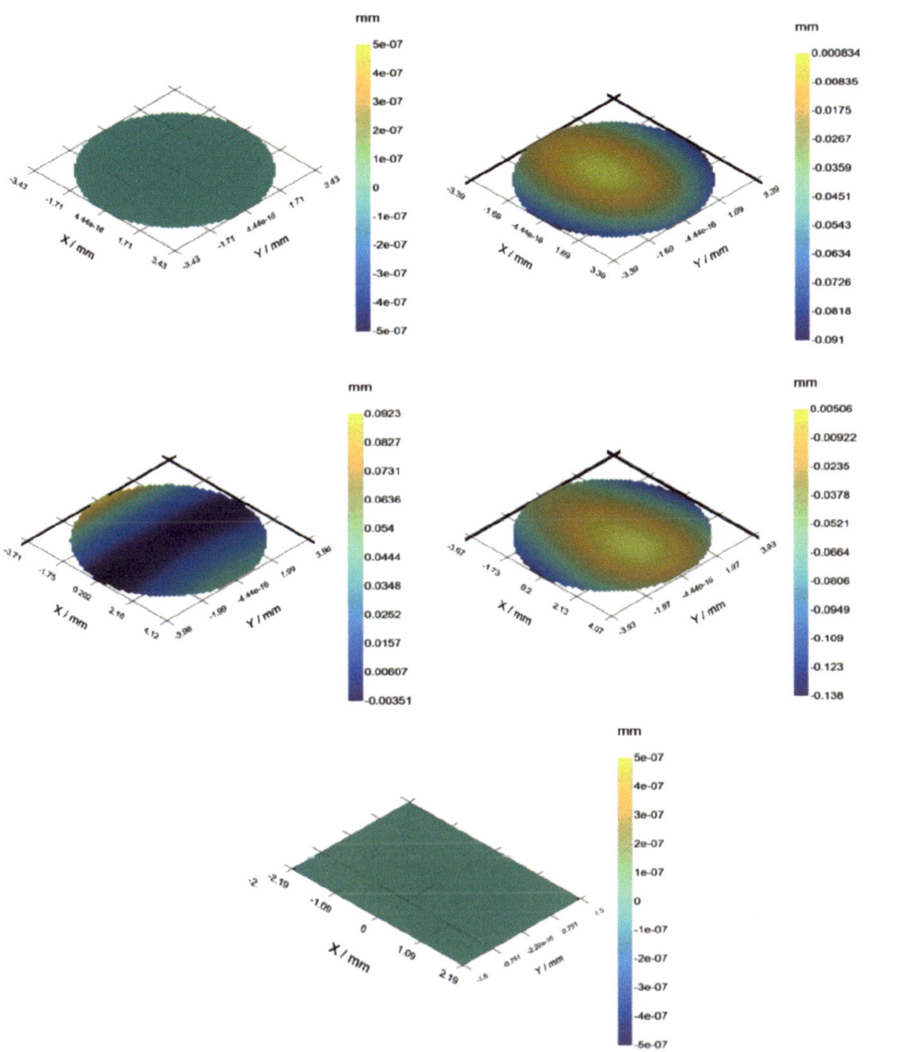

FIGURE 9.8
`Surface Form` of `10xe2`, *left to right stop:* mirrors number 1, 2, 3, and sensor.

finally the sensor. Observe that the area covered by rays gradually diminishes across the mirrors.

In figure 9.8 is plotted the `Surface Form` of each of the elements in the system. The top row comprises the stop and the first mirror, the second row consists of the second and third mirrors, and lastly, there's the sensor. Notice that the three mirrors are freeform and the aperture stop and sensor are flat.

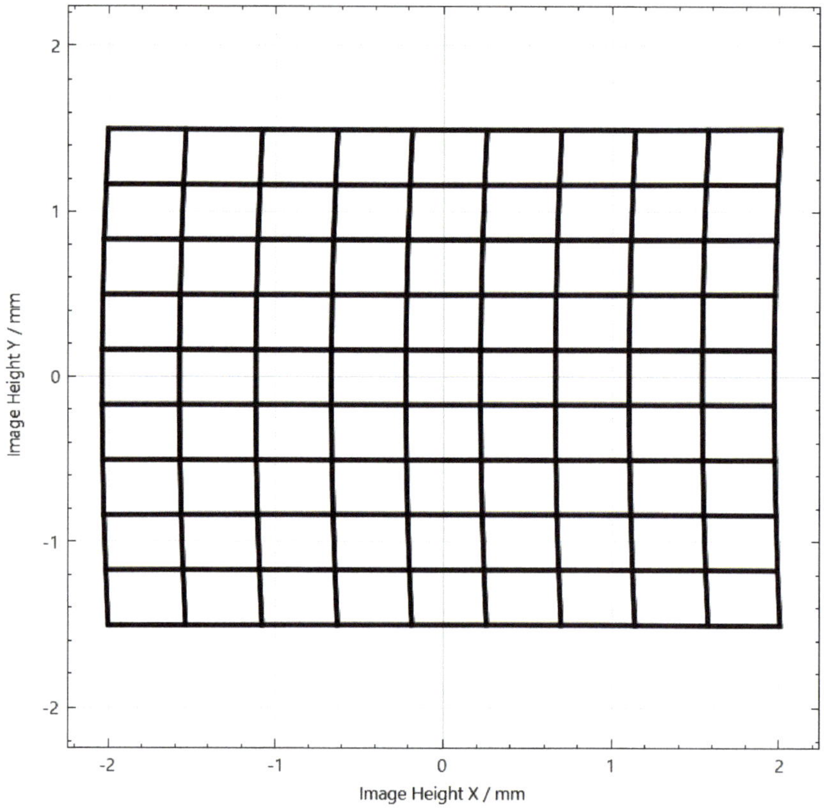

FIGURE 9.9
`Distortion 2D` of `10xe2`.

The system suffers from some slight distortion as can be seen in figure 9.9. The reader may improve this with re-optimization.

The `Distortion 2D` on the grid is not symmetrical due to the freeform shape of the system.

The `FTT MTF Plot` should have a really good shape since the spots are inside the Airy disk, figure 9.6. So, we plot `FTT MTF Plot` and the result is as expected the system is diffracted-limited. See figure 9.10.

Let's assume now that our light source is polarized. So in the source, we choose a linear polarization with an angle equal to zero. We want to see how much the polarization state changes along the optical system. So we plot the `Poincare Sphere` in figure 9.11, for the on-axis field. On the left side of figure 9.11 is the initial polarization state, at the aperture stop. So, at this point, it is just a single point in the `Poincare Sphere`, In the right size is the polarization at the image plane, now the polarization is not a point small region.

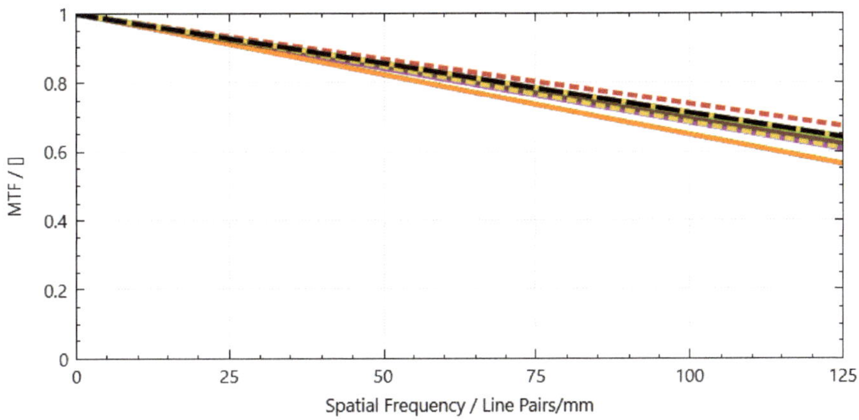

FIGURE 9.10
`FTT MTF Plot` of `10xe2`. The system is diffracted-limited.

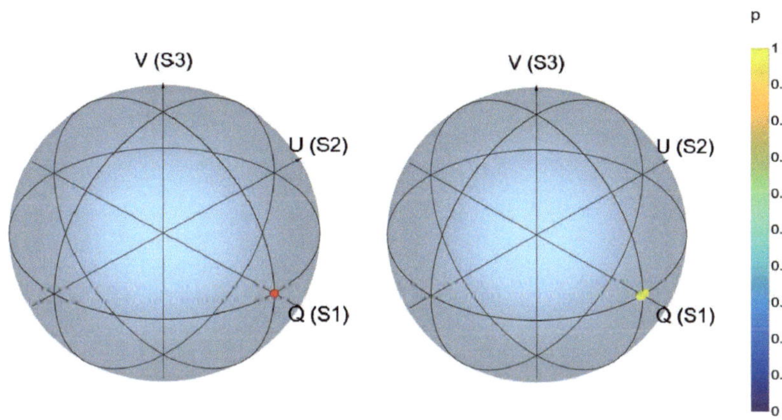

FIGURE 9.11
`Poincare Sphere` of `10xe2`. *Left* at stop. *Right* at the image plane.

This means that the polarization changes across the system but it does not change too much, at least for the on-axis field.

We can confirm this with the `Polarization Map` presented in figure 9.12. It shows the polarization state of the on-axis field along the system, the stop of the three mirrors, and the image plane. The polarization slightly changes across the system. This is helpful to avoid aberration polarization a topic widely studied in [32], but out of the scope of this book.

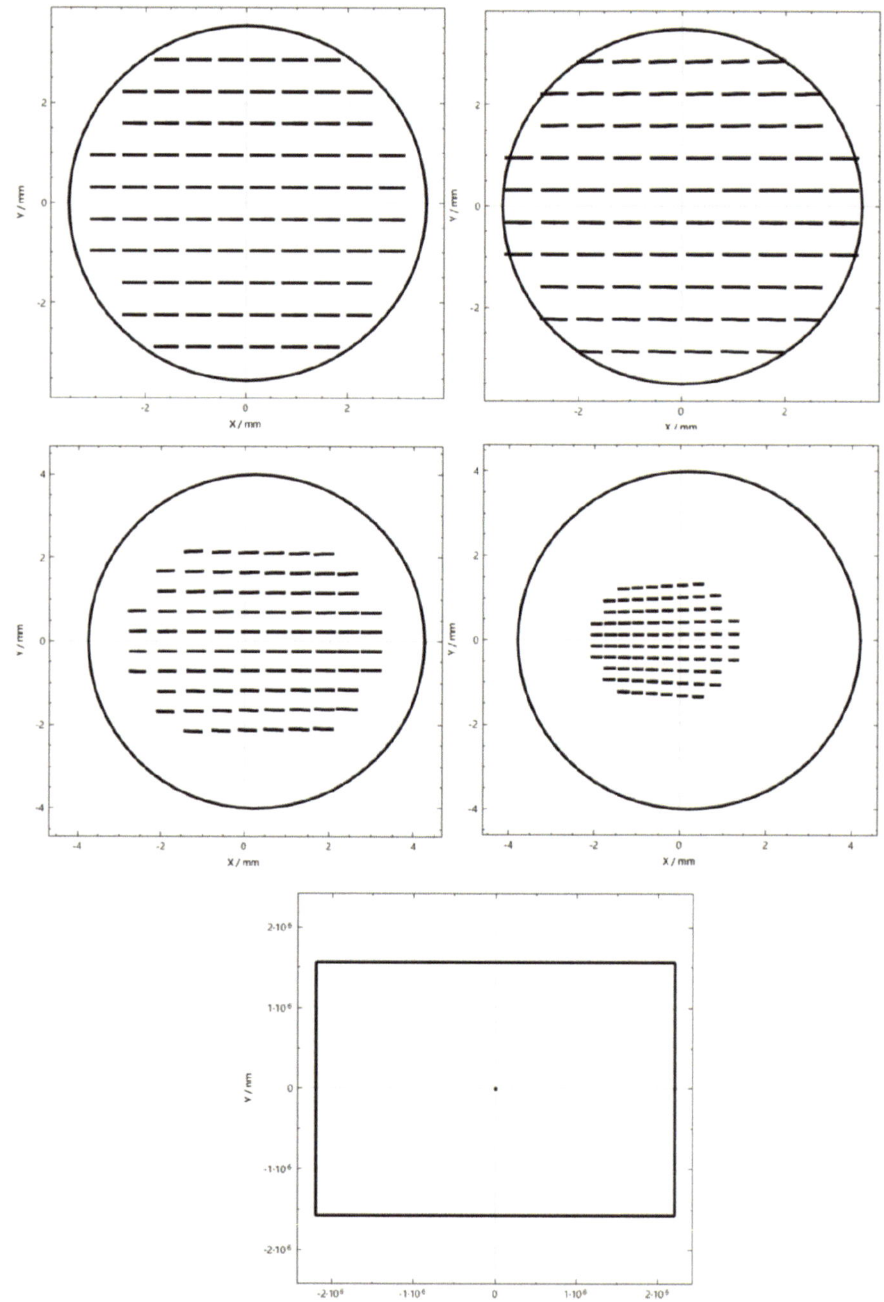

FIGURE 9.12
Polarization Map of 10xe2.

9.4 Summary

In this chapter, we studied the step by step of how to design a freeform camera. The camera was made of three off-axis freeform mirrors. The steps were placing the initial considerations of the system, designing the merit function, optimizing, and analyzing the result.

The crucial point in this process was setting the merit function. Place the corner fields in the corner of the sensor. The rest was just to analyze the performance of the system.

As a tip, I consider that the reader should pay more attention to the design of the merit function and also where and how the mirrors are placed in supplementary file `10xe2`.

10

Examples of Optical Systems

> Divide each difficulty into as many parts as is feasible and necessary to resolve it
>
> Rene Descartes

10.1 Examples of Optical Systems

In this chapter, we are going to explore and analyze several optical systems. Imaging optical systems encompass various devices that capture, manipulate, and display images using optical principles.

We are going to present the optical system from the most basic and increase its level of sophistication. In each section of the chapter, it is presented an example of an optical system, we describe its main function and plot some of its aberrations in `Quadoa Optical CAD`.

10.2 Singlet

A singlet lens comprises just one lens element, making it the most fundamental type of lens. Usually crafted from a single piece of glass or transparent material with curved surfaces, singlet lenses find common use in uncomplicated optical setups necessitating minimal focusing or magnification, like magnifying glasses, reading glasses, or viewfinders. So as you made spec, they have limited capability to correct optical aberrations, such as chromatic and spherical aberrations, which can compromise image quality. However, the singlet lens can be diffraction-limited for an on-axis field, if they have an aspherical surface.

DOI: 10.1201/9781003506355-10

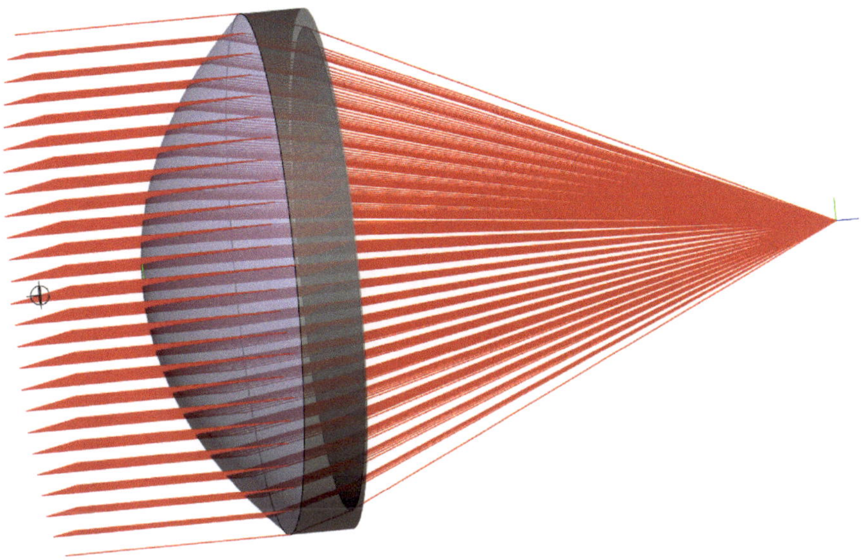

FIGURE 10.1
Ratrying of a singlet aspherical lens for an on-axis field. The first surface of the lens is aspherical and the second is flat.

In figure 10.1, we present an aspherical singlet lens. The first surface of the lens is `Aspherical Form` and the second is flat. The radius of the first surface is $R = 29.54$mm, the conic constant is $k = -0.69$mm and the apsherical coefficients are $A_4 = 6.5568e^{-07}$, $A_6 = 1.778e^{-10}$, $A_8 = 2.988e^{-14}$. This example can be found in `C:\Program Files\Quadoa\examples\Objectives\Asphere`.

The singlet presented in figure 10.1 consists of a singlet element made by `C79-80`, a fused silica. This aspherical lens has been designed for a primary wavelength of 285nm. The stop is defined in the sequence of the lens, and it is placed at the first surface of the lens.

This particular example is diffracted-limited, one of the best ways to see this is to compute the PSF, point spread function.

Remember that the PSF describes the distribution of light intensity around a point source as it appears in the image produced by an optical system [8, 23]. It indicates how much the point source is spread out or blurred in the image. PSF has significant implications for image quality and resolution [5, 58]. A smaller and more compact PSF indicates better resolution and less blurring, while a larger and more spread-out PSF indicates poorer resolution and more significant blurring [15, 13]. The shape of the PSF depends on various factors including the optical design of the system, aberrations present in the system,

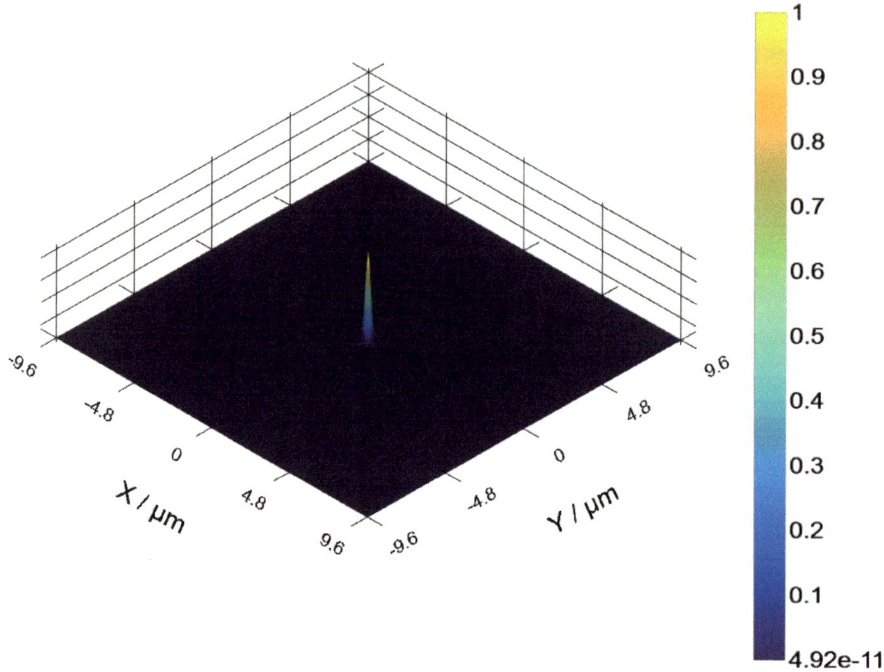

FIGURE 10.2
`FFT PSF Plot` of the example presented in file which is path is `C:\Program Files\Quadoa\examples\Objectives\Asphere`. The singlet is diffracted-limited. The plot shows a single bright peak at the center of the image plane which rapidly goes to zero outside its center of intensity.

and the wavelength of light being used. Common PSF shapes include Gaussian, Airy disk (for diffraction-limited systems), and more complex shapes for systems with aberrations [37]. Since our lens its diffracted-limited lets compute the PSF of the on-axis field and confirm if it looks like a Gaussian function.

In figure 10.2 we present the `FFT PSF Plot` of the singlet lens `C:\Program Files\Quadoa\examples\Objectives`. From the figure, it is clear that the on-axis field is diffracted-limited. The intensity peak is placed in the center of the PSF and it rapidly goes to zero around the central peak. The intensity units of the plot are normalized

The result presented in figure 10.2 is the PSF but based on the `Fast Fourier Transform`. Remeber that there several ways to compute the PSF of an optical system. The difference between the PSF depends on how acquired the diffraction pattern is computed. The most acquired diffraction pattern for scalar diffraction theory is computed with Huygens integral. The Huygens integral (seen in Chapter 2) is based on the Huygens-Fresnel principle. This

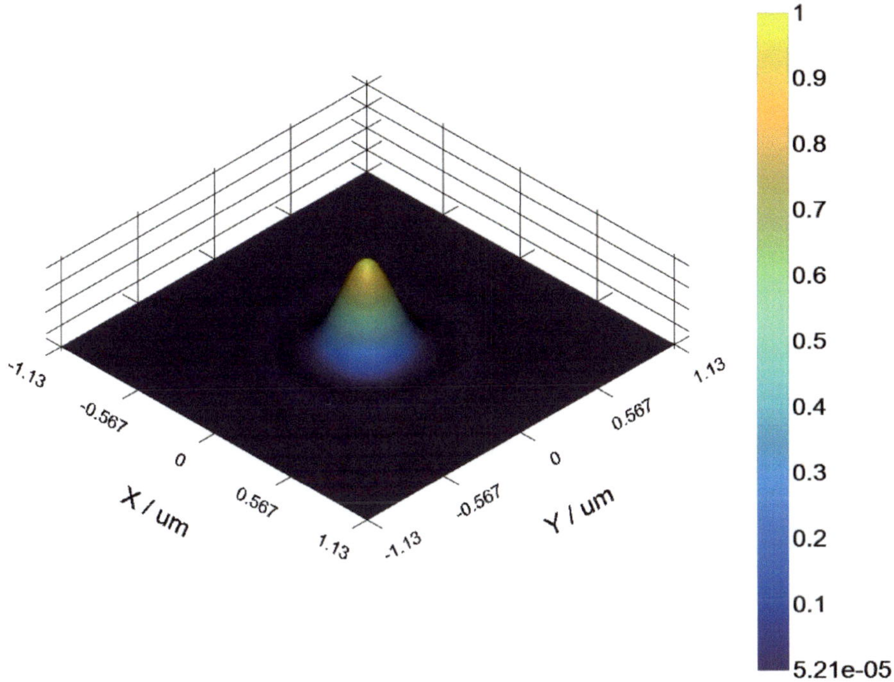

FIGURE 10.3
`Huygens PSF Plot` of the example presented in following path `C:\Program Files\Quadoa\examples\Objectives\Asphere`. The intensity of the PSF goes to zero for a radius beyond 0.567μm from the center intensity peak.

principle, also known as the Huygens principle, is a fundamental concept in optics that describes how light waves propagate through a medium. The Huygens principle states that every point on a wavefront serves as a source of secondary spherical wavelets, and the new wavefront at a subsequent time is the result of the superposition of these wavelets.

So back on track, let's compute the `Huygens PSF Plot` which used the Huygens integral [37]. The plot is presented in figure 10.3. Notice that the system still diffracted limited and beyond a radius of 0.567μm the intensity goes to zero. Observer that the intensity units of the plot are normalized.

10.3 Tessar

The next example is the Tessar camera. The Tessar lens is a classic photographic lens design known for its simplicity, compactness, and excellent optical

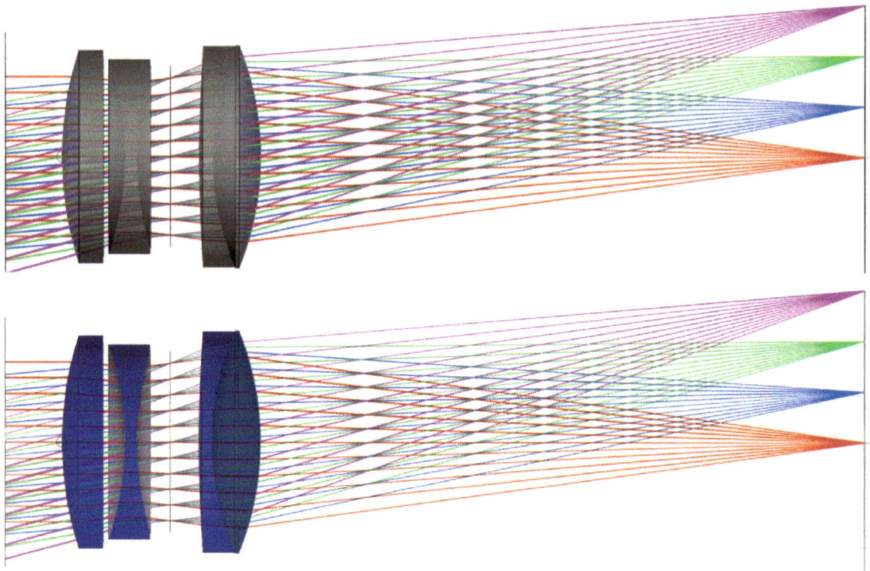

FIGURE 10.4
Ray-tracing of Tessar example.

performance. It was developed by the German physicist and optician Dr. Paul Rudolph in 1902 while working for the Carl Zeiss company [9, 31, 30]. Its multiple versions consist of three of four lens elements, depending on the version. The Tessar lens has a symmetrical optical layout, with two positive lens elements positioned between two negative lens elements, in the case of the four-element version. For the three-element configuration two positive lens elements are positioned between one negative singlet lens element. This symmetrical design helps to correct various aberrations, resulting in sharp and well-corrected images.

The three-element version is also called the Cooke Triplet. However, in this book, we refer to it as the Tessar because the example that we are going to study in this section is called Tessar can be found in `C:\Program Files\Quadoa\examples\Objectives\Tessar`. The raytracing is presented in figure 10.4.

The design presented in figure 10.4 is made by three spherical lenses. The materials of the elements are `N-SK16`, `N-BAK2`, and `N-SF10`, all glasses that can be found in the `SCHOTT Catalog`. The system has four `Fields` with 0°, 5°, 10°, 15° angle of incident. The `Spot Diagram Plot` of this file is presented in figure 10.5. From the spot diagram, we can see that the of-axis fields suffer from coma and the field at 15° angle of the incident also suffers from astigmatism, due to the loss of symmetry of its spot diagram.

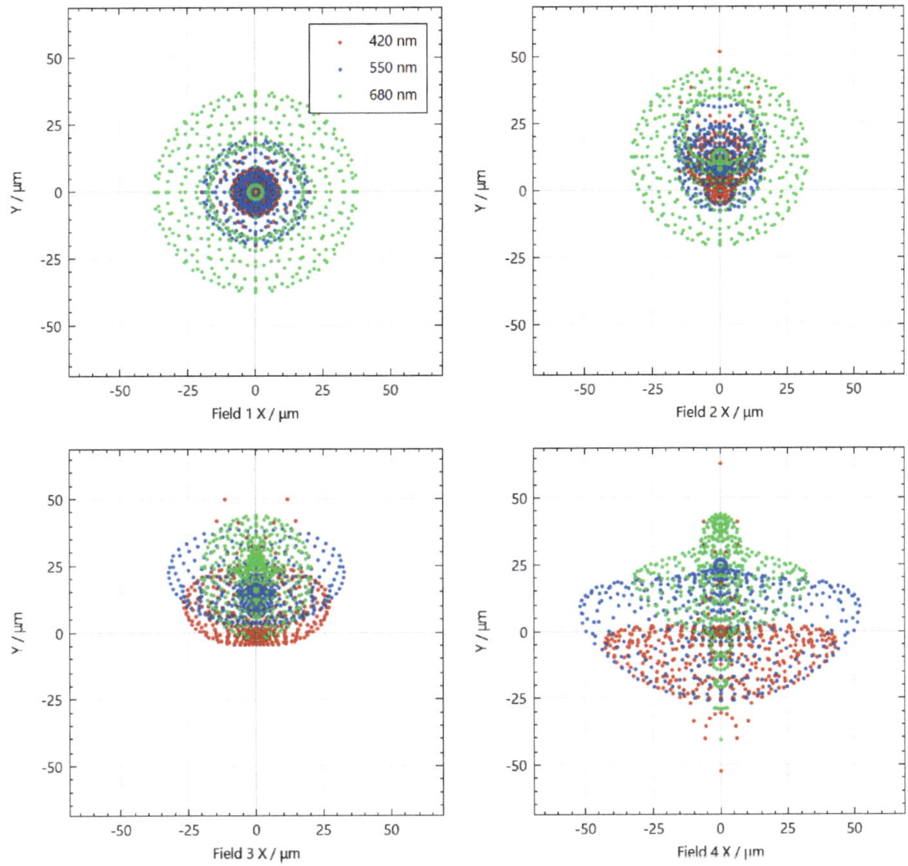

FIGURE 10.5
`Spot Diagram Plot` of the Tessar lens. In the *top left* is the spot diagram of the on-axis field. *top right* is the spot diagram of the off-axis field with 5° angle of incidence. *bottom left* is the spot diagram of the off-axis field with 10° angle of incidence. *bottom right* is the spot diagram of the off-axis field with 15° angle of incidence.

Observe that from the `Spot Diagram Plot`, the four fields are not diffracted-limited. Since the spot is much bigger than the Airy disk, also plotted in figure 10.5. However, it is small compared to the spot diagram, so it is not easy to recognize.

Also, it is important to remark that the spot diagram differs in its position and shape concerning the wavelength. So let's plot the `Lateral Color Plot` and the `Defocus` or also called the `Lateral Aberration Plot`. Both plots are presented in figure 10.6. The primary wavelength is 420nm and in the lateral color plot, it does not suffer from dedication because it is the reference.

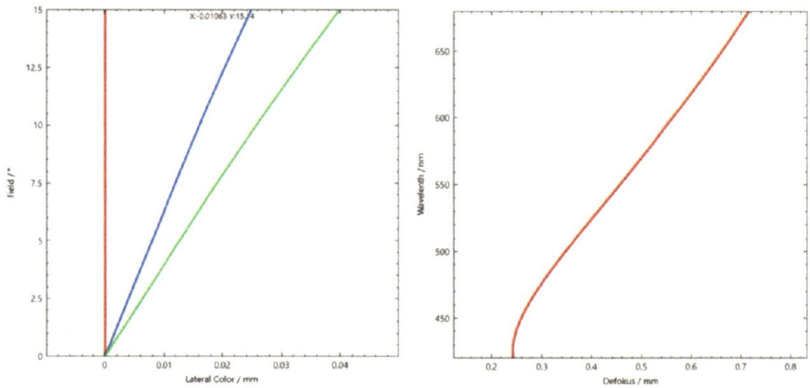

FIGURE 10.6
Left: `Lateral Color Plot`. The lateral color increases with the field of view. *Right:* `Defokus` of the Tessar. The defocus increases of the on-axis field with the increase of the wavelength.

But notice that the lateral color increases with the field of view. Just as we saw in the spot diagram. on the other side, the defocus plot shows that the defocus increases of the on-axis field with the increase of the wavelength. This can be confirmed on the *right* side of figure 10.6.

We can consider that the performance of a Tessar and Cook lens is good for such simple systems, four and three spherical elements respectively.

The Tessar lens design, along with the Cook triplet, has a long and storied history in photography, and it has been used in many iconic cameras throughout the 20th century. It has been employed in both film and digital cameras, and variations of the Tessar design are still in use today.

Let's talk a little bit more about the Cook triplet. The Cooke Triplet is a classic lens design known for its simple yet effective optical formula and high image quality. It was developed by Harold Dennis Taylor, the Chief Optical Designer at Cooke Optics Ltd., in the late 19th century. The Cooke Triplet is known for its compact size and lightweight construction, making it suitable for use in portable cameras and lenses. The Cooke Triplet design is versatile and can be used for a wide range of photographic applications, including portrait, landscape, and general-purpose photography. It offers good contrast, resolution, and color reproduction across the image frame.

10.4 Camera

The next example is an optical camera. The example can be found in `C:\Program Files\Quadoa\examples\Objectives\Camera_Lens`. See figure

FIGURE 10.7
Ray-tracing of a camera. This particular example can be found after successfully installing `Quadoa Optical CAD` in `C:\Program Files\Quadoa\examples\Objectives\Camera_Lens`.

10.7. The system is composed of six lenses with a `Sphere Form`. The materials of lenses are glasses from the `SCHOTT Catalog` such as `N-BK7` and `N-F2`. The five `Fields` in the system have angles of incident given by 0°, 10°, −10°, 20°, −20°, respectively.

The `STOP` is placed on the first surface of the fifth lens. This made some vignetting in the off-axis field. This helps to reduce aberrations in the off-axis fields but also affects the amount of light that goes to the sensor.

This camera has been designed to work for the following wavelengths: 500nm, 515nm, and 530nm. The primary wavelength is 500nm. Since we have several fields and several wavelengths, it maybe useful to plot the spot diagram for each wavelength. `The Field vs. Wavelength Spot Diagram Plot` depicts the 2D ray distribution of rays on the image surface within a sequence. It organizes the rays into a grid, where each subplot represents the rays for one field/wavelength combination. The wavelengths are arranged from left to right, and the fields are arranged from top to bottom. `The Field vs. Wavelength Spot Diagram Plot` of the design of figure 10.7 is presented in figure 10.8.

From figure 10.8, we see the asymmetric presented for the pairs of `Fields` composed by 10°, −10°, and 20°, −20°. This result is expected since the system is radially symmetric.

The `Transversal Ray Fan Plot` in the $y-$ and $x-$ axis is also plotted in figure 10.9. The color code of the plot is red, blue, and green for the wavelengths: 500nm, 515nm, and 530nm, respectively. The transverse aberrations are defined by the axial distance between the position where a ray intersects the image plane and the reference image point. So, from this definition, it is

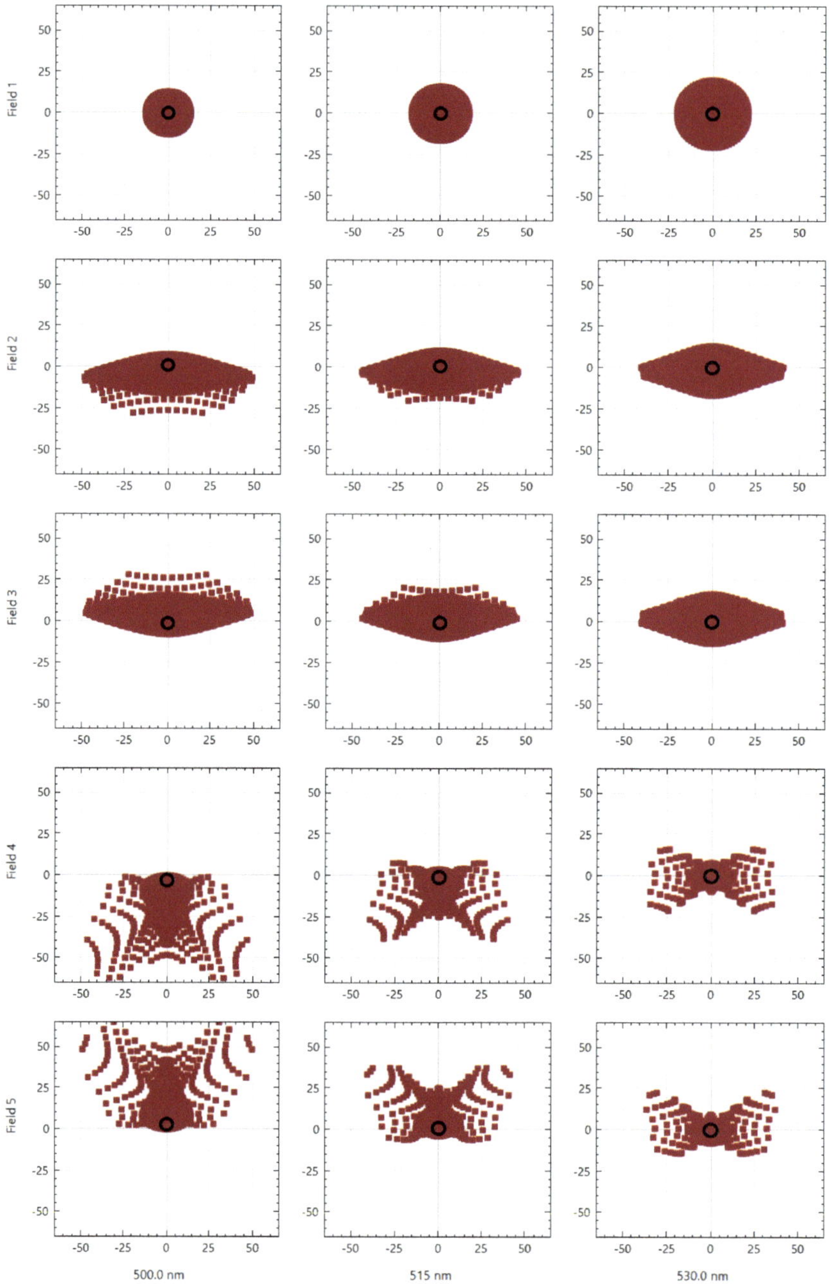

FIGURE 10.8
`The Field vs. Wavelength Spot Diagram Plot` of the design of figure 10.7.

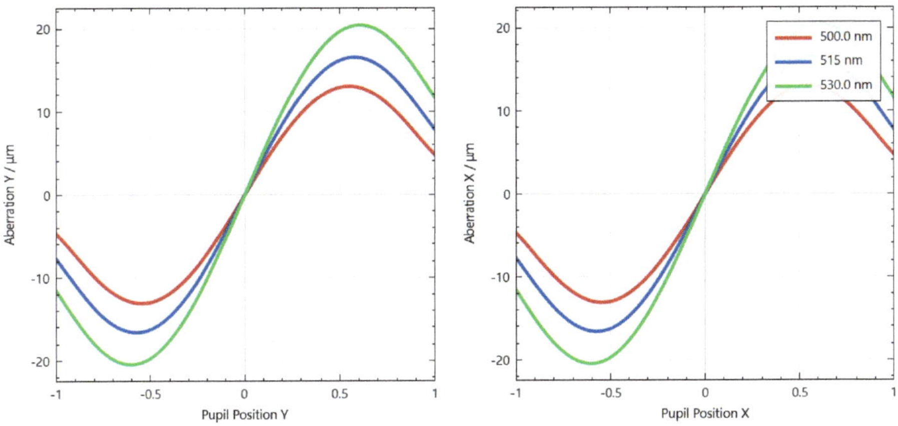

FIGURE 10.9
The **Transversal Ray Fan Plot** of the design of figure 10.7. *left* is the transversal aberration along the y axes across the pupil of the sequence. *right* is the transversal aberration along the x axes across the pupil of the sequence. The color-code of the plot is red, blue, and green for the wavelengths: 500nm, 515nm, and 530nm, respectively.

clear that the **Transversal Ray Fan Plot** will also be symmetric for the pairs of **Fields** composed by 10°, −10°, and 20°, −20°. Therefore for a radially symmetric system is not necessary to have symmetric pairs of **Fields** like 10°, −10°, or like 20°, −20°. They don't give extra information but just double the computations done by the computer. This is an important tip for lens designers.

Also, it is the same for other plots, like MTF, PSF, distortion etc.

Remember that in the transverse aberration plot, the origin of the coordinate system usually corresponds to the optical axis of the system. If the system is radially symmetric there is no traverse aberration in the optical axis. Thus both plots in figure 10.9 start from zero.

Also from figure 10.9, we can see that in the plots, the transversal aberration along the x and y axes across the pupil differs with the wavelength. Just like the spot diagram varies in both position and shape with respect to the wavelength. See figure 10.9. This means the optical system presented in `C:\Program Files\Quadoa\examples\Objectives\Camera_Lens` suffers from chromatic aberration the combination of materials and radius of lenses is not enough to balance the chromatic aberrations.

Let's illustrate the **Lateral Color Plot**. The primary wavelength is 550nm, and in the lateral color plot, it remains unaffected because it serves as the reference. However, it's important to observe that the lateral color escalates with the field of view. But at some point around 15° it is reduced again and then increases but in the other direction. This observation is corroborated in the right panel of figure 10.10.

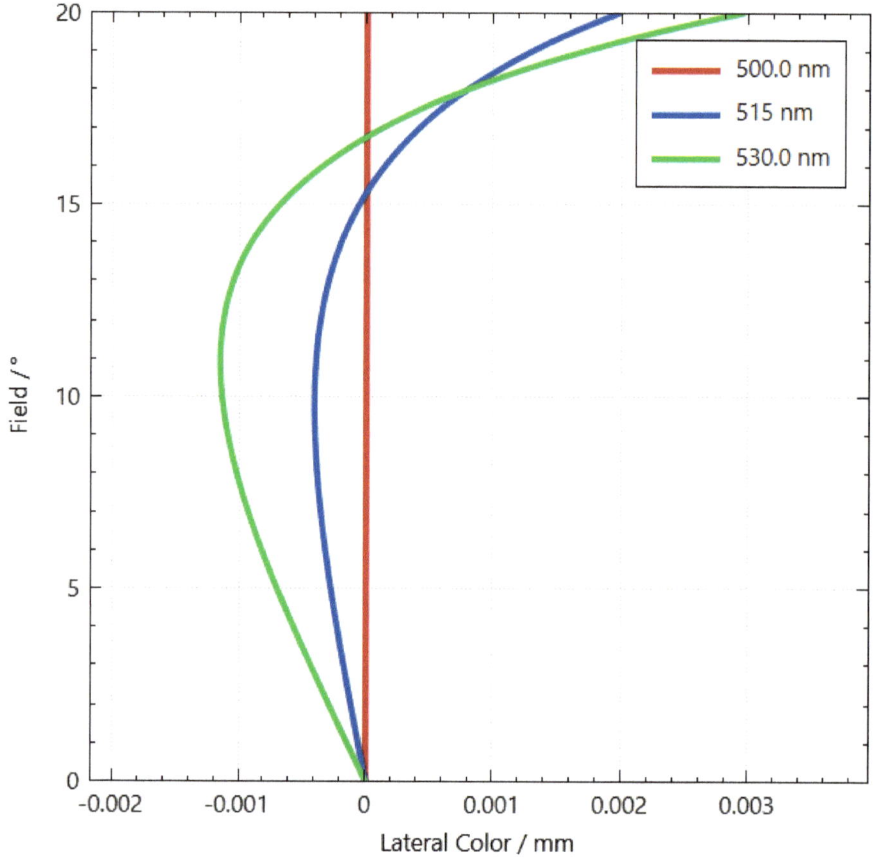

FIGURE 10.10
`Lateral Color Plot` of the design is presented in figure 10.7. The primary wavelength is 550nm, serving as the reference in the lateral color plot, and remains unaffected. It is just a vertical line. However, it's crucial to note that the lateral color increases with the field of view. Interestingly, around 15°, it experiences a reduction before resuming an increase in the opposite direction. The color code of the plot is red, blue, and green for the wavelengths: 500nm, 515nm, and 530nm, respectively.

10.5 Wide-angle Camera

The next example is an optical camera with a wide field of view, which is in `C:\Program Files\Quadoa\examples\Objectives\Camera_Lens_75deg_FOV`, see figure 10.11. Wide-angle cameras are photographic devices with lenses that have a broader field of view compared to standard or telephoto lenses. These

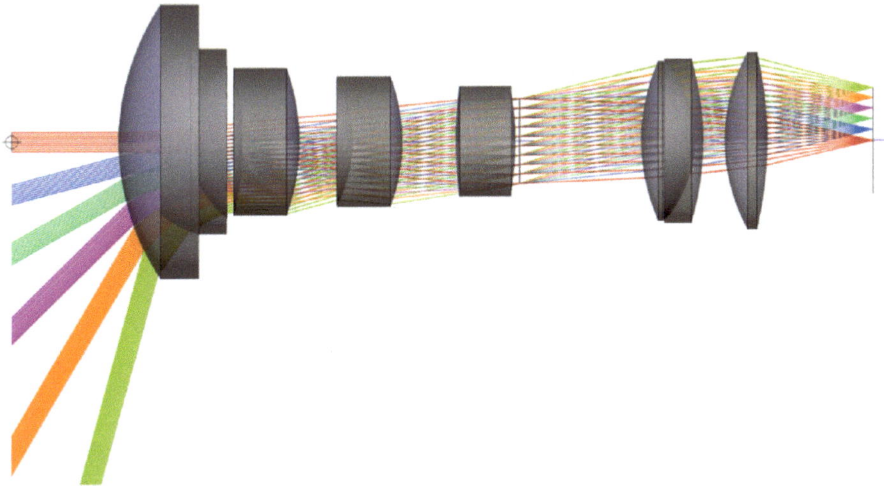

FIGURE 10.11
The ray-tracing of Wide-angle camera.

cameras are designed to capture more of the scene in a single frame, making them ideal for landscape photography, architectural photography, interior photography, and capturing group shots in tight spaces. So, The particular of this camera is its field of view which is extremely large.

The six `Fields` for this design are 0°, 15°, 30°, 45°, 60°, 75°. So the total field of view is 150°. The camera consists of seven elements with a `Sphere Form`, all glasses from the `SCHOTT Catalog`. These materials are `N-LASF44`, `N-LASF41`, `N-SF66`, `N-PK51`, `P-SF68`, `N-FK51A`, `N-LASF31A`, respectively for the seven lenses. The `STOP` has its own surface, surface 12 in the system. The ray-tracing is presented in figure 10.11.

The main problem of this type of design is the distortion. Wide-angle lenses introduce a significant amount of distortion, particularly near the edges of the frame. Barrel distortion, where straight lines appear curved outward, is a common characteristic of wide-angle lenses. Therefore it is crucial, to measure the distortion. In figure 10.12, we show `Distortion 1D Plot` for the design from figure 10.12. On the *left*, we plot `Distortion 1D Plot` with $f\tan\theta$ and in the *right* $f\theta$. Notice that both plots are very different. Therefore, the distortion definition is highly important. There is a wide discussion of which definition should be used. But generally, it is to use $f\theta$ for wide-angle cameras. However, modern lens designs and software correction techniques can minimize distortion to some extent.

Also, it is important to remark that it is very hard to get diffracted-limited performance for the off-axis fields as can be in the `OPD Fan Plot`, figure 10.13.

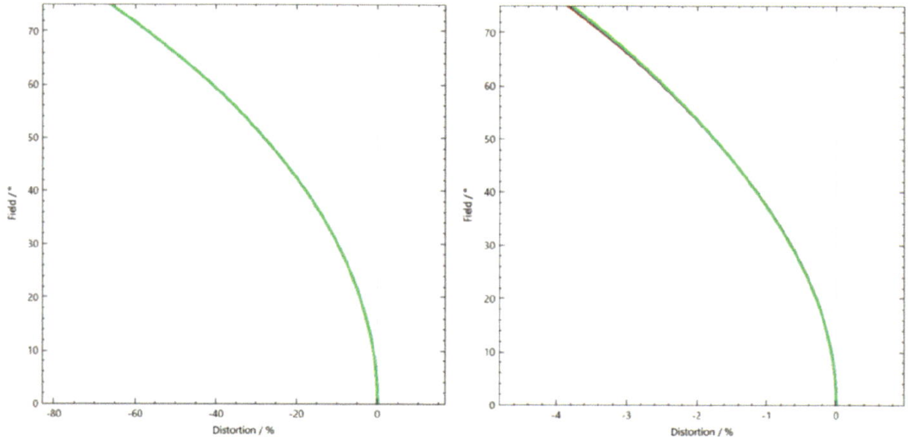

FIGURE 10.12
`Distortion 1D Plot` on the *left* the distortion is given by $f\tan\theta$ and on the *right* it is given by $f\theta$.

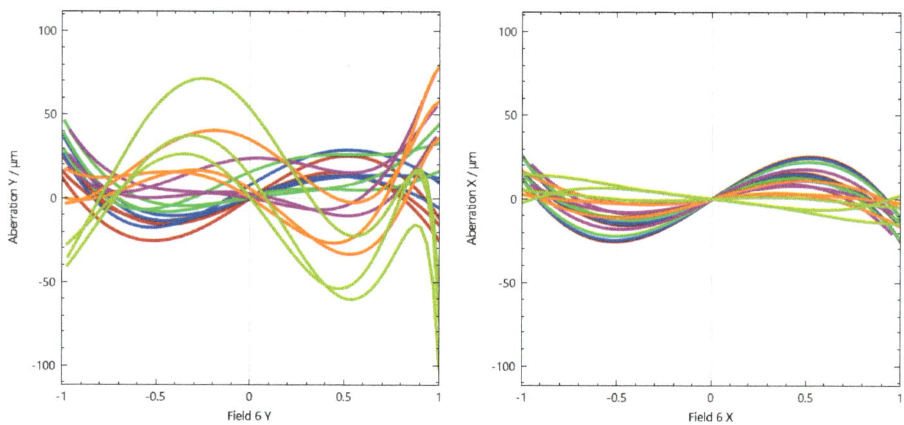

FIGURE 10.13
`OPD Fan Plot` in the right of the figure is the OPD along the y dimension and in the left is the OPD along the x.

10.6 Microscope Objective

Microscope objectives are a fundamental component of optical microscopes, responsible for magnifying and resolving the details of specimens under observation. They are designed to gather light from the sample and focus it into an

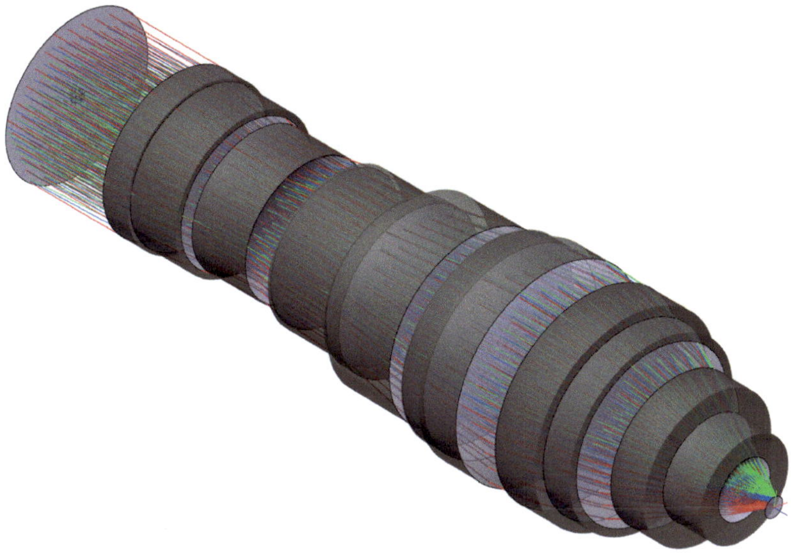

FIGURE 10.14
Ray-tracing of a microscope objective.

image, enabling users to study objects at high magnification with enhanced resolution.

Microscope objectives are characterized by their magnification power, which determines how much larger the specimen appears compared to its actual size. Objectives with different magnifications. Magnification is the ratio between the object size and image size, For example, 4x, 10x, 40x, and 100x are commonly used to observe specimens at various levels of detail. In this notation, 4x means the image is 4 times bigger than the object [76].

In figure 10.14 we present an example of a microscope. This example is in `C:\Program Files\Quadoa\examples\Objectives\Microscope\Objective`. The numerical aperture for this example is 0.548. The numerical aperture of a microscope objective measures its ability to gather light and resolve fine details in the specimen. Higher NA objectives offer greater resolution and are suitable for observing smaller details. Because the Higher NA is smaller the Airy disk in the PSF [23].

The horizontal cut of the ray-tracing of the microscope in figure 10.14 is displayed in figure 10.15. The working distance distance for this microscope is 2mm. The working distance of a microscope objective refers to the distance between the front lens of the objective and the surface of the specimen when the image is in focus. Objectives with longer working distances provide more space between the objective and the specimen, making them suitable

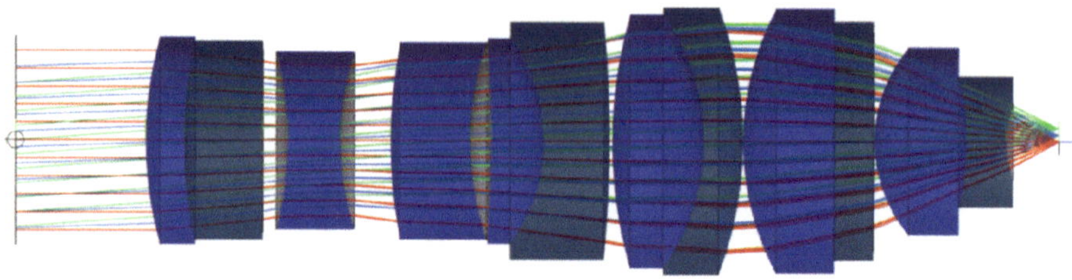

FIGURE 10.15
The horizontal cut of the ray-tracing of a microscope objective. The wavelengths of this microscope object are 550nm, 650nm, and 750nm. This example has three fields with 0°, 2°, 4° degrees, respectively.

for observing thick or irregular samples and for accommodating additional accessories like slides or coverslips.

The materials in this example are not real materials but theoretical materials described by the refraction index and the Abbe number. Notice that this example uses an `Assembly`, which is a way to `Quadoa Optical CAD` use to group several elements. In this case, the `Assembly` contains a singlet lens and a doublet. The rest of the elements of the microscope objective are singlet lenses or doublets.

Microscopes in general have a small field of view. This example has three `Fields` with 0°, 2°, 4° degrees of incident angle, respectively. The reason why microscopes have so a small field of view is because they have a high numerical aperture. The bigger the numerical aperture the harder to get a diffracted-limited system for an off-axis field with a large field of view. For example, photographic cameras have a much bigger field of view but the numerical aperture is much smaller than a microscope. Therefore the price to pay to have a great resolution is the field of view.

The wavelengths of this microscope object are 550nm, 650nm, and 750nm. For this example, the polarization is not taken on count, but for several types of microscopy, polarization takes a big role. Polarization microscopy is a technique used to study the optical properties of materials by observing how they interact with polarized light. It involves the use of polarizing filters to control the orientation of light waves as they pass through a sample, allowing for the visualization and analysis of various optical properties such as birefringence, optical activity, and stress in materials. Polarization microscopy is out of the scope of this book but good references are given at the end of the chapter.

Let's start analyzing this microscope objective. We shall start first with the spot diagram. It gives us a first instance of how good the quality we may expect at the image plane. In figure 10.16, it is displayed the `Spot Diagram Plot`. From the figure, we can guarantee that for the on-axis field, the microscope is diffracted-limited. The spots are around the Air disk for the

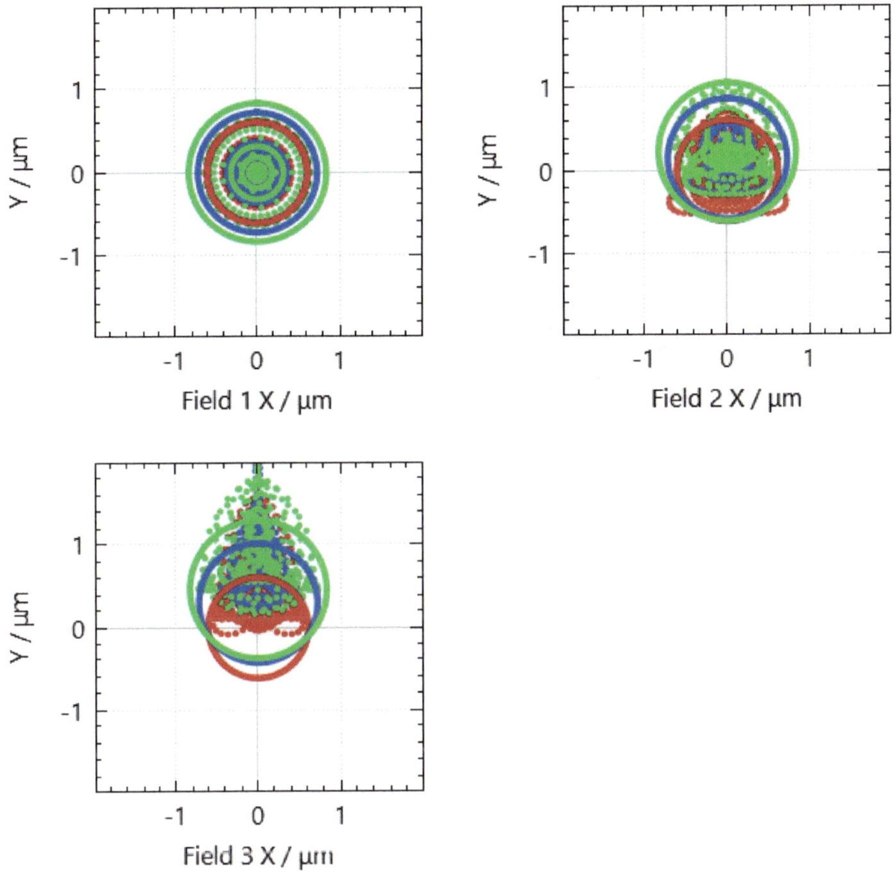

FIGURE 10.16
`Spot Diagram Plot` of a microscope objective. The color code of the wavelength is 550nm for red, 650nm for blue, and 750nm for green. *top left* on-axis field, *top right* and *bottom* fields with 2° and 4° degrees of incident angle, respectively.

three different wavelengths. The color code of the wavelength is 550nm for red, 650nm for blue, and 750nm for green. The off-axis fields are also close to being diffracted-limited. Naturally, the field with less performance, due to color shift and coma, is the marginal field with incident angles of 4°.

The `Spot Diagram Plot` of the field with 2° degrees of incident angle is very interesting, the shape of its like a trefoil. A trefoil is a geometrical shape characterized by three overlapping loops, resembling the form of a three-leaf clover or a knot with three loops. When the `Spot Diagram Plot` has the shape of a trefoil we can consider that the image suffers from astigmatism

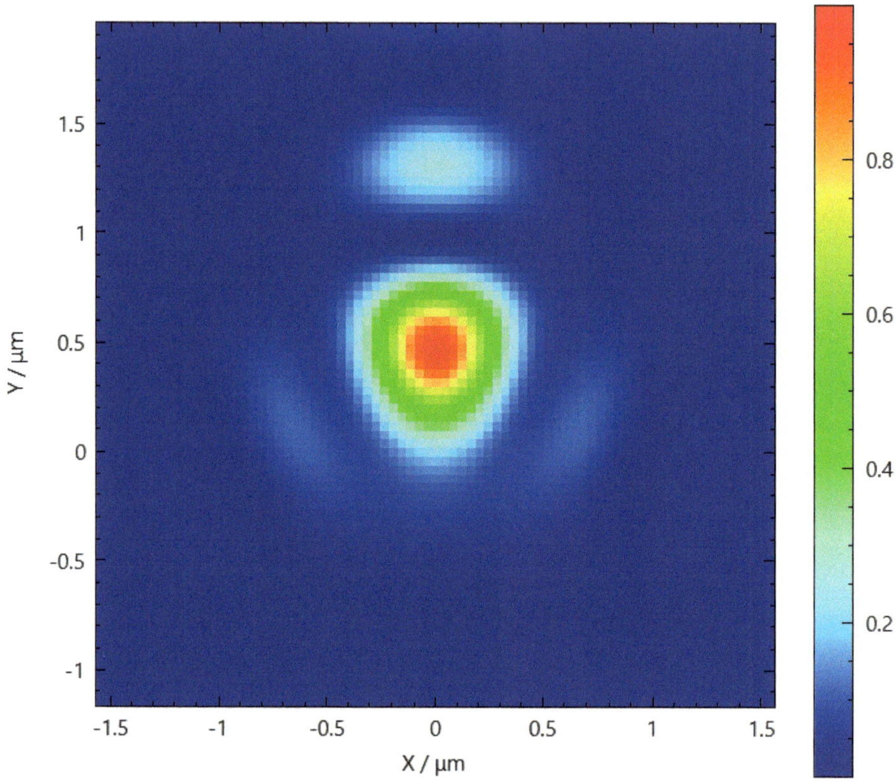

FIGURE 10.17
`Huygens PSF Plot` of the second field of microscope object presented in figure 10.14. The shape of the PSF is the shape of a trefoil. A trefoil is a geometrical shape characterized by three overlapping loops, resembling the form of a three-leaf clover or a knot with three loops.

and coma. Let's compute the `Huygens PSF Plot` for this particular field. It is displayed in figure 10.17. As expected the form of the trefoil appears once more in this plot. With a single peak of energy at the center and three smaller regions with less intensity surrounding it.

From the spot diagram, we notice the color shift of the third field, the marginal field with incident angles of 4°. So it would be interesting to plot the `Lateral Aberration Plot`, also called `Defokus`. This plot is presented in figure 10.18. Where the vertical axis is the wavelength and the horizontal axis is the defocus present by each of the wavelengths. The primary wavelength of this example is 550nm. The range of the defocus is low around 0.006mm.

The next and most important test we can do is the microscope objective in question, the one presented in figure 10.14, the MTF. For the purpose of

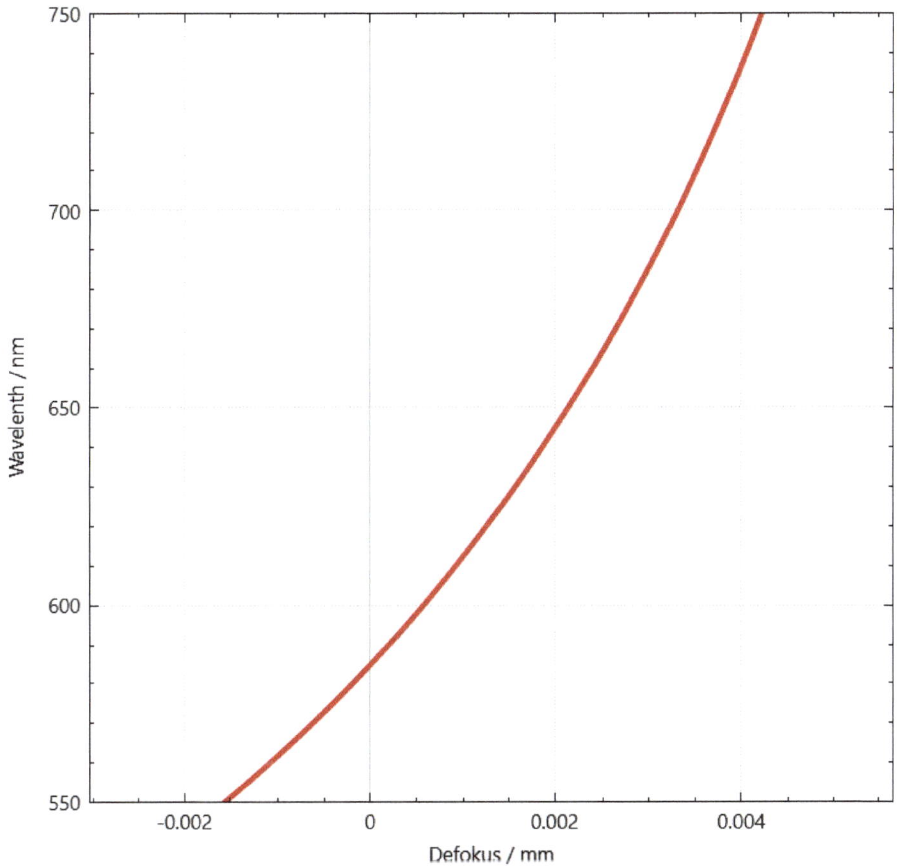

FIGURE 10.18
`Lateral Aberration Plot` of the microscope object of figure 10.14. The vertical axis is the wavelength and the horizontal axis is the defocus present by each of the wavelengths.

speed, we use the MTF based on the Fast Fourier transform, rather than the Huygens integral. Thus, the **`FFT MTF Plot`** is plotted in figure 10.19. There is very interesting information that we can extract from the **`FFT MTF Plot`** in figure 10.19:

- The First thing to notice from the plot, is the Dashed line. The Dashed line represents the performance of an ideal stigmatic optical system. This ideal system has its MTF to be 1 at 0 spatial frequency per millimeter. Then it starts to slowly going down until it reaches zero at some point after 500 lines per millimeter. This dashed line is the best possible performance we

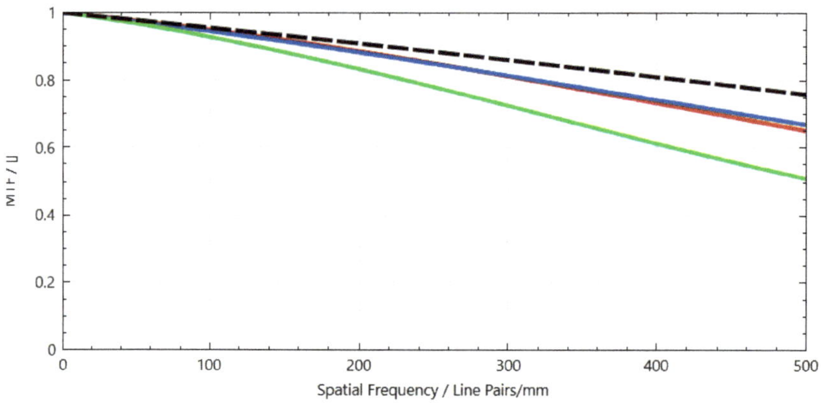

FIGURE 10.19
`FFT MTF` of the design of figure 10.15.

can get. However, the actual performance is given by the other three lines, which have the same color code as the spot diagram.

- The performance of the on-axis field and the second field, lines red and blue, are very similar. This performance can be considered acceptable since the dashed line is no so far away from it them. But there is still space for improvement.

- The green line, which corresponds to the marginal field, has the lowest performance, this can be expected from the size of the spot predicted by the spot diagram of figure 10.19. The bigger the spot the lower the MTF performance.

- Finally compare the MTF performance of the microscope study in this section with the camera studied in Chapter 4. The MTF has much better performance since the numerical aperture is bigger, and the PSF is smaller.

10.7 Cassagrain Telescope

The Cassegrain telescope is a type of reflecting telescope that was invented by the French astronomer Laurent Cassegrain in the 17th century [77]. It is widely used in both amateur and professional astronomy due to its compact design and excellent image quality. The ray-tracying of the Cassagrain Telescope is presented in figure 10.20, from two different points of view. The design,

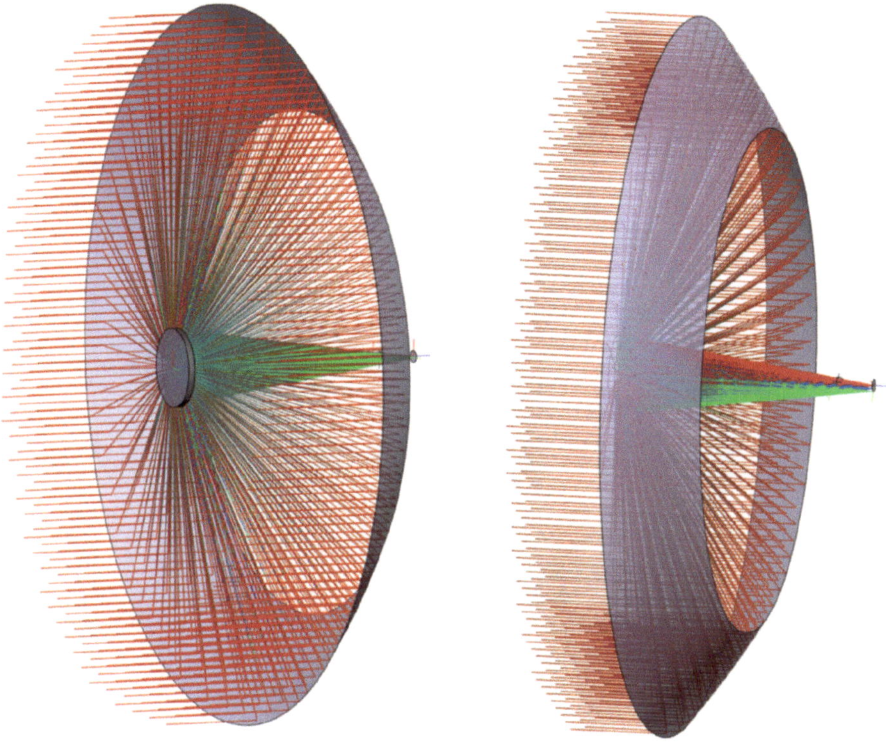

FIGURE 10.20
The Cassegrain telescope is a type of reflecting telescope. This figure shows the ray-tracing of a Cassegrain telescope from two different points of view.

shown in the figure 10.20, is composed of two mirrors. The file is in `C:\Program Files\Quadoa\examples\Objectives\Microscope\Cassagrain_Anular_Telescope`.

This telescope has there `Fields` with a tiny angle of incident, the first field is on-axis. The second and third fields are off-axis and their respective angle of incidence is 0.075° and 0.15°, respectively. Since we are working with a system only composed of mirrors the wavelength is not relevant from the point of view of the materials of the system, it can be only relevant for the sensor. But that topic is outside the scope of the book,

The primary mirror has an annular aperture and its shape is parabolic. The secondary mirror has a circular aperture, and its shape is hyperbolic [12, 58]. Let's take a look on the form the of the shape of the surface of the primary mirror, the parabolic one. In figure 10.21, it is displayed the `Surface Form` of it. From the figure, it is clear to see the annular aperture of the primary mirror. The inner radius of the annular aperture is 350mm and the

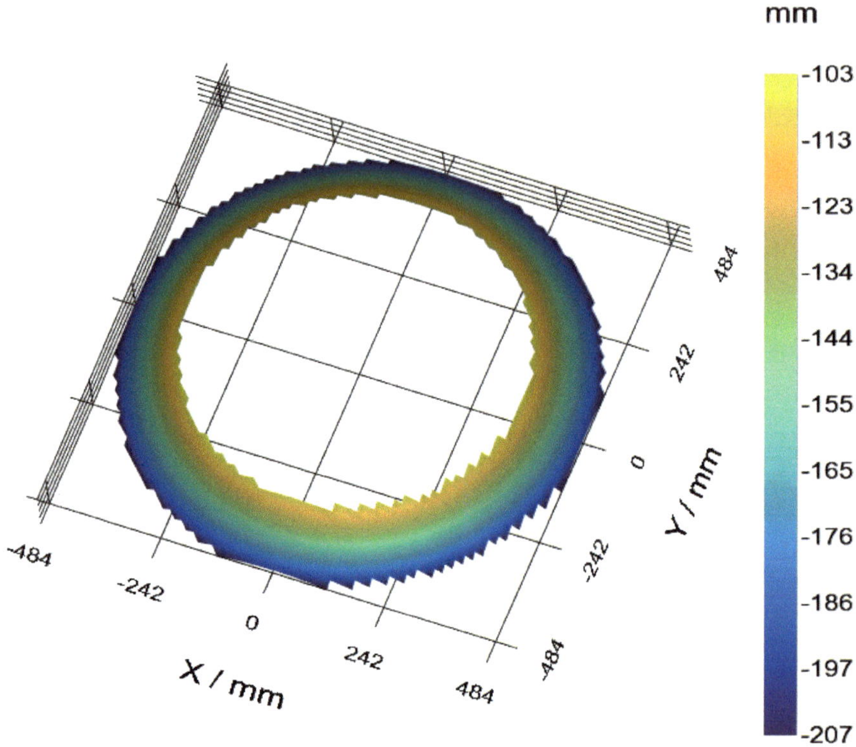

FIGURE 10.21
`Surface Form`: The parabolic primary mirror has an annular aperture. The inner radius of the annular aperture is 350mm and the outer radius is 500mm. The diagonal bar depicted in the figure illustrates the mirror's height level within the local coordinate system of the surface.

outer radius is 500mm. The diagonal bar presented in the figure shows the level of the height of the mirror in the surface local coordinate system. Where the zero is the vertex of the parabola. But since the mirror has an annular aperture the vertex is not plotted. Thus the values of the height go from -103mm to -207mm since the mirror can be seen as convex in the way how it is placed in the local coordinate system.

This combination of conic surfaces is such that the system is stigmatic for the on-axis field. Remember for chapter 2, that Stigmatism pertains to the image-forming characteristic of an optical system, wherein it concentrates a singular point source from one phase of optics space into a solitary point within image space. A pair of such points is termed a stigmatic pair within the optical system. Therefore that means that the spot diagram of the on-axis field

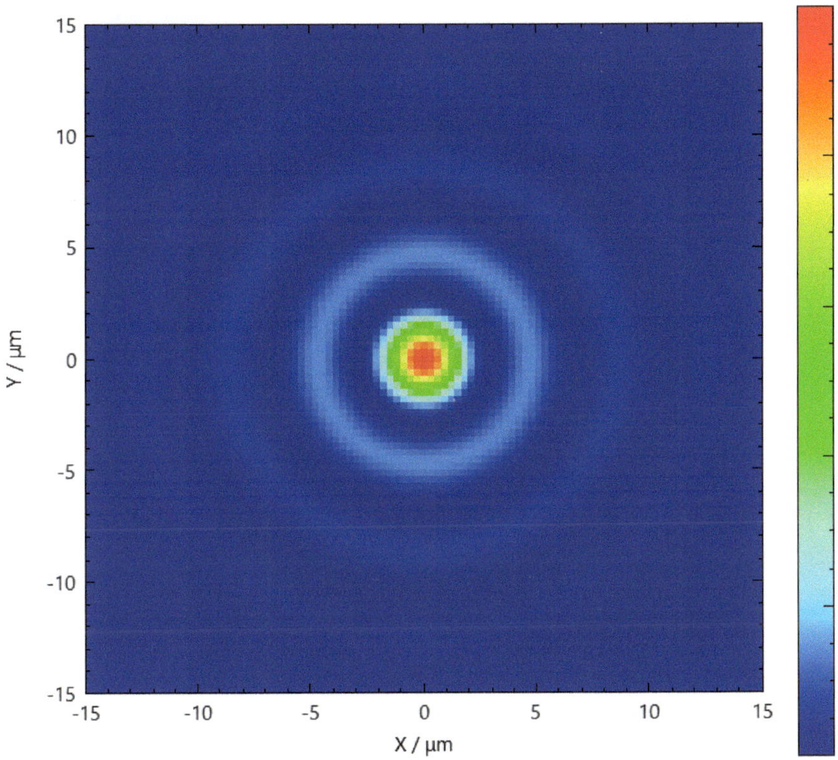

FIGURE 10.22
`Huygens PSF Plot` of the on-axis field of the Cassagrain Telescope. The Cassagrain Telescope is stigmatic for the on-axis field, therefore it is also diffracted-limited for this field.

is a single point. This is only true by ignoring the wave nature of light and only using the postulates of geometrical optics. But if we plot the `Huygens PSF` we cannot get a single point but the Bessel function point spread function as studied in chapter 2. The `Huygens PSF` for this telescope is present in figure 10.22. From the plot, we confirm that the system is diffracted-limited.

10.8 Nasmyth Telescope

James Nasmyth was a Scottish engineer born in 1808 and known for his contributions to the field of mechanical engineering and astronomy [78]. Nasmyth's

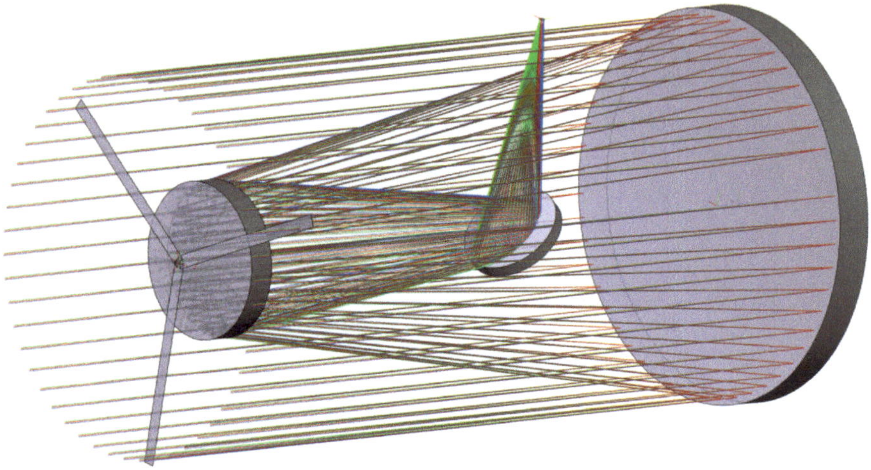

FIGURE 10.23
Rat-tracing of the Nasmyth telescope. The Nasmyth telescope is a type of reflecting telescope that incorporates a design feature known as the Nasmyth focus. Nasmyth focus is an additional focal plane located outside of the primary focal plane of the telescope.

significant advancements in astronomy and is known for his detailed observations and sketches of celestial objects, particularly the moon. His observations were highly regarded by astronomers of his time and contributed to our understanding of the lunar surface. Also, James Nasmyth is known for the Nasmyth telescope.

The Nasmyth telescope is a type of reflecting telescope that incorporates a design feature known as the Nasmyth focus. It was developed by Nasmyth in the 19th century as a way to facilitate astronomical observations and instrument mounting.

The defining feature of the Nasmyth telescope is the Nasmyth focus, which is an additional focal plane located outside of the primary focal plane of the telescope. This focus is created by redirecting the light path from the primary mirror to a fixed location on the telescope mount, where instruments can be attached for observations. This additional focus can be seen in figure 10.23 where also ray-tracing of the telescope is displayed.

So from the figure, it is clear that the image plane is outside the axis where the first and second mirrors are aligned. The third mirror works as a folding mirror and it is just flat. The first and second mirrors are conic. In this case, the on-axis system is not stigmatic because it is such that tries to compensate for the aberration of the off-axis fields. The example has three `Fields`, with the following incident angles 0°, −0.1°, and 0.1°. The `Spot Diagram Plot` is shown in figure 10.24.

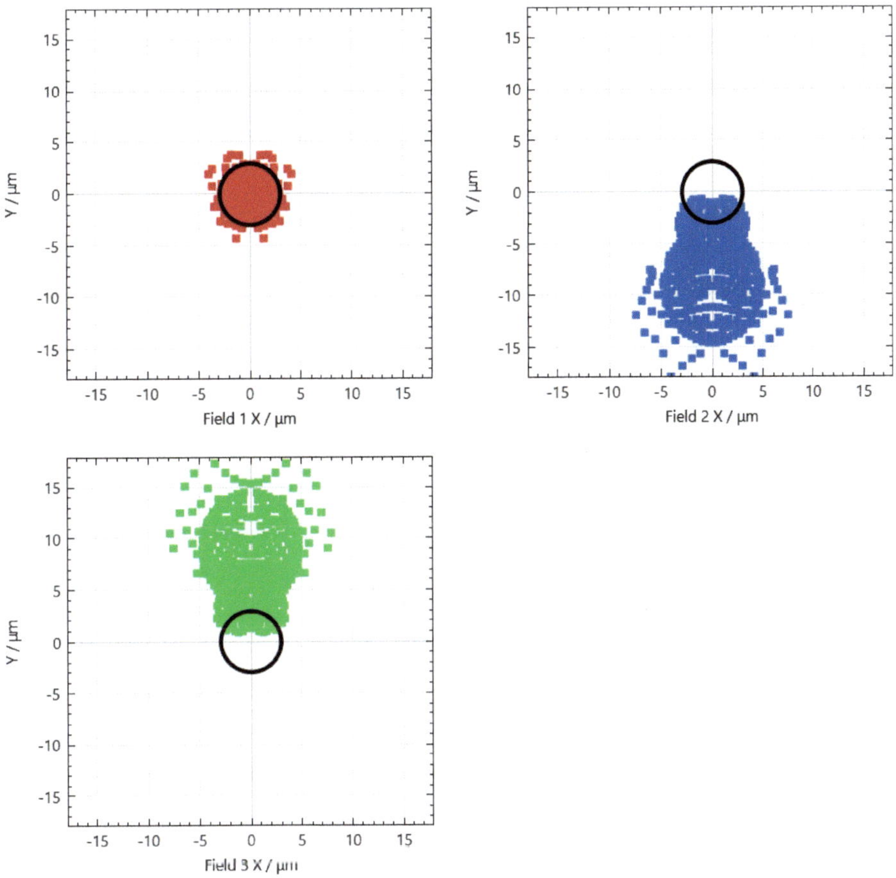

FIGURE 10.24
`Spot Diagram Plot` of the Nasmyth telescope. The fields have the following incident angles 0°, −0.1°, and 0.1°. The color code is red, blue, and green for the fields, for the 0°, −0.1°, and 0.1°, respectively.

The color-code is red, blue, and green for the fields, for the 0°, −0.1°, and 0.1°, respectively. Notice that the off-axis field with −0.1°, and 0.1° have also very similar patterns, this is due to the symmetries preserved with the folding mirror and the rest of the elements of the system being radially symmetric.

Since the field of view of the system is very small the distortion is also very small. But this is just an initial guest, so let's compute the `Distortion 2D` of the Nasmyth telescope with `Quadoa Optical CAD`. The result is presented in figure 10.25 the rectangular grid stays with no major distortion across the image height in the $x-$ and $y-$ axis.

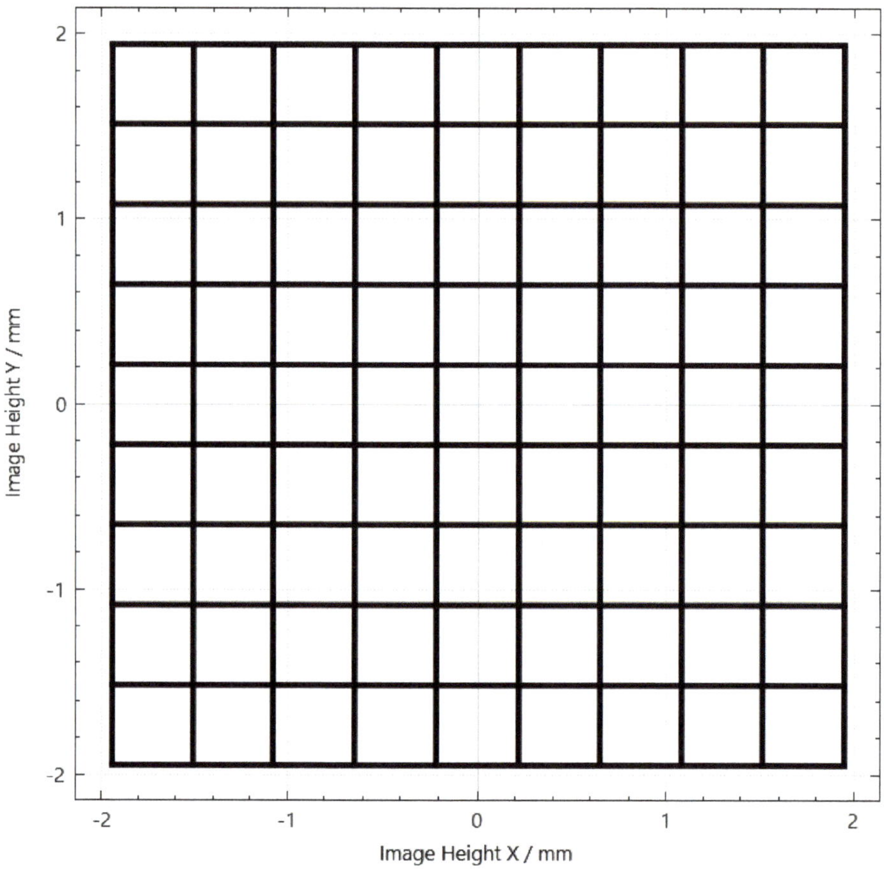

FIGURE 10.25
`Distortion 2D` of the Nasmyth telescope computed in `Quadoa Optical CAD`. There is no major distortion across the image height in the x and y axis.

The next plot is the `FTT MTF Plot`, which is displayed in figure 10.26. From the figure, it is clear that the on-axis field is mostly diffracted-limited. For both off-axis fields, the performance drops signidfically compared to the on-axis system, this can be seen from the spot diagram. Due to the symmetry of the system, the performance of both off-axis is mostly similar.

Now compare the `FTT MTF Plot` of the Nasmyth telescope with the `FTT MTF Plot` of the microscope is presented in figure 10.26. The resolution of a microscope objective is better than the telescope. Since the microscope has a bigger numerical aperture, thus its PSF is smaller. Which means a better resolution or more lines per millimeter can be detected.

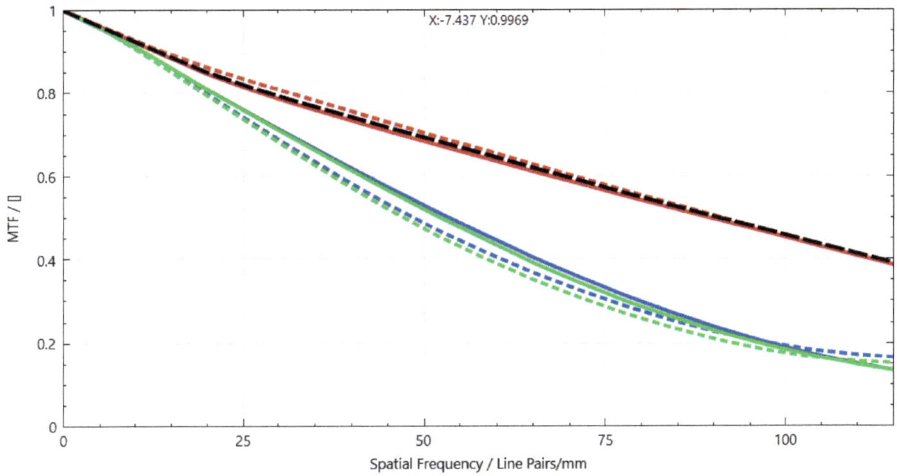

FIGURE 10.26
`FTT MTF Plot` of Nasmyth telescope. The on-axis field is primarily constrained by diffraction, showcasing mostly diffracted-limited behavior. As for the off-axis fields, their performance notably diminishes compared to the on-axis configuration.

10.9 James Web Telescope

The James Webb Space Telescope (JWST) is a large, space-based observatory that is set to be one of the most powerful telescopes ever built [79]. Named after James E. Webb, who was NASA's administrator during the Apollo program, the JWST is a joint project of NASA, the European Space Agency (ESA), and the Canadian Space Agency (CSA).

The James Webb Space Telescope represents a major technological advancement in space-based astronomy and is expected to revolutionize our understanding of the cosmos. It is a flagship mission for NASA, ESA, and CSA, and has the potential to make groundbreaking discoveries across many areas of astrophysics [80, 81, 82]. The ray-tracing of this flag ship is in figure 10.27.

The JWST is composed of four mirrors. The JWST features a segmented primary mirror composed of 18 hexagonal mirror segments, which together have a total diameter of 6.5 meters (21.3 feet). This large mirror will allow the telescope to collect and focus faint infrared light from distant celestial objects.

Let's compute the `FFT PSF Plot` of the James Webb Space Telescope for the on-axis field, the result is displayed in figure 10.28. Notice that the pattern of the `FFT PSF Plot` has some complex shapes related to hexagons. This pattern comes from the fact that the aperture of the segmented elements of

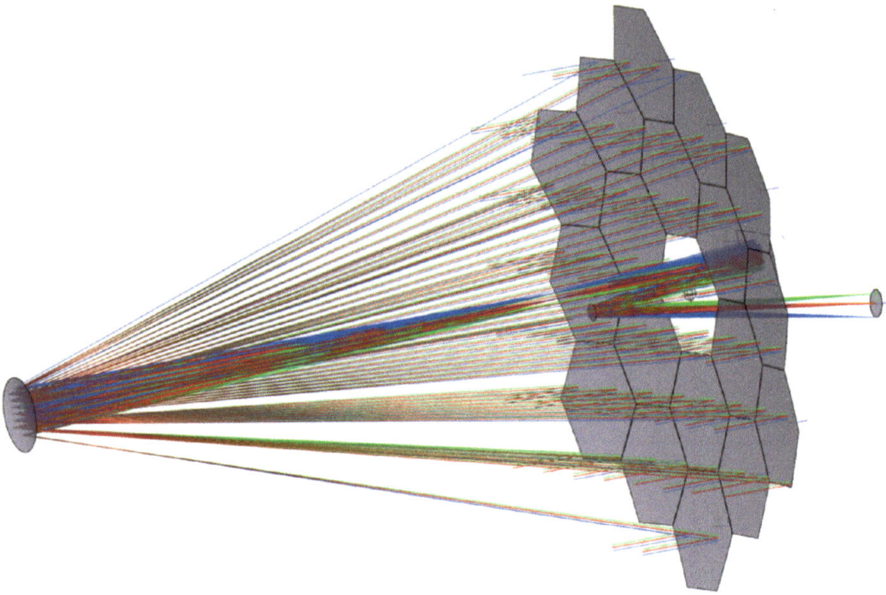

FIGURE 10.27
Ray-tracing of the James Webb Space Telescope. The JWST is composed by four mirrors. The JWST features a segmented primary mirror composed of 18 hexagonal mirror segments.

the primary has a hexagonal shape. The aperture affects the computation of the PSF since it delimits the integration area [32].

The shape of the aperture of one of the 18 hexagonal mirror segments is presented in figure 10.29. From the height color plot, we can confirm that is one of the segments in the bottom right.

`FFT MTF Plot` is shown in figure 10.30. The performance of the three `Fields` is mostly diffracted-limited the incident angles are 0.18°, 0.24° and 0.14° and their color code is red, blue and green. Notice that the `FFT MTF Plot` drops almost to zero at 20 lines per millimeter. This is due to the low numerical aperture, NA=0.0216865, and the high f-number, which is equal to 23.0566.

10.10 Freeform Mirror Telescope

The next example is a very interesting telescope. It is a telescope made of three freeform mirrors [68, 69, 70, 71]. The description of the mirror

FIGURE 10.28
`FFT PSF Plot` of the James Webb Space Telescope for the on-axis field.

surfaces is given by `Zernike Polynomial Form`. The ray-tracing is displayed in figure 10.31. The file is in `C:\Program Files\Quadoa\examples\Objectives\Microscope\Free_Form_Telescope`.

Since the system is not radially symmetric it is convenient to describe the incident angles of the fields with two values. he angles respect to the x and y, respectively. In our notation will be (tilt n x, tilt in y). This example has the following `Fields` $(0°, 0°)$, $(0°, 1.5°)$, $(0°, -1.5°)$, $(1.5°, 0°)$, $(-1.5°, 0°)$. The color code is red, blue, green orange, and purple, respectively.

The `Spot Diagram Plot` of our freeform telescope is presented in figure 10.32. It is presented stop of each of the fields. Notice the patterns of the fields are not symmetric. Even for the on-axis field, the one with $(0°, 0°)$, the spot is not symmetric. The symmetric has been broken. With the broken symmetric other aberrations appear, but the idea is that with the freeform surface the system has more degrees of freedom that can be optimized.

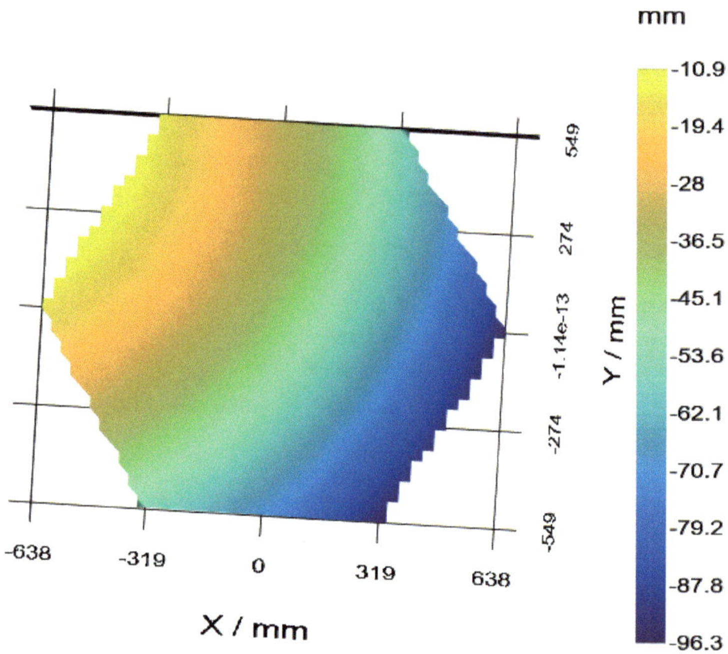

FIGURE 10.29
`Surface Form` of one of the elements of the primary mirror the James Webb Space Telescope.

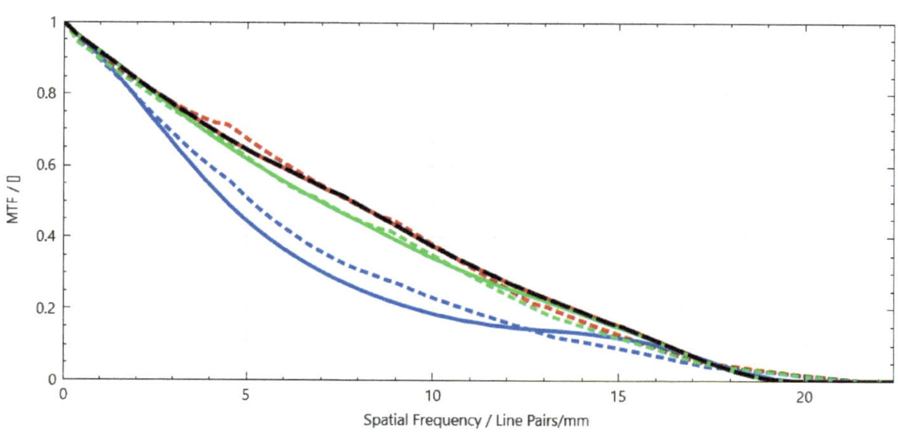

FIGURE 10.30
`FFT MTF Plot` of the James Webb Space Telescope.

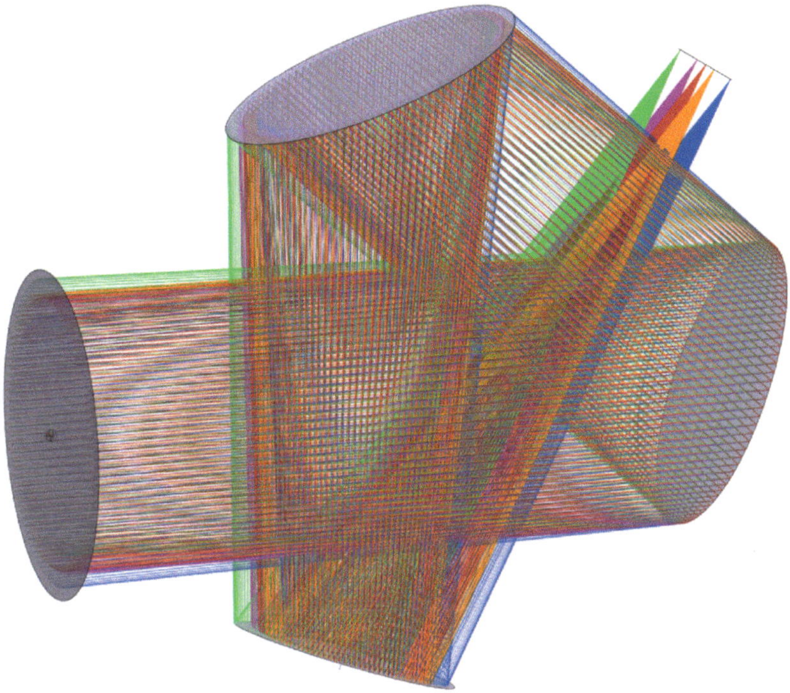

FIGURE 10.31
Ray tracing of a freeform telescope. The telescope has three freeform mirrors described by `Zernike Polynomial Form`. The fields have the following tilts $(0°, 0°)$, $(0°, 1.5°)$, $(0°, -1.5°)$, $(1.5°, 0°)$, $(-1.5°, 0°)$ and their colors are red, blue, green orange and purple, respectively.

The `FFT PSF Plot` is plotted in figure 10.33. For the on-axis field, red field. The distribution of intensity is not radially symmetric. The shape resembles the shape presented in the spot diagram. The plot intensity normalized, the peak of intensity one is not at the centroid of the point spread function. This means the on-axis field from off-axis aberration like coma and astigmatism.

The range where the point spread function is distributed in the plot is quite big. The x-local axis is around 60μm, and also 60μm for the y-local axis. The Airy disk has a radius around 7μm.

Notice that the freeform telescope of figure 10.31 has three freeform mirrors with spherical apertures. The apertures in the freeform system can be crucial to block rays and for vignetting, this is something the design should be aware of to avoid stray light. In figure 10.34, we plot the `Footprint Diagram` of the freeform telescope of figure 10.31. The first three plots are the floor prints

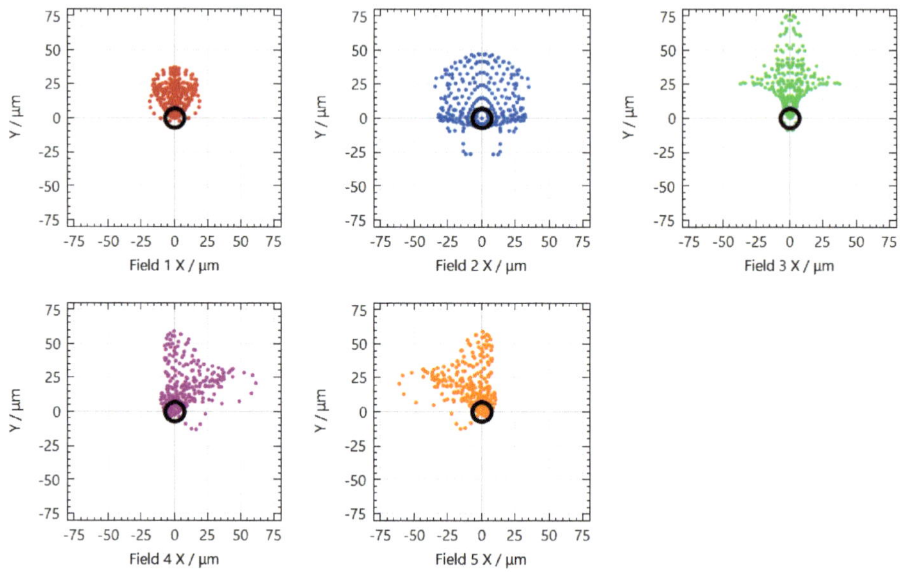

FIGURE 10.32
`Spot Diagram Plot` of the freefrom telescope of figure 10.31. The fields have the following tilts $(0°, 0°)$, $(0°, 1.5°)$, $(0°, -1.5°)$, $(1.5°, 0°)$, $(-1.5°, 0°)$ and their colors are red, blue, green orange and purple, respectively. The patterns of the fields are not symmetric, due to the freeform shape and off-axis positions of the mirrors.

of the mirrors. The last one, bottom left, is the image plane. Notice that the aperture of the third mirror is not fully covered. Additional fields should be added in other to see how the entire aperture is covered by the rays of the additional fields. On the other hand, the apertures of the first and second mirrors are mostly covered by the rays of the given fields.

The `Surface Form` of the three mirrors is displayed in figure 10.35. From right to left, the first, second, and third plots represented the first, second, and third mirrors. At the center, the mirrors seem to be aspherical but near the borders, the freeform features start to appear.

10.11 Smart Phone Camera

The optics of smartphone cameras play a crucial role in determining the image quality, versatility, and overall performance of these devices. Despite their compact size, smartphone cameras incorporate sophisticated optical

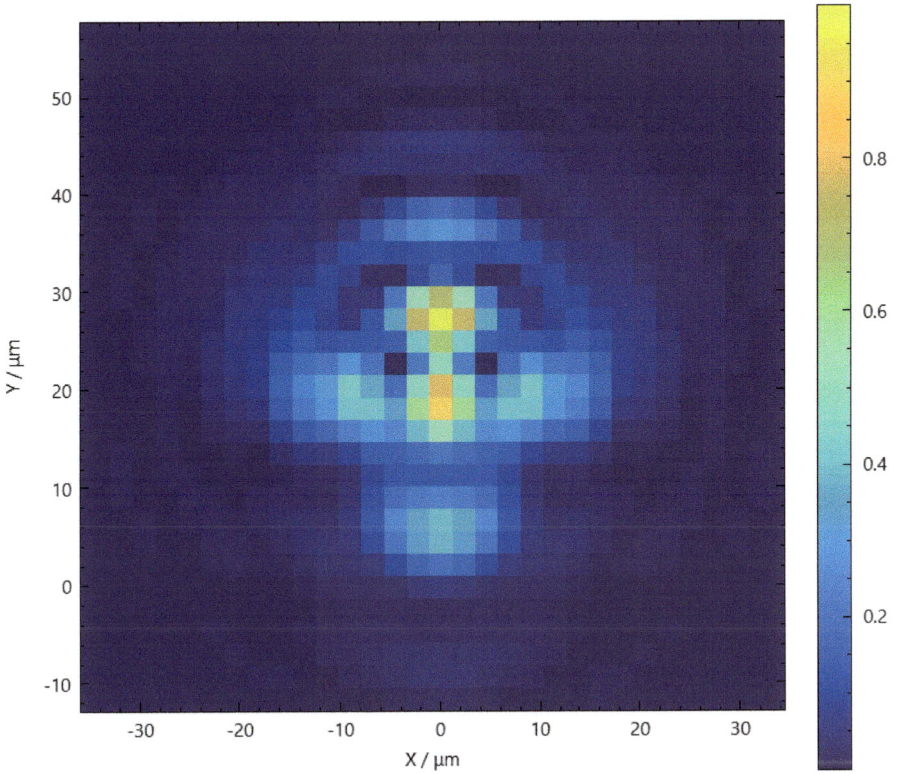

FIGURE 10.33
`FFT PSF Plot` of the on-axis field of the freeform telescope of figure 10.31. The plot intensity normalized. The distribution of intensity is not radially symmetric due to the broken symmetry of the system.

components and technologies to capture high-quality photos and videos. Our next example is a smartphone camera. This camera is also called the main, camera because nowadays most phones include several cameras, But the main camera is the one with the most image quality. Most main cellphone cameras share several characteristics which are the following [83]:

- The dimensions of image sensor pixels range from $1.0\mu m$ to $2.0\mu m$.
- Typically, camera lenses incorporate six or seven elements made of plastic optical materials. These materials are well-suited for crafting the intricate surfaces necessary for correcting aberrations commonly found in mobile phone camera lenses.
- The shape of the surface of the elements is the `Asphere Form`.
- The f-number ranges from 1.5 to 3.

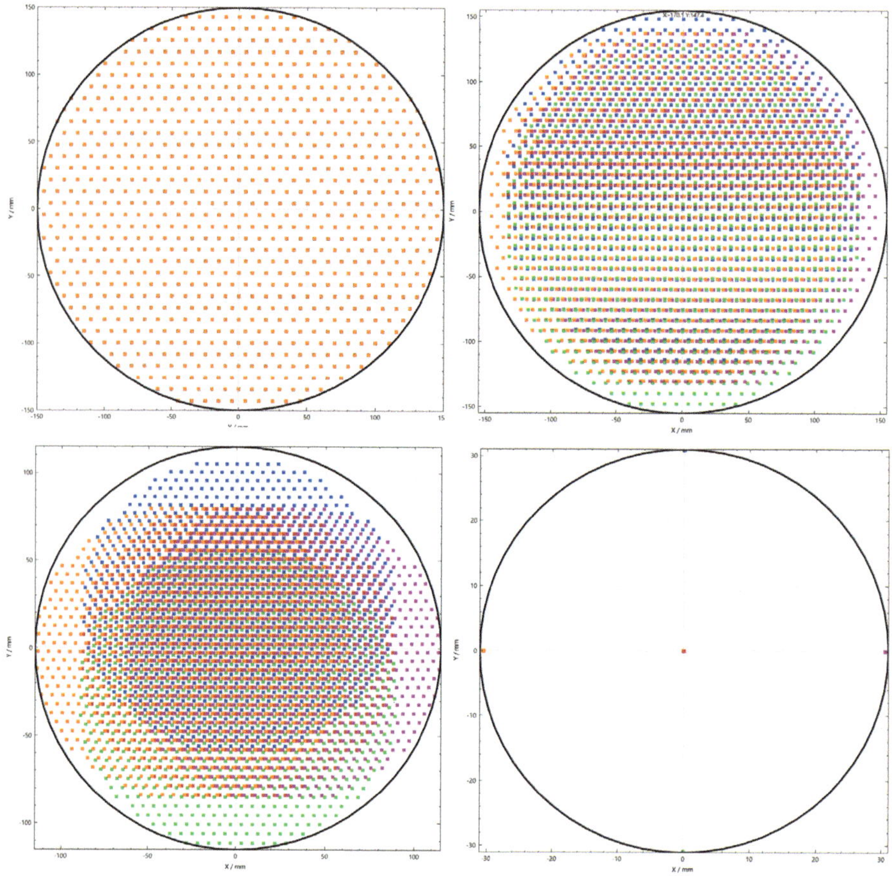

FIGURE 10.34
`Footprint Diagram` of the freeform telescope of figure 10.31.

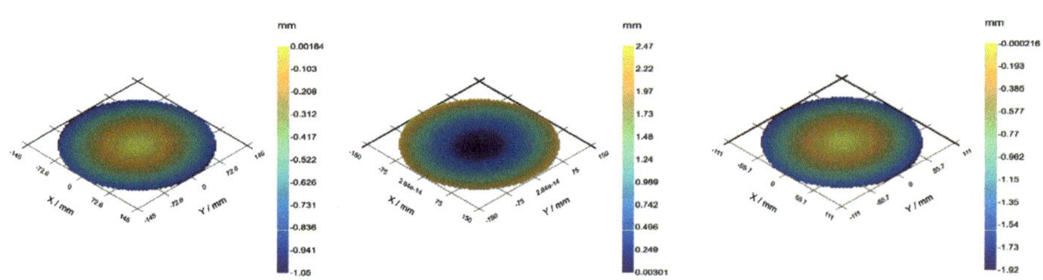

FIGURE 10.35
`Surface Form` of the freeform telescope of figure 10.31.

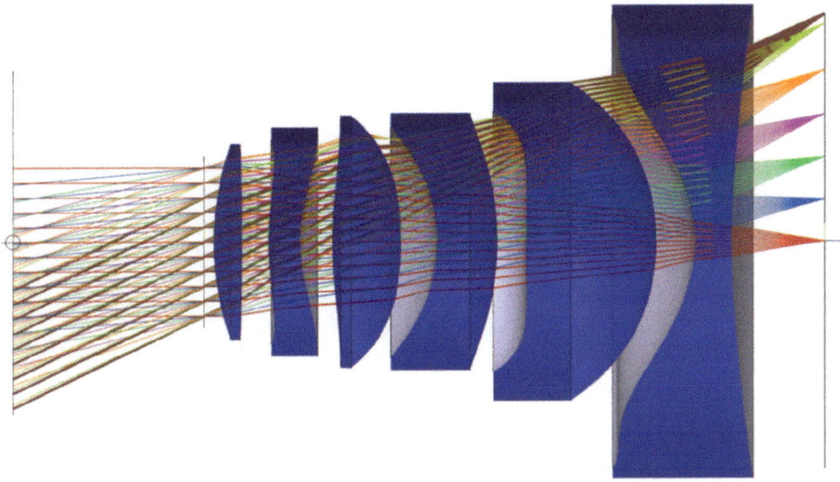

FIGURE 10.36
An example of a main camera.

- The field of view (FOV) provided by mobile phone cameras is expansive, with typical values ranging from 70° to 85°.

An example of a main camera is presented in figure 10.36. Another view of ray-traying is presented in figure 10.36. This design does not come from a patent, it is just an illustrative example.

This particular design has a field of view of 70°. The design contemplates seven fields. The first field is an on-axis field. The rest increases the angle of the incident until reaches the half-of-view angle of 35° with the last field. In other means the seventh field has an angle of 35°. The rest have some angle between zero and 35°, which gradually increases with the number of the field.

It is made of six plastic elements. The materials are `PMMA` and `COC` plastics. There is a vast catalog of plastic used in cellphone cameras, but for this illustrative example, we focus on the two mentioned materials, which are also widely used in main cameras.

The total track length is a major aspect in the design of smartphone cameras due the space restrictions. For this example is total track length is 6.64mm. The effective focal length is 5.067 mm, a back focal length of 0.279 mm, and an f-number of 2.8. The image height measured 3.96 mm, slightly exceeding the diagonal size of the image sensor, which is 7.906 mm.

The stop surface is placed before the first lens. Usually in cellphone main cameras that happens too. But it is not a general rule. There are cases where the stop is placed between the first and second lens or the second and third lens.

FIGURE 10.37
Ray-tracing of smartphone main camera. The total track length is 6.64mm. The effective focal length stands at 5.067 mm, with a back focal length of 0.279 mm, and an f-number of 2.8. The image height is measured at 3.96 mm, and the diagonal size of the image sensor is 7.906 mm.

The `Spot Diagram Plot` is presented in figure 10.38. The first four fields are mostly diffracted and limited. The fields from five to seven suffer from a coma. Usually, these fields are very hard to optimize such a good performance. The main reason is that this optimization is more related to the numerical approach than the theoretical one since the theory of lens design is mostly based on the third aberration theory. The third aberration theory is mostly based on paraxial approximations. These fields have incident angles much bigger than the paraxial regime.

Another important aspect of main cellphone cameras is distortion. Usually, the acceptance distortion level is around 2%. Let us compute the `Distortion 1D Plot` of this main camera, figure 10.36, with `Quadoa Optical CAD`. The result is presented in figure 10.39.

From the figure, it is clear that the distortion increases in negative values as the field of view increases. The maximum distribution presented is below 3%. It is around 2.7%, which for several vendors could not be acceptable. So the design presented in figure 10.36 should be re-optimized to decrease the distortion. However, the design of figure 10.36 can be considered as a good starting point.

Now, let's check the `Petzval Curvature 1D Plot` of the design of figure 10.36. With `Quadoa Optical CAD` we compute it and display it in figure

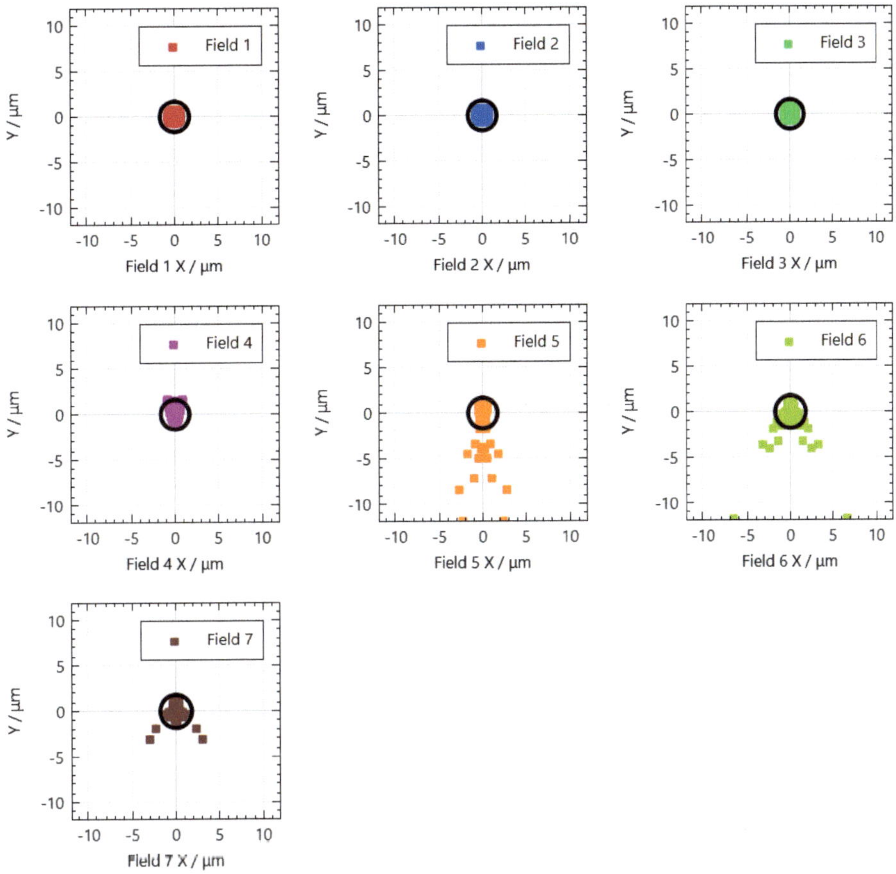

FIGURE 10.38
`Spot Diagram Plot` of the design presented in figure 10.36.

10.40. The plot shows continuous and dashed lines, the first one corresponds to the sagittal plane, and the second one to the tangential plane. The range of both is between 0.001mm which is considered acceptable for the design of main cameras in cellphones

10.12 Spectrometer

The last example we are going to study in this chapter is a spectrometer. The fundamental idea behind a spectrometer is to separated the different wavelengths from a light source [84]. This can be very useful in several areas

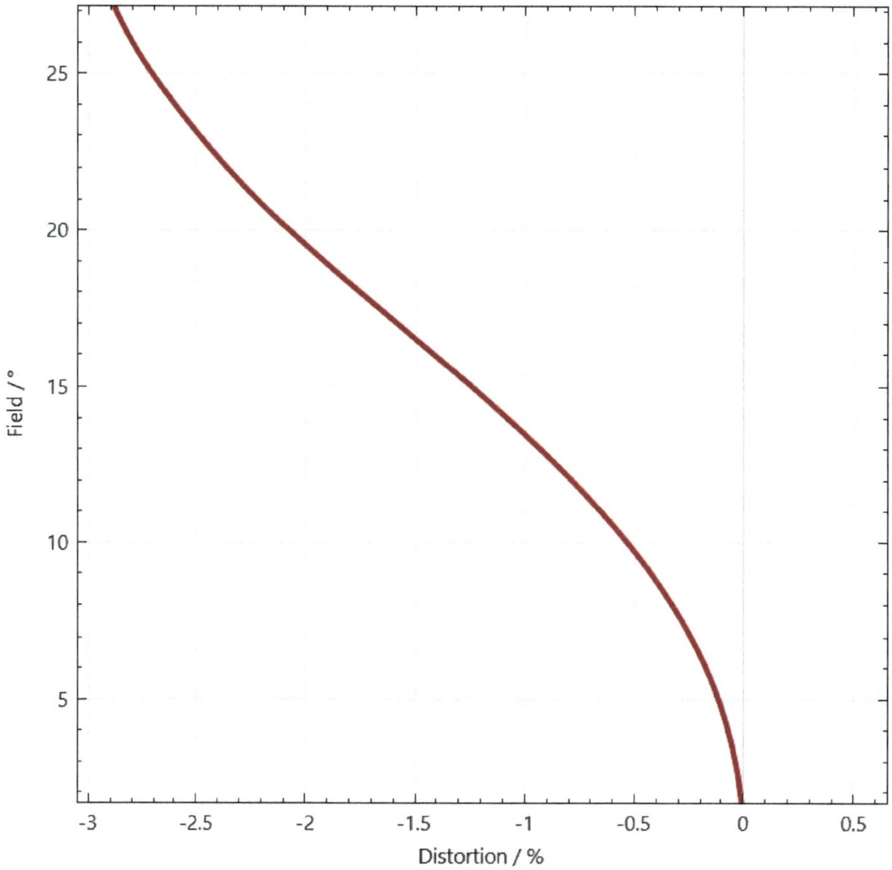

FIGURE 10.39
`Distortion 1D Plot` of the design is presented in figure 10.36. The maximum observed distribution is around 2.7%.

of science and engineering [85, 86, 87]. In figure 10.41 we present the example of a spectrometer. This example can be found in:

`C:\Program Files\Quadoa\examples\Objectives\Spektrometers`.

It consists of three elements. The first element is not contained in an `Assembly`. But, the second and third elements are in a group together as an `Assembly`. The reason is that the second and third elements are off-axis. The `Assembly` is tilted by $-35°$ respect to the global coordinate system, optical axis.

The first element is a flat glass lens made of `N-BK7`, with the feature that the first surface is a grading phase with 1000 lines per millimeter. The second element is a lens, with `Cylindric Form` in its both surfaces, with a radius of 28mm and 55mm, respectively. The third and last element is a

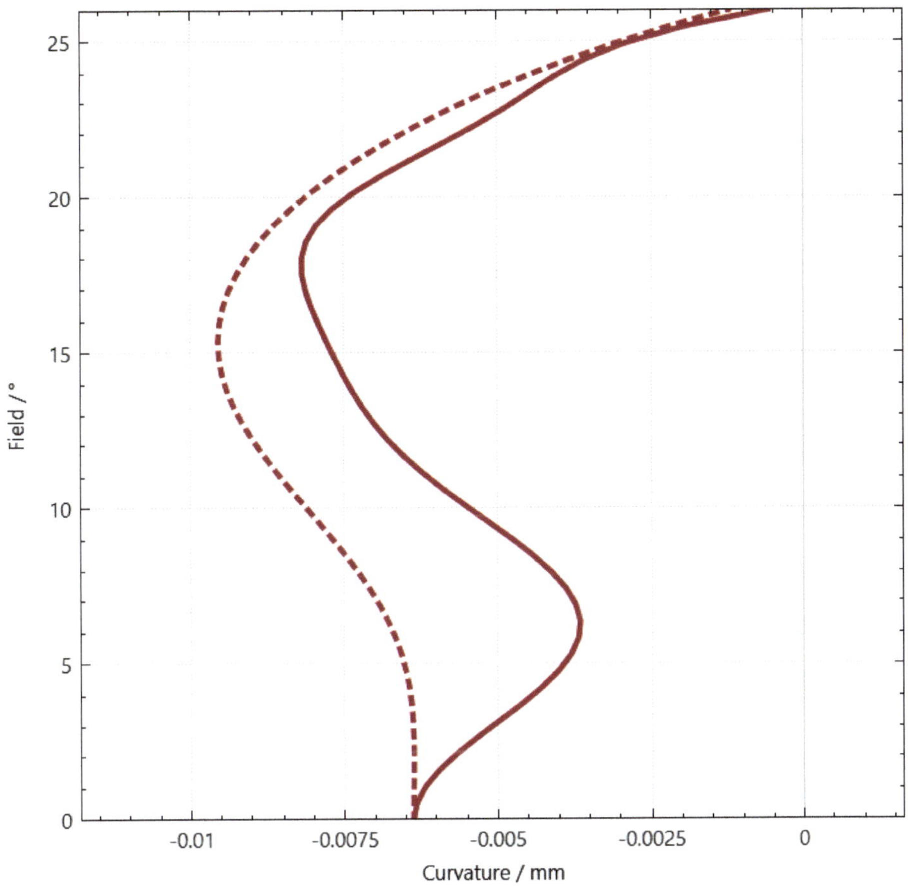

FIGURE 10.40
`Petzval Curvature 1D Plot` of the design presented in figure 10.36. The range of field curvature is between 0.001mm, which is considered acceptable.

lens with `Cylindric Form` and `Biconic Form` surfaces. The radius of the first surface of the third element is -40mm. The radius of the last surface of the third element is 0 and -26mm, in the x and y axis respectively. The conic constants of the last surface of the third element are 0 and -1.9mm, in the x and y axis respectively. The material of both lens is `N-BK7`.

The source is composed of an on-axis field with seven wavelengths. The wavelengths are 400nm, 450nm, 500nm, 550nm, 600nm, 650nm, and 700nm. The wavelengths are separated once they touch the grading, due to diffraction. Then the different diffraction orders are focused by the two lenses in the image plane.

A side view of the ray-tracing of the spectrometer of figure 10.41 is displayed figure 10.42. From the figure, it is clear that this simple, but

FIGURE 10.41
Ray-tracying of a basic spectrometer.

illustrative example, of a spectrometer separates the light and focuses it in the image plane.

We can explore the image plane generated by this illustrative example with `Footprint Diagram`. A tool provided by `Quadoa Optical CAD`. The footprint of the spectrometer of figure 10.41 is shown in figure 10.43. From the figure, its clear the spectrometer successfully separated the seven different wavelengths and their diffraction orders. At the top of the `Footprint Diagram` we can see the order of the 700nm wavelength, and it goes down in descendent order until the 400nm wavelength.

10.13 Summary

In this chapter, we have seen several designs and analyzed them with `Quadoa Optical CAD`. The different tools provided by `Quadoa Optical CAD` help us to analyze and make a possible plant for potential optimization in the current design. Also, we see the difference between several optical designs, like f-numbers and field of view. These differences come from the purpose of the designs.

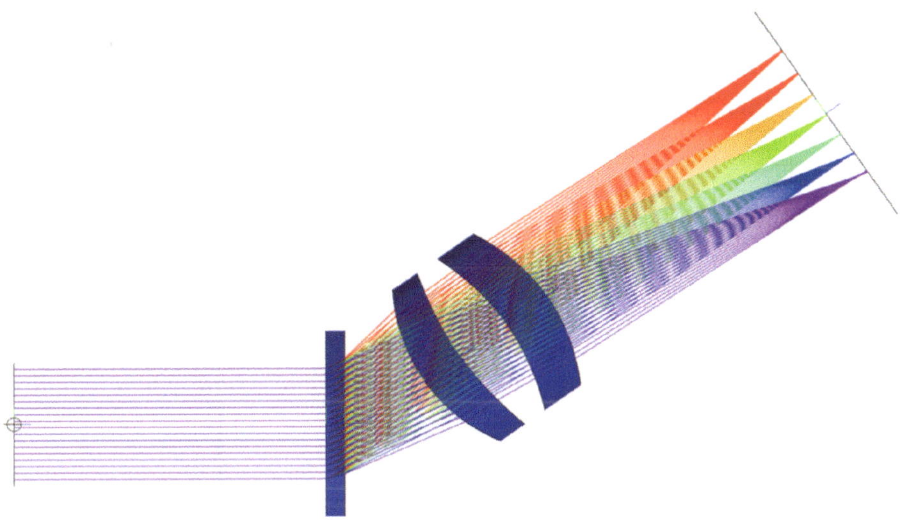

FIGURE 10.42
Side view of the ray-tracying of the spectrometer of figure 10.41. It consists of three elements. The first element is a flat glass lens with grading phase and two lenses. The source is composed of an on-axis field with several wavelengths. The wavelengths are separated once they cross the grading. Then lenses focus on the different diffraction orders in the image plane. The wavelengths are 400nm, 450nm, 500nm, 550nm, 600nm, 650nm, and 700nm.

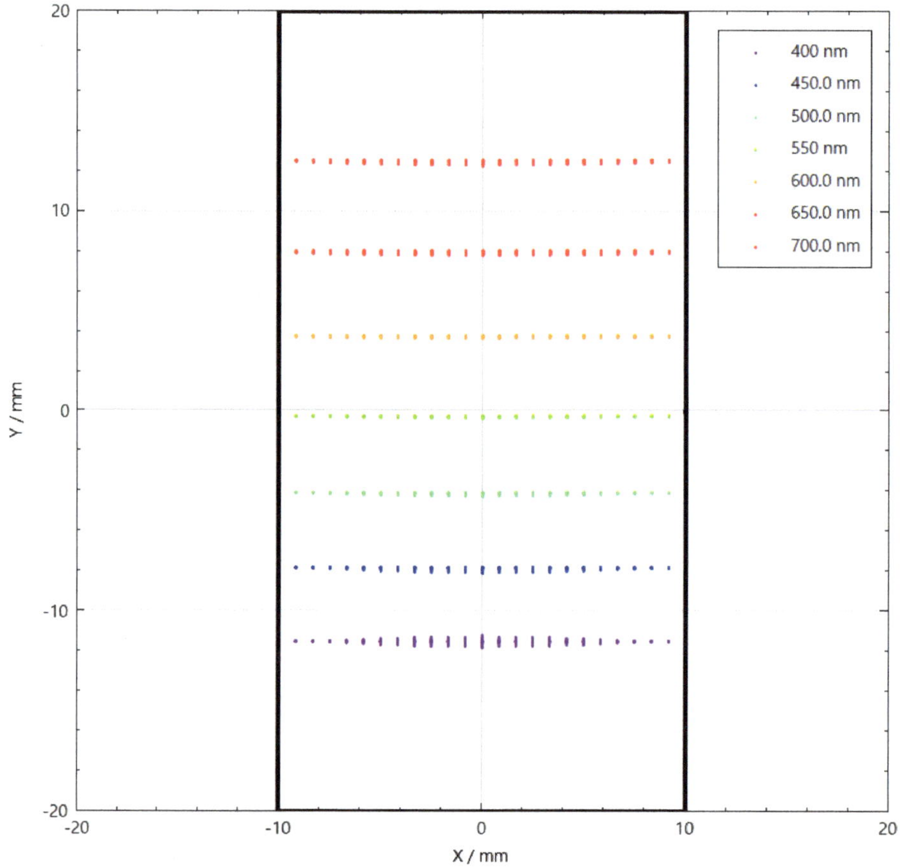

FIGURE 10.43
The `Footprint Diagram` of the spectrometer of figure 10.41. The wavelengths are 400nm, 450nm, 500nm, 550nm, 600nm, 650nm, and 700nm.

11

Multiple Sequences and Configurations

> Music is the arithmetic of sounds as optics is the geometry of light.
>
> Claude Debussy

11.1 Multiple Sequential Raytracing

In chapter six, before our first example, we review the concept of sequence in `Quadoa optical CAD`. We saw that in `Quadoa optical CAD` a sequence acts as the foundational element utilized to conduct a sequential ray trace within the system. It delineates the sequential path along which the rays are traced through the system.

One of the most important properties of `Quadoa optical CAD` is that one model can handle multiple sequences. Contrary to other optical design software, it is easily possible to have more than one sequence in a single model [18]. Each sequence can have its own path, and properties given by source, wavelengths, fields, ray distributions etc.

This property enables `Quadoa optical CAD` to be a powerful tool for designing optical systems with more than one path and configuration. A configuration in optical design can be a set of parameters that compose an optical system for example the position of lenses, the field of view of the system, the position of an object/image, etc. So when an optical system has several configurations, this set of parameters differs from at least one value with other configurations.

In optical engineering, there are many cases where an optical design should consider additional sequences. The most popular example is the interferometer. The idea of this divide is to split a source of light and then merge it again at some point, to highlight some properties of a system. Another interesting example is a virtual reality lens. Next, we are going to study it.

DOI: 10.1201/9781003506355-11

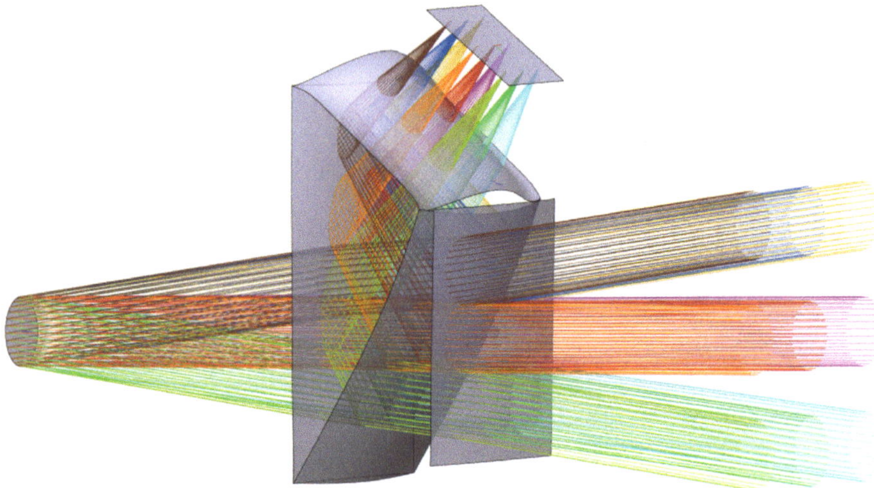

FIGURE 11.1
Ray-tracing of FF_Prism_With_Corrector_Merch. The design has one configuration and two sequences. Both sequences depart from the same surface but end on different surfaces due to a Beam Splitter Coating that splits them. The Beam Splitter Coating is placed at the second surface of a Custom Element made of four freeform surfaces.

11.2 Example 1: Virtual Reality Lens

Our first example is an augmented reality optical system. An augmented reality optical system is capable of overlaying computer-generated imagery onto a person's real-world view. It has been hailed as a groundbreaking technology, poised to reshape our interaction with digital information [77, 88, 89, 90, 91]. A virtual reality lens is presented in figure 11.1. The file can be found in:
C:\Program Files \Quadoa\examples\VR\FF_Prism_With_Corrector_Merch
Here, this design will be called FF_Prism_With_Corrector_Merch.

This particular design has only one configuration but it has two sequences. The system consists of the first surface, which is considered to be the eye. Collimated rays emerge from the eye in both sequences. Then the ray goes inside a Custom Element which is a freeform prism with four surfaces made by N-BK7. The sequences split at the second surface of the Custom Element. This surface has a Beam Splitter Coating. The first sequence then suffers a total internal reflection at the first surface, then passes the third surface, and ends in the display. The second sequence passes the fourth surface. This configuration makes the eye believe that there are additional elements generated by the image display.

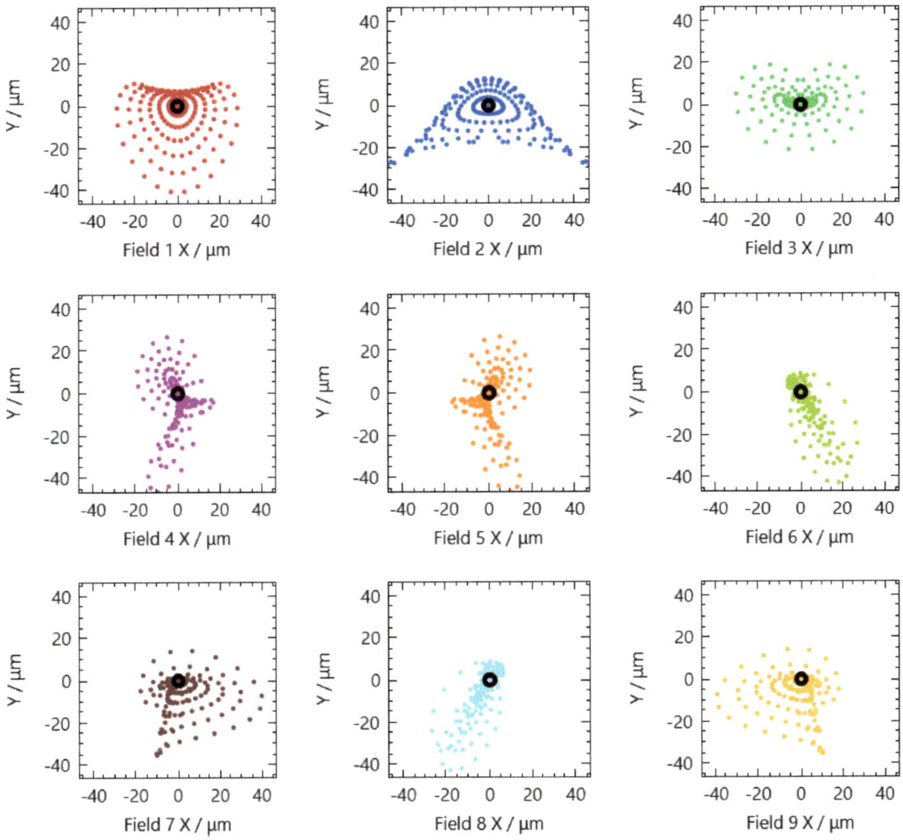

FIGURE 11.2
`Spot Diagram Plot` of the first sequences of the `Multiconfig_Model`. This sequence ends in the surface called `Display` in the mentioned file.

In the design, the sequences are inverted concerning the real life. So, it is assumed that the eye receives collimated beams from two sources. One source is the collimated rays from the second sequence and the other source is the rays from the display. The shape of the surface of the prism is such that the fields of both sequences merge in the eye.

The shape of the fourth surface is described by `XYZ Polynomial Form`. The field of view of the system is $16° \times 16°$. The wavelength of the source in this particular example is 550nm.

We can study the performance of both sequences using the tools provided by `Quadoa optical CAD`. Let us start with the first sequence, with the `Spot Diagram Plot`. The `Spot Diagram Plot` is plotted in figure 11.2.

The `Spot Diagram Plot` shows that all the fields suffer from several high-order aberrations due to the freeform nature of the `Custom Element`.

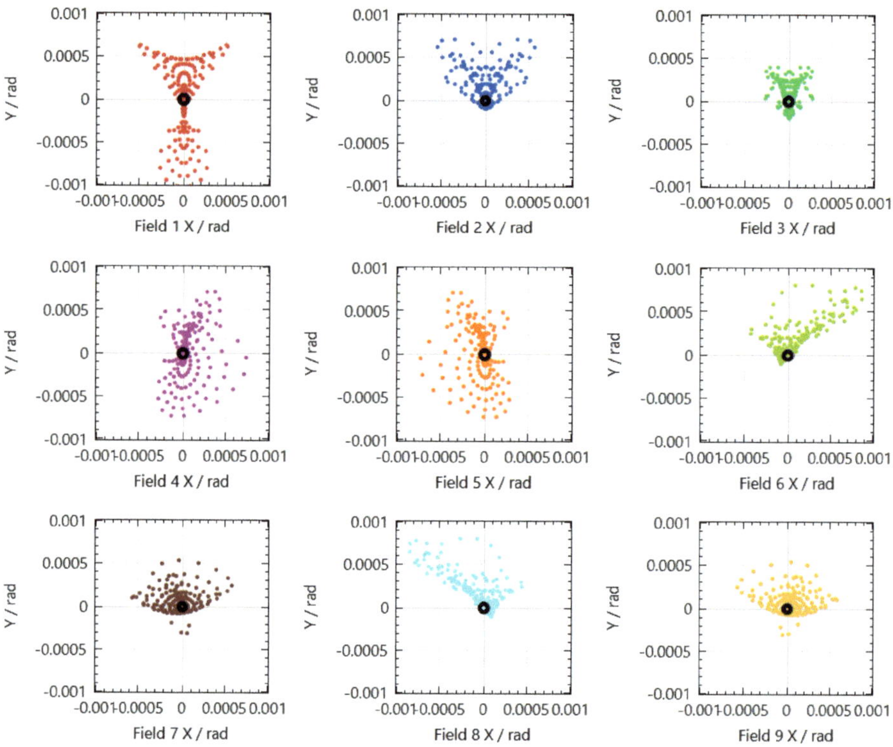

FIGURE 11.3
`Spot Diagram Plot` of the second sequences of the `Multiconfig_Model`. This sequence ends in the surface called `SURFACE 1` in the mentioned file. This surface is not displayed because, in the real world, it is the object space. But, in this design, the sequence order is the inverter and it is the image space.

The spots for all the fields are several times larger than their respective Airy disk. The design has space for optimization.

The `Spot Diagram Plot` of the second sequence is plotted in figure 11.3. The fields in this sequence must be collimated at the output. Thus, the `Spot Diagram` measures, for a given field, the angles of deviation of a ray concerning the chief ray. The units are in radians and the destination is around 0.001 radians in the worst-case scenario. Also, the spot patterns presented in the figure have freeform shapes due to the nature of the design in the `Multiconfig_Model`.

The key in this design is the feature of `User Sequence`. This made the design, optimization, and analysis process much easier than with other optical design software. Because the user sets the path of the sequence and because `Quadoa optical CAD` can work with several sequences at the same time.

11.3 Multiple Configurations

In optical design, there are several scenarios where multiple configurations are needed for a single optical design. The most common cases are the focus-able lens, the zoom lens, and the thermal stabilization [12, 13, 58]. But in general, there are many other scenarios with multiple configurations or multiple sequences.

11.3.1 Focus-able Lens

A focus-able lens is a lens optimized for several configurations related to the object position. These configurations are mostly based on the position object or source [5].

If it is a point object/source, the distance between this source and the first element of an optical system is extremely important for the performance of the optical system. A focus-able lens is optimized for several positions of the point object sources. Also, a focus-able lens can be a lens optimized for a plane wave source and a point object source with a finite distance to the first surface of the lens [5].

This optimization is achieved during the merit function but also by changing the configuration (in this case the position of the object) of an element of the optical system. For example, some lenses are moving within the system or the image plane is moving. That usually is considered in cell phone cameras [5, 83].

11.3.2 Zoom Lens

A Zoom lens is a lens that is optimized for several configurations related to different fields of view. Usually in this case, the source is considered to be a plane wavefront but the field of view changes between each configuration. The image plane stays fixed. Therefore the effective focal length changes also in each configuration [9, 31, 30].

Like the Focus-able lens, the optimization process for the zoom lens is accomplished not only through the merit function but also by modifying the configuration of an element within the optical system. For instance, certain lenses may be adjusted within the system, or the position of the image plane may be altered [5].

11.3.3 Thermal Stabilization

The thermal stabilization considers that the performance of an optical system changes at different temperatures. To compensate for this change and keep a relatively good performance at different temperatures. The thermal stabilization applied the movement of some of the elements of the optical system [5, 15, 18].

11.3.4 Lookup Parameters and Lookup Tables

In Quadoa optical CAD the Multi-Config Lookup Table serves to provide various discrete states. This functionality is particularly useful for modeling systems with moving assemblies, commonly utilized in the development of focusable lenses, zoom lenses, or thermal stabilization. By accommodating several discrete states simultaneously, the entire system can be optimized comprehensively.

You can access the primary functionality through the menu bar located at the top of the Multi-Config Lookup Table interface.

- **Add Config:** To add a new configuration to the Multi-Config Lookup Table, simply click on the icon. The configurations are added as columns in the Multi-Config Lookup Table.
- **Delete Config:** Click on the icon to remove the selected configuration from the table.
- **Add Multiconfig Parameter:** Click on the icon to add a new Multiconfig Parameter to the Multi-Config Lookup Table. Each column within the Multi-Config Lookup Table contains the lookup values for individual configurations. Additionally, the lookup parameter can be designated as a variable, allowing for flexibility in its state. This variable state can be activated independently for each configuration. Consequently, when the current configuration changes or when specific configurations are needed, such as for analysis windows or optimization tasks, the values of parameters referring to the Multi-Config Lookup Table will automatically update to reflect the current configuration.
- **Add Parameter:** Click on the icon to add a new parameter to the Multi-Config Lookup Table. Unlike the Multi-Config Lookup Table, the Simple Parameter holds only a single value that remains constant regardless of the current configuration.
- **Add Math Expression:** Click on the icon to add a new Math Expression to the Multi-Config Lookup Table. The Math Expression feature enables the computation of simple mathematical expressions. The expression to be computed can be configured in the second row of the Math Expression and remains consistent across all configurations. However, if a parameter used within the Math Expression is a Multi-Config Lookup Table, its value will vary based on the current

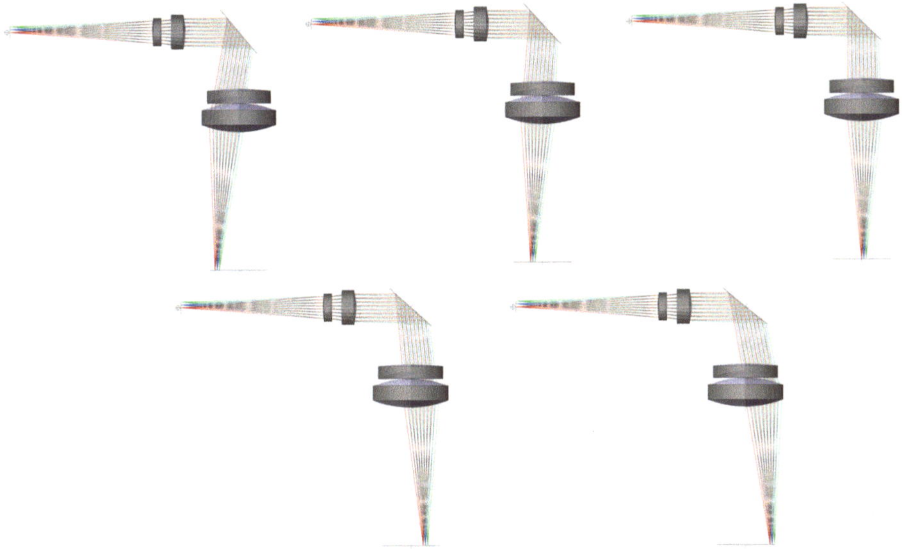

FIGURE 11.4
The five configurations of `Multiconfig_Model`. The system consists of two doublets and a mirror between them. The tilt angle of the mirror changes on every configuration. The change of the tilt angle of the mirror changes where the image is placed and its performance.

configuration, consequently affecting the value of the `Math Expression` f(x) accordingly. If the expression includes an unknown parameter or is not valid, the corresponding row will be highlighted in red, and an error message will be displayed.

11.4 Example 2: Multiple Configurations

Our second example is a simple optical system that consists of five different configurations. See figure 11.4. The tilt angle of the mirror between the doublets changes in each of the configurations.

This example is installed also with `Quadoa optical CAD` and it is located at:
`C:\Program Files \Quadoa\examples \Basic Features\Multiconfig_Model`

Here, we are going to call this design just as `Multiconfig_Model`.

The changing of the tilt angle of the mirror, changes where the image is placed as well as the performance of the system. The system consists of three

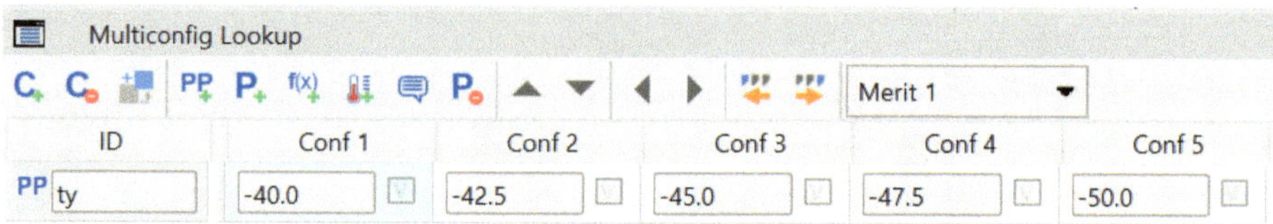

FIGURE 11.5
The multiple configuration parameters of the `Multiconfig_Model`. The parameter `ty` changes on each configuration from -40° to -50°, in steps of -2.5°.

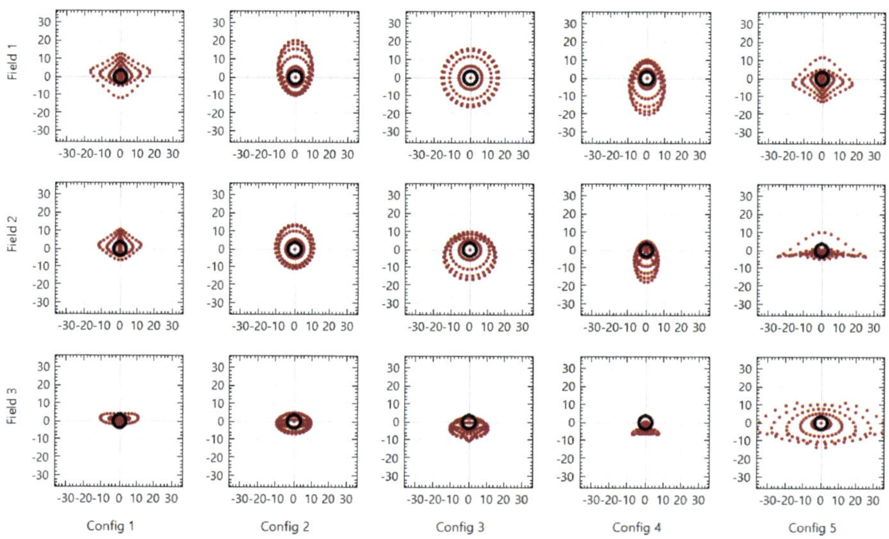

FIGURE 11.6
`Spot Diagram Config vs. Field` of `Multiconfig_Model`. The rows are the fields and the columns are the configurations.

fields with a different decenter in y of 0mm, 1mm, and 2mm, respectively. The source has only one wavelength, which is 632.82nm. The multiple configuration parameters of the `Multiconfig_Model` can be found in `Multiconfig Lookup`, which is shown at figure 11.5.

The parameters are the rows, in this case only one row. The configurations are the columns. It's clear that the tilt angle changes from -40° to -50°, in steps of -2.5°. This variation is under the `ID`/name of `ty`.

The spot diagram for each field and each configuration can be analyzed with `Spot Diagram Config vs. Field`. The results are displayed in figure 11.6. The rows are the fields and the columns are the configurations. From

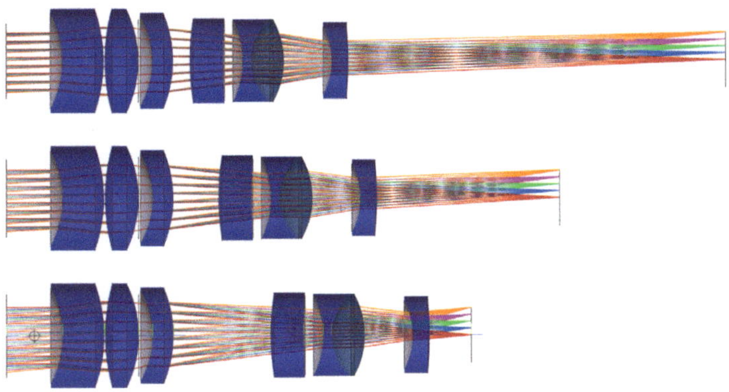

FIGURE 11.7
Ray-tracing of three configurations of `Zoom_Lens`.

the figure, it is clear that each configuration has a different performance. But, nearly all of the possible spots are close to being diffracted-limited since the Airy disk has a similar size to the respective spot.

11.5 Example 3: Zoom Lens

Our final example is a zoom camera with three different configurations. The file can be found in:
`C:\Program Files \Quadoa\examples \Objectives\Zoom_Lens`

The ray-tracing of the three different configurations is shown in figure 11.7 and 11.8. The half field of view of the system is 20° and it is constant along the configurations. The system is composed of two assemblies. The source has three different wavelengths which are 450nm, 550nm, and 650nm.

From figure 11.9, we know that the system has three configurations and two multi-configuration parameters, `z_lens` and `z_cam`, respectively. `z_lens` represents distances between the first assembly and the second assembly. `z_cam` represents distances between the second assembly and the image plane. This information is displayed in figure 11.10.

We plotted `Spot Diagram Config vs. Field`, the result is presented in figure 11.11. From it, we see that the performance of the on-axis field does not change too much among the configurations. But the field with the incident angle of 20° changes a lot. This is due to the nature of the problem of having several configurations in a single design. When the task is to design a

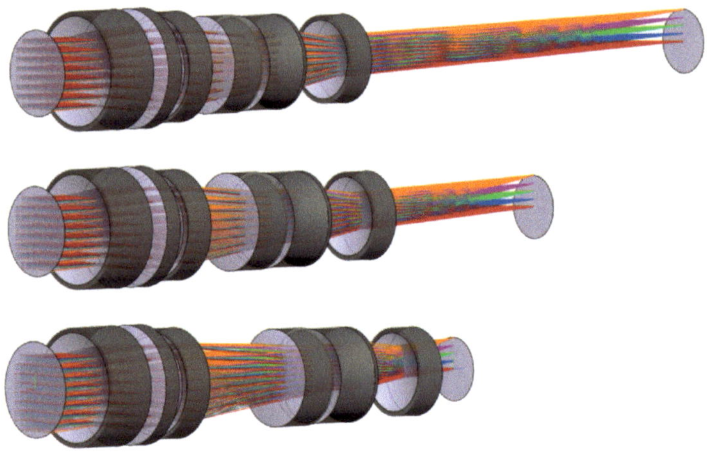

FIGURE 11.8
Side-view of ray-tracing of `Zoom_Lens`.

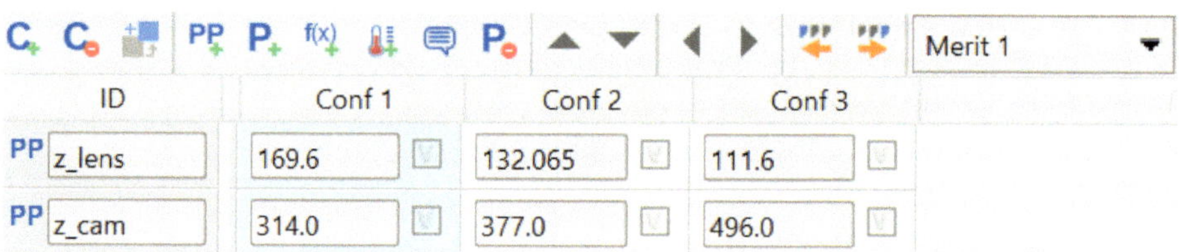

FIGURE 11.9
Multiple configuration parameters of `Zoom_Lens`.

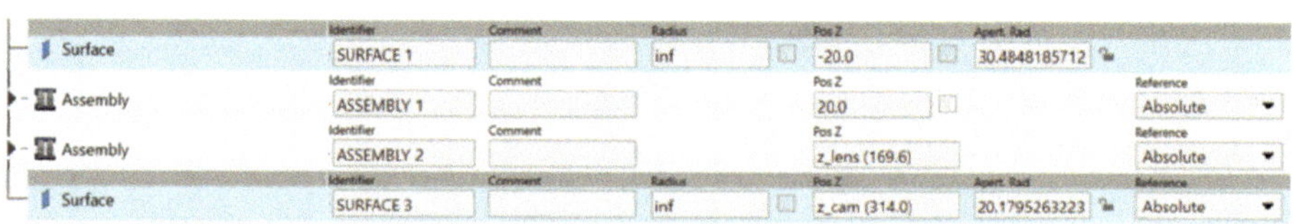

FIGURE 11.10
`Optical Design` of `Zoom_Lens`.

multi-configuration optical system, it is recommended, first to optimize the main configuration. Then, other configurations configurations must be added and optimized. The design process is challenging. It required more computation capabilities than single-configuration optical systems.

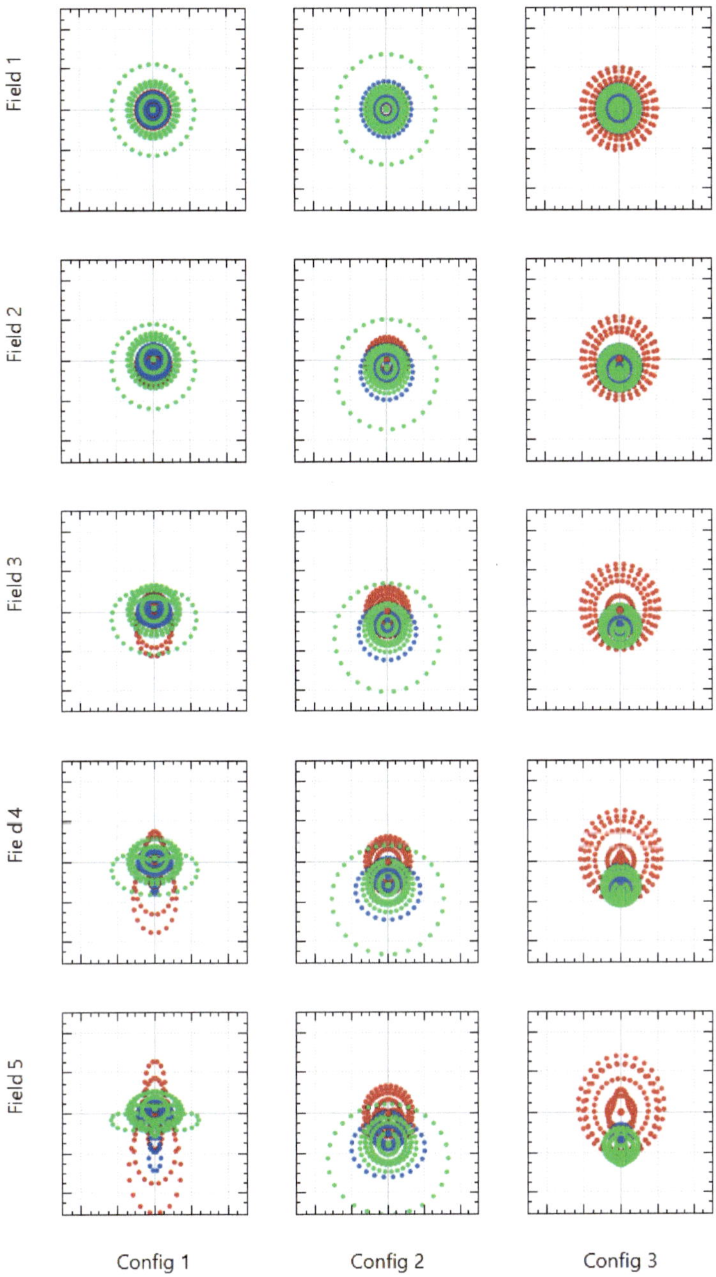

FIGURE 11.11
Spot Diagram Config vs. Field of Zoom_Lens.

11.6 Summary

In this chapter, we learned how to work with multiple sequences and configurations in `Quadoa optical CAD`. We show the concepts of focus-able lens and a zoom lens. Also, we studied other scenarios where multiconfiguration is needed like thermal stability.

Finally, we show three illustrative examples. The first example has only one configuration, but two sequences. The second example is a system with several configurations but one single sequence. Finally, the last example was the most popular in multiconfiguration a zoom lens.

All the examples were explained in detail, highlighting the key features of the design. The files are given in the supplementary material.

Part III

Final Considerations of an Imagining Optical System

12

Ghost Images

The shortest distance between two points is a straight line.

Archimedes

12.1 Introduction

Ghosting or ghost images occur when scattered rays from surfaces, edges, and lens housing inadvertently focus on the sensor, leading to the formation of unwanted images [5, 12, 13, 58]. In this chapter are going to study more on them.

But it is important to mention, that given the infrequency of ghosting in visible image lenses, it's advisable to exclude considerations of ghosting during the optimization phase of the lens design process. Instead, reserve this topic for final adjustments. As the design nears completion, it's crucial to thoroughly examine for any potential ghost images.

12.2 Ghost Analysis in `Quadoa Optical CAD`

`Quadoa Optical CAD` has several tools to make a proper ghost analysis of an optical system. In the following list are the main important tools provided by the software [18].

- **Generate Ghosts:** This button initiates the generation of `Ghost Reflexes` from a selected `Sequence`. This action will open the `Ghost Sequence Wizard`. `Ghost Sequence Wizard` will be studied in detail in the next subsection.

DOI: 10.1201/9781003506355-12

- **Diffractive Ghosts:** This button is employed to generate Diffractive Ghost Reflexes on surfaces with a Phase Function from a selected Sequence. It will open the Diffractive Ghost Sequence Wizard.

- **To User Sequence:** This button is utilized to transform a Ghost Sequence into a User Sequence. To convert a Ghost Sequence into a User Sequence, you must first select the desired Ghost Sequence in the System Setup Editor. If no Ghost Sequence is selected, an error message will be displayed. Upon selection of the Ghost Sequence in the System Setup Editor, it will be converted into a User Sequence.

- **To Ghost Sequence:** This button is employed to transform a User Sequence into a Ghost Sequence. To convert a User Sequence into a Ghost Sequence, you must first select the desired User Sequence in the System Setup Editor. If no User Sequence is selected, an error message will be displayed. Upon selection of the User Sequence in the System Setup Editor, it will be converted into a Ghost Sequence and will be included in the ghost analysis alongside other Ghost Sequences.

- **Remove all Ghosts:** By pressing this button, it deletes all Ghost Sequences from the Sequence List in the System Setup Editor.

- **Hide All Ghosts:** This button is utilized to conceal all Ghost Sequences. The Ghost Reflexes will be hidden in the 3D View and displayed in gray in the System Setup Editor. To hide or reveal individual Ghost Sequences, you can use the right-click menu on the corresponding Sequence.

- **Show All Ghosts:** This button is employed to reveal all Ghost Sequences. The Ghost Reflexes will be displayed in the 3D View and reinstated with color in the System Setup Editor. To conceal or reveal individual Ghost Sequences, you can utilize the right-click menu on the respective Sequence.

12.2.1 Ghost Wizard

The Ghost Wizard Dialog is employed to generate a set of Sequences, with each sequence representing one ghost reflection path in the optical system.

To generate ghost sequences using the Ghost Wizard Dialog, follow these steps: Select the Sequence for which to generate Ghost Reflexes from the drop-down menu in the Ghost Wizard Dialog. From the Ghost Order drop-down menu, choose between First Order and Second Order.

- **First Order:** will generate ghosts with a single reflection on one of the surfaces of Sequence.

- **Second Order:** will generate ghosts with two reflections on the surfaces of the `Sequences`.

Press `OK` to generate the ghost reflex sequences. For each ghost reflex, a corresponding `Ghost Sequence` will be created in the `Sequences List` within the `System Setup Editor`. If the ghost rays are not hidden, they will be displayed in the `3D View`.

To minimize the ray count, choose `Minimal` on the `Ray Count parameter`. This setting will generate ghosts with a minimal xy-fan `Distribution`.

12.2.2 Diffractive Ghost Wizard

The `Diffractive Ghost Wizard Dialog` is employed to generate a set of `Sequences` for systems containing at least one diffractive surface with at least one `Phase Property Item` is assigned to it.

`Ghosts` are characterized as the diffraction orders of the diffractive element that deviate from the design order. To generate diffractive ghosts, select the Sequence for which to generate them in the wizard. The `Maximum Diffractive Order specifies` the highest order of diffraction to generate. For instance, if +1 is the design order for the diffractive element, the following orders will be generated: -3, -2, -1, 0, +2, +3. Note that +1 is not included as it represents the design order.

The `Minimal` option in the diffractive ghost wizard works like `Ghost Wizard`.

12.2.3 Ghost Foot Print Diagram

The `Ghost Spot Diagram` plots the x- and y-coordinates of any `Sequence` marked as a ghost sequence within the local coordinate system of the last surface of the ghost sequences. In cases where there are different sets of ghost sequences, such as single and double bounces, ending at different surfaces, you can choose the set of ghosts to be plotted by selecting the surface to be analyzed via the `Surface` parameter. Only the ghosts where the selected surface is the image surface of the sequence will be included in the plot.

12.2.4 Ghost Irradiance Plot

The `Ghost Irradiance Plot` serves to illustrate the effects of ghost reflections on the image surface. The intensity on the image plane is summed for any `Sequence` marked as a `Ghost Sequence`, sharing the selected image surface.

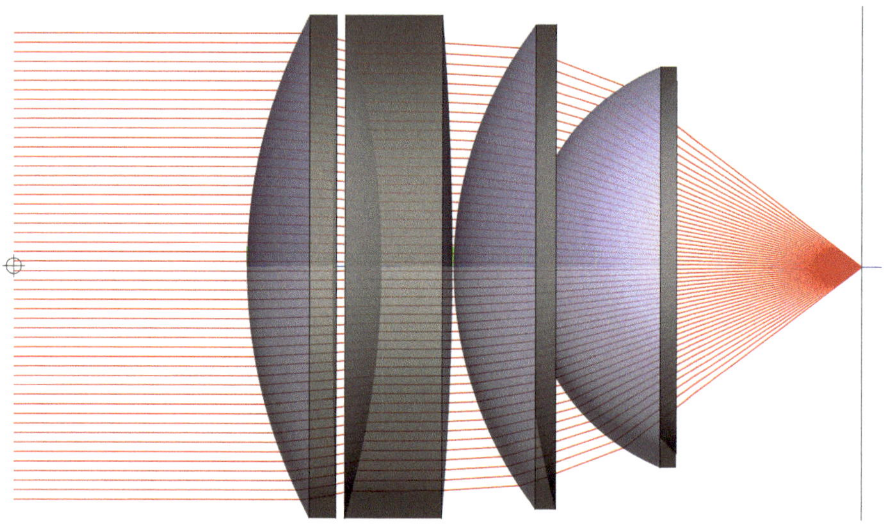

FIGURE 12.1
Ray-tracing of Objective_57mm_Ghosts. The optical system consists of four lenses. The first is made of N-BK7 and the next three are made of N-SF6. The system considers an on-axis field with a wavelength of 533nm.

12.3 Example of a System With Ghost Images

The following example illustrates an optical system suffering from ghost images. The example can be found in:
C:\Program\Files\Quadoa\examples\Objectives\Objective_57mm_Ghosts

The design will be called Objective_57mm_Ghosts.

The ray-tracing of the design is placed in figure 12.1. For simplicity, we are only considering the on-axis field for the ray-tracing. The wavelength of the field is 533nm. The system consists of four lenses. The first is made of N-BK7 and next three are made of N-SF6.

If we go to the Sequence tab, and go to Generate Ghosts the Ghost Wizard Dialog will be open. Ghost Wizard Dialog is used to generate a set of Sequences. Each tot eh sequence generated by Quadoa Optical CAD represents one ghost reflection path in the optical system.

So let's generate ghost sequences for Objective_57mm_Ghosts. using the Ghost Wizard Dialog.

We start Select the Sequence for which to generate Ghost Reflexes from the drop-down menu in the Ghost Wizard Dialog. For our case we choose Auto Sequence 1, the only sequence on the file so far.

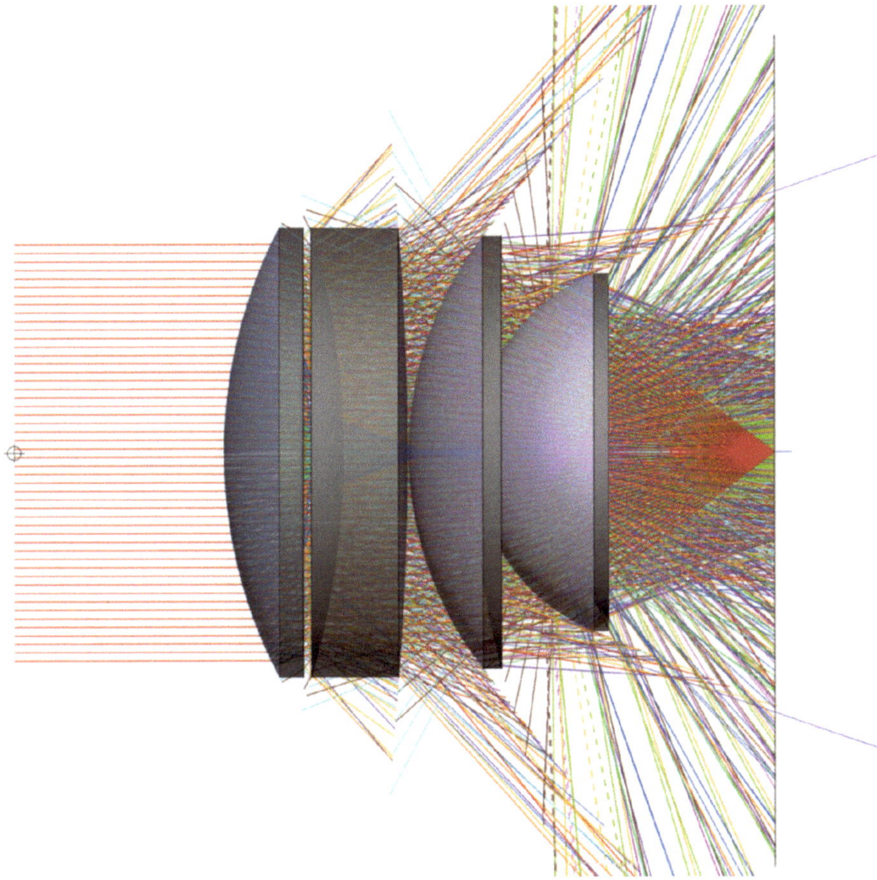

FIGURE 12.2
Ray-tracing of Objective_57mm_Ghosts after adding **Second Order** ghost sequences. The second-order ghost sequences consider two reflections on the surfaces of the Sequences.

Then, from the Ghost Order drop-down menu, choose between Second Order, instead of First Order.

Finally, we press OK to generate the ghost reflex sequences for Auto Sequence 1. Remember that for each ghost reflex, a corresponding ghost Sequence will be created in the Sequences List within the System Setup Editor. The generated ghost reflex sequences will be displayed and the result should be similar to the one presented in figure 12.2. Figure 12.2 shows the ray-tracing of Objective_57mm_Ghosts after adding Second Order ghost sequences.

Figure 12.2 shows that ghost sequences can have chaotic behavior. The ghost sequence can be hidden with the bottom Hide all Ghosts, or

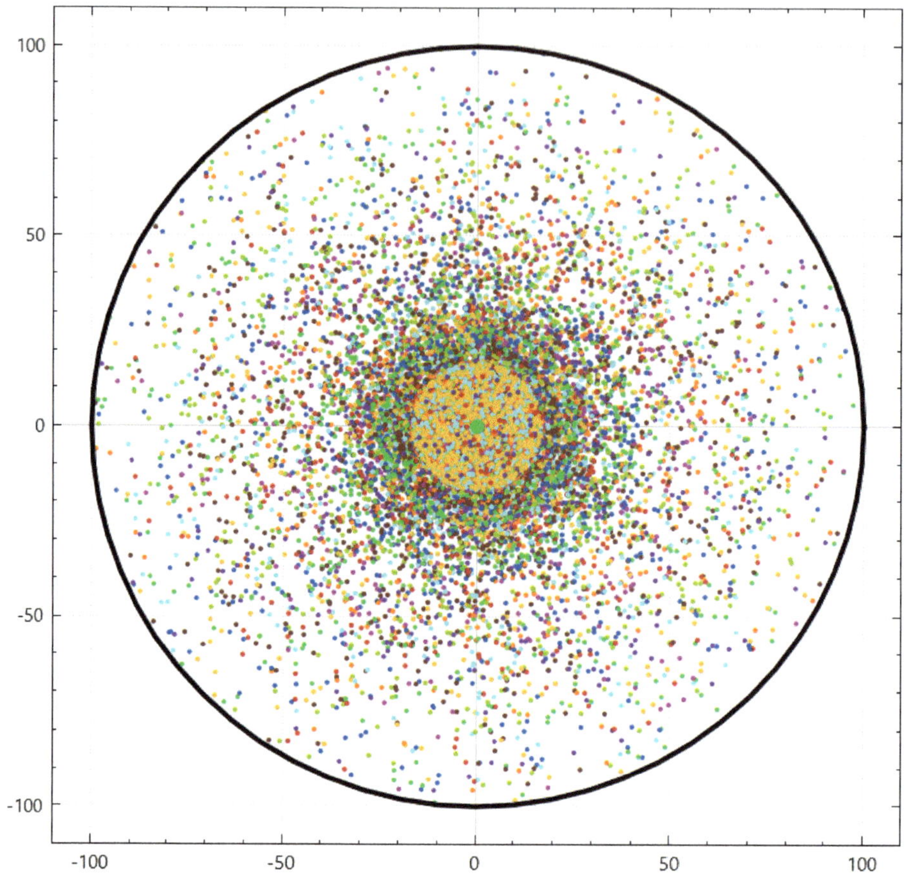

FIGURE 12.3
The `Ghost Footprint Diagram` of `Objective_57mm_Ghosts`.

displayed once more with `Show all Ghosts`. Finally, they can be removed with `Remove all Ghosts`.

The most important thing about Ghost images is that they should not form an image in the image sensor. For example, from the ray tracing displayed in figure 12.2, we can see that most of the rays from ghost sequences do not end even in the image sensor. With `Ghost Footprint Diagram` we can see which of the ghost rays strike in the image plane. The `Ghost Footprint Diagram` of `Objective_57mm_Ghosts` is displayed in figure 12.3. As long as the ghost rays do not form an image point the sensor there is not a big trouble with them. The tolerance for the potential image point generated intensity of the ghost rays depends on the wavelength of the light, the sensitivity of the sensor, and the coatings presented in the lenses. The irradiance of this potential point can be analyzed with the `Ghost Irradiance Plot`.

12.4 Summary

Ghosting or ghost images arise when light rays are reflected or scattered from mainly optical surfaces, and converge on the sensor, resulting in the creation of undesired images. In this chapter, we explore the tools provided by `Quadoa Optical CAD` to analyze the possible ghost images in an optical system.

13

Tolerance Analysis and Technical Drawings

> I have had my results for a long time: but I do not yet know how I am to arrive at them.
>
> Carl Friedrich Gauss

13.1 Tolerance Analysis

Tolerance analysis in lens design involves assessing the impact of manufacturing tolerances and variations on the performance of optical systems, particularly lenses. It aims to predict how variations in parameters such as surface shapes, refractive indices, thicknesses, and alignment errors may affect the optical performance of a lens system [5, 12, 13, 58].

Tolerance analysis typically involves several steps:

- **Identifying Tolerances:** The first step is to identify the relevant parameters and dimensions of the optical system that are subject to manufacturing tolerances. These can include parameters such as lens curvatures, radii, thicknesses, refractive indices, and alignment tolerances.
- **Defining Tolerance Limits:** For each identified parameter, tolerance limits are defined based on manufacturing capabilities and performance requirements. These limits specify the acceptable range of variation for each parameter.
- **Modeling Variations:** Next, variations or deviations from the nominal values of the parameters are modeled using mathematical or computational techniques. Monte-Carlo simulations, ray tracing software, or other numerical methods may be employed to model the effects of parameter variations on system performance.

DOI: 10.1201/9781003506355-13

- **Analyzing Performance:** The modeled variations are then used to assess the impact on system performance metrics such as image quality, resolution, aberrations, and other optical properties. This analysis helps identify which parameters have the greatest influence on system performance and may require tighter tolerances or additional compensation.

- **Optimizing Design:** Based on the results of the tolerance analysis, optical designers may need to refine the design to minimize sensitivity to manufacturing variations. This could involve adjusting the lens prescription, modifying the optical layout, incorporating compensators or tolerance rings, or selecting more robust materials or manufacturing processes.

- **Verification and Validation:** Finally, the optimized design is verified and validated through prototyping, testing, and evaluation to ensure that it meets performance requirements under real-world manufacturing conditions.

Tolerance analysis is a critical aspect of lens design, helping to ensure that optical systems perform reliably and consistently despite variations in manufacturing processes. By understanding and mitigating the effects of manufacturing tolerances, optical designers can optimize lens designs for improved performance, reliability, and manufacturability. In this chapter, we are going to study the tolerance analysis with more detail in `Quadoa optical CAD`. First let's focus on the error distributions.

13.1.1 Error Distributions

In optical design tolerance analysis, error distribution refers to the statistical distribution of errors or deviations from ideal optical parameters within the optical system. These errors can arise from various sources such as manufacturing imperfections, material variations, or environmental factors [18].

Understanding the error distribution is crucial for assessing the impact of tolerances on the performance of the optical system. Different types of error distributions may be considered, depending on the specific parameters being analyzed and the characteristics of the system. Some common error distributions include:

- **Gaussian (Normal) Distribution:** This is one of the most widely used distributions in tolerance analysis. In a Gaussian distribution, errors are symmetrically distributed around a mean value, with most errors clustering close to the mean and fewer errors occurring further away. Many manufacturing processes naturally produce errors that follow a Gaussian distribution.

- **Uniform Distribution:** In a uniform distribution, errors are equally likely to occur within a specified range. This distribution is often used when there is no prior knowledge about the error distribution, or when errors are expected to be uniformly distributed within certain limits.

- **Triangular Distribution:** In a triangular distribution, errors are more likely to occur near the center of the range and less likely at the extremes. This distribution is often used when there is some knowledge or expectation about the most probable error value.
- **Exponential Distribution:** In an exponential distribution, errors are more likely to occur at smaller values and decrease exponentially as the error magnitude increases. This distribution is often used when errors are more likely to be small but can occasionally be large.
- **Custom Distribution:** In some cases, the error distribution may be specified based on empirical data, theoretical models, or specific requirements of the optical system.

By modeling the error distribution, optical designers can perform tolerance analysis to assess the impact of tolerances on system performance, optimize system designs to minimize sensitivity to errors, and ensure that the optical system meets performance requirements in real-world conditions. The mentioned errors distributions are considered by `Quadoa optical CAD`.

13.1.2 Element Surface Decentring

One of the most common errors in manufacturing a lens is element surface decentring, which is also considered by `Quadoa Optical CAD`. Surface decentering refers to the displacement or offset of an optical element's surface from its ideal or centered position within the optical system. When an optical element is decentered, it is shifted laterally or axially relative to the optical axis, causing the incoming light rays to deviate from their intended paths.

Surface decentering can occur intentionally as part of the optical design process to achieve specific optical effects or to correct aberrations. However, in many cases, surface decentering is unintentional and can result from manufacturing imperfections, assembly errors, or mechanical stresses.

The effects of surface decentering on optical performance depend on the specific design of the optical system and the degree of decentering. In some cases, surface decentering can introduce optical aberrations such as astigmatism, coma, or image distortion, leading to degraded image quality or reduced system performance. However, surface decentering can also be used strategically to control aberrations or optimize system performance in certain applications.

Optical designers often perform tolerance analysis to assess the impact of surface decentering and other manufacturing tolerances on system performance. By considering the effects of decentering and other tolerances during the design process, designers can optimize optical systems to minimize sensitivity to manufacturing variations and ensure that performance requirements are met in real-world conditions.

13.1.3 Radius of Tolerance

In lens design, the radius of tolerances refers to the allowable variation or deviation from the nominal radius of curvature for a lens surface. The radius of curvature is a critical parameter in determining the optical properties of a lens, such as its focal length, imaging characteristics, and aberrations [5].

Tolerances on the radius of curvature are specified to ensure that manufactured lenses meet performance requirements within acceptable limits. These tolerances define the allowable range of variation from the nominal radius of curvature, taking into account manufacturing capabilities, material properties, and performance specifications [5, 92].

For example, if a lens design specifies a nominal radius of curvature of 100 mm with a tolerance of ± 0.1 mm, means that the actual radius of curvature of the manufactured lens can vary within the range of 99.9 mm to 100.1 mm. Tolerances on the radius of curvature are typically specified in terms of absolute values (e.g., ± 0.1 mm) or as a percentage of the nominal radius (e.g., $\pm 1\%$).

Tolerances on the radius of curvature are critical in lens design because even small deviations from the nominal value can affect the optical performance of the lens. For example, variations in the radius of curvature can lead to changes in the lens's focal length, image magnification, and aberrations such as spherical aberration and coma.

Optical designers use tolerance analysis techniques to assess the impact of radius of curvature tolerances on lens performance and to optimize lens designs to minimize sensitivity to these variations. This may involve adjusting the lens prescription, selecting appropriate manufacturing processes, or incorporating compensating elements into the design to ensure that performance requirements are met within specified tolerances [5].

In interferometry, a common technique used to measure optical surfaces, fringes are used to quantify surface irregularities or deviations from an ideal shape. The number of fringes corresponds to the difference in optical path length between adjacent regions of the surface, which in turn relates to the curvature of the surface [5, 92].

For example, in a spherical lens, the number of fringes associated with a given radius of curvature tolerance will depend on the wavelength of light used for measurement and the sensitivity of the interferometric setup. A higher number of fringes indicates a smaller deviation from the nominal curvature, while a lower number of fringes indicates a larger deviation.

The tolerance for the radius of an optical element's surface can be defined using either fringe units or as a percentage of the radius. These two methods are interconnected through a nonlinear relationship

$$\text{Fringes} = \frac{\Delta \text{Sag}}{(\lambda/2)} = \frac{1}{(\lambda/2)8} \frac{D^2}{r} \frac{\Delta r}{r} \tag{13.1}$$

In the equation, ΔSag represents the tolerance for the sag of the optical element, λ denotes the wavelength of the test laser, D stands for the diameter of the element, r represents the radius of the element, and Δr is the tolerance in radius. Fringes can be directly quantified through measurement using a test plate. For instance, if the surface irregularity amounts to 1 wave, a minimum of two fringes is required for accurate measurement. It is important to note that the radius tolerance cannot be measured directly; instead, it must be calculated from the fringes using equation (13.1).

When dealing with percentage radius tolerance $\Delta r/r$, fringes may drop below 1, rendering them unmeasurable, especially when the radius r is exceedingly large. Hence, employing fringes to define radius tolerance proves more practical, although the radius percentage tolerance remains prevalent. In the case of a flat surface, where the radius is infinite, fringes are necessary to specify tolerance.

Therefore, the radius of tolerance in lens design specifies the allowable variation in the radius of curvature for a lens surface, while the number of fringes relates to the interference patterns observed when measuring optical surfaces or evaluating optical system performance. Both parameters are important considerations in ensuring the quality and performance of optical components and systems [5, 92].

13.1.4 Total Indicator Ronout

Total Indicator Runout (TIR) in lens design refers to a measurement used to assess the axial and radial alignment of a lens element's surface relative to its intended position or axis of rotation. TIR is a critical parameter in ensuring the accuracy and quality of optical components, particularly in precision optical systems where tight tolerances are required.

When measuring TIR, a gauge or measuring device is placed in contact with the surface of the lens element. As the lens rotates, the gauge detects any deviation in the surface profile from a perfectly cylindrical or spherical shape. TIR is then calculated as the difference between the highest and lowest points of deviation along the surface profile.

In lens design, TIR is typically specified as a maximum allowable deviation from the ideal surface profile, often expressed in terms of microns or fractions of the lens's design wavelength. Exceeding the specified TIR tolerance can lead to optical aberrations, image degradation, and reduced system performance.

Optical designers use TIR measurements to verify the quality of manufactured lens elements and to ensure that they meet performance requirements. TIR measurements are also used during the manufacturing process to monitor and control the quality of lens surfaces, allowing adjustments to be made to machining or polishing parameters as needed.

13.1.5 Glass Homogeneity

Glass homogeneity in optical design tolerance analysis refers to the uniformity of optical properties within a volume of optical material, typically glass. In other words, it assesses how consistent the refractive index, dispersion characteristics, and other optical properties are throughout the glass material.

In practical terms, glass homogeneity is important because variations in optical properties can introduce optical aberrations and degrade the performance of optical systems. For example, suppose different regions of a lens element have slightly different refractive indices. In that case, it can lead to variations in the focusing behavior of the lens and result in image distortion or blurring. `Quadoa optical CAD` can model the glass homogeneity with parameters related to the Abbe number and the refractive index, we are going to see them in detail later in this chapter for the moment let's continue with the general overview of tolerance analysis.

13.1.6 Compensators

In tolerance analysis, a compensator refers to a mechanism or feature designed to adjust for variations or deviations from specified dimensions or tolerances in order to maintain desired performance or functionality within a system or component.

For example, in mechanical engineering, compensators can include features such as springs, shims, or adjustable components that can be used to accommodate dimensional variations or misalignments in assembled parts. By incorporating compensators into a design, engineers aim to ensure that the overall system operates within acceptable limits despite variations in individual components.

Compensators are particularly important in systems where tight tolerances are required for proper function or where variations in manufacturing processes can result in dimensional discrepancies. They help to mitigate the effects of tolerances on overall system performance, improving reliability and consistency.

13.1.7 Monte-Carlo Analysis

Monte-Carlo analysis, also known as Monte-Carlo simulation, is a computational technique used in optical design to assess the performance of optical systems under various conditions, taking into account uncertainties and random factors [9, 30, 31].

In optical design, Monte-Carlo analysis involves simulating multiple instances or realizations of an optical system with randomized parameters such as surface shapes, refractive indices, or positioning errors. Each instance represents a possible configuration of the system that could occur due to manufacturing tolerances, material variations, or environmental factors.

For each simulated instance, the performance of the optical system is evaluated using ray tracing or other numerical methods. By running a large number of simulations, Monte-Carlo analysis provides statistical information about the distribution of system performance metrics, such as image quality, spot size, or system throughput, considering the uncertainties in the system's parameters.

Monte-Carlo analysis helps optical designers to:

- Assess the impact of manufacturing tolerances and variations on system performance.
- Identify critical parameters that have the greatest influence on system performance.
- Optimize system designs to minimize sensitivity to uncertainties and improve robustness.
- Estimate the overall yield and reliability of optical systems in real-world conditions.

13.2 Zernike Polynomials

Zernike Polynomials can also be used to simulate variations in lens/optical surface parameters curvature, and surface defects. This is done by stacking a Zernike Polynomial over a given optical surface. The Zernike Polynomial represents some deviation in the optical surface due to a manufacturing error. These errors, or surface deviations, can significantly impact the performance of optical systems by introducing aberrations that degrade image quality.

As we have seen previously, Zernike polynomials are a series of orthogonal polynomials defined on the unit disk. This makes them particularly useful for describing surface deviations since each term of the Zernike polynomial describes a particular error since Zernike polynomials are orthogonal.

There are several benefits of using Zernike polynomials for this task. The first one describes the phenomena of surface deviations in a compact representation. A relatively small number of Zernike coefficients can accurately describe complex surface errors. Also, Zernike polynomials are widely used and accepted in the optical industry, facilitating communication and standardization. Finally, we can acknowledge the precision of Zernike polynomials for this task. They allow precise quantification and correction of specific aberrations for the wavefront and deviation in the optical surfaces.

Finally, Zernike coefficients can be used to specify tolerances for surface quality. For example, specifications might limit the allowable magnitude of certain Zernike terms to ensure the surface meets performance criteria. This can help to determine how sensitive the optical system is to specific surface errors, aiding in the design of robust systems.

13.3 Tolerance Analysis in `Quadoa Optical CAD`

The `Tolerancing Workbench` within `Quadoa optical CAD` empowers users to assess the as-built performance of an optical system in real-world conditions. It also aids in identifying critical parameters for the alignment and accuracy of optical components. Moreover, this data can be utilized to optimize the system for improved as-built performance even with looser tolerances. Ultimately, this contributes to the development of an optical system with enhanced as-built performance and reduced production costs.

The disparity between the as-built performance and the nominal design performance stems from various uncertainties inherent in the fabrication process of the optical system. For surfaces, this could manifest as deviations from the intended form, introduced during processes like polishing or diamond turning. In the case of lens elements, variations in the index of the glass substrate may alter light refraction, or uncertainties in alignment during polishing may result in wedge errors. Additionally, errors in the manufacturing process of mounting mechanics, as well as uncertainties in alignment between optical and mechanical components, can lead to inaccuracies in the position and orientation of individual lenses or entire assemblies.

13.3.1 Setting up Tolerances

The initial step for any tolerance analysis, regardless of the specific analysis to be conducted, is always to establish the tolerances for the system. These tolerances are categorized into three groups:

- **Surface Tolerance:** This category encompasses tolerances that can be assigned to individual surfaces, such as tilt or decenter, as well as curvature or form deviations.
- **Element Tolerance:** This category includes tolerances that can be assigned to individual elements, such as position and orientation, index, or thickness tolerances.
- **Assembly Tolerance:** This category comprises tolerances that can be assigned to entire assemblies, including position and orientation tolerances.
- **Multiconfig Tolerance:** This category encompasses tolerances that may be assigned to `Multiconfig Parameters`.

13.3.2 Overview on Tolerance Parameters

Tolerance parameters can be inserted into the tolerance containers to specify the values for each tolerance.

All tolerances share the first four parameters in common. Depending on the statistical distribution, additional parameters may vary. The statistical distributions are:

- **Rect**: The random value will fall within the interval `[-Range, Range]`, with equal probability for any value within that range. The parameter Value determines the extent of this distribution. In sensitivity analysis, the perturbation parameter is assumed to be `+Range` in the positive direction and `-Range` in the negative direction.

- **Rect Asymetric:** The value will reside within the interval `[Min, Max]`, with an equal probability for any value within that range. In sensitivity analysis, the perturbation parameter is assumed to be `Max` in the positive direction and `Min` in the negative direction.

- **Triangular :** the random value will fall within the interval `[Min, Max]`, where the probability is 0.0 at both `Min` and `Max`, and increases linearly towards `Peak` from both sides. In sensitivity analysis, the perturbation parameter is assumed to be `Max` in the positive direction and `Min` in the negative direction. Additionally, for sensitivity analysis, the perturbation is taken as 3σ in both the positive and negative directions.

- **Normal:** The random value will be generated assuming a Gaussian normal distribution for the tolerance. The parameter `Value` specifies the standard deviation of the distribution.

- **Parabolic:** The random value will be generated based on a parabolic distribution, with the vertex of the parabola positioned at the midpoint between `Min` and `Max`, where the chance is 0%, and the highest chance is at the `Min` and `Max` values.

- **Extreme:** The random value will be precisely set to either `Min` or `Max` with a 50% chance. In sensitivity analysis, the perturbation is `Min` in the negative direction and `Max` in the positive direction.

The following list provides an overview of the available tolerance parameters:

- **Position:** The `Position Tolerance` specifies the allowable deviation of an element or assembly's position in the x, y, and z directions within the object's reference coordinate system.

- **Position Z:** The `Position Z Tolerance` is employed to specify the permissible deviation in the z-position of individual surfaces, such as the image surface or a dummy surface.

- **Angle:** The `Angle Tolerance` specifies the allowable deviation in the angular orientation of an element or assembly around the x, y, and z-axes within the object's reference coordinate system.

- **Surface Tilt:** The `Surface Tilt Tolerance` specifies the allowable deviation in the tilt of a surface around the x and y axes within the object's reference coordinate system.

- **Surface Decenter:** The `Surface Decenter Tolerance` specifies the allowable deviation in the decentering of a surface in the x and y directions within the object's reference coordinate system.

- **General:** The `General Surface Tolerance` specifies the allowable deviation for any parameter of a surface feature that has been added to the surface.

- **Thickness:** The `Thickness Tolerance` specifies the allowable deviation in the thickness of a lens.

- **Index:** The `Index Tolerance` specifies the allowable deviation in the refractive index of the glass from which an element is made.

- **Abbe:** The `Abbe Tolerance` specifies the allowable deviation in the `Abbe Vd` of the glass from which an element is made.

- **Radius:** The `Radius Tolerance` specifies the allowable deviation in the radius of a surface. This tolerance is applicable only to surfaces with a spherical or aspherical form.

- **Zernike:** The `Zernike Tolerance` specifies the tolerance of the surface form using a `Zernike Polynomial` to describe the sag deviation. This tolerance can be applied to any surface. The norm radius for the `Zernike Polynomial` will be automatically calculated from the surface aperture. If no aperture is defined for a surface, the radius of the floating aperture will be used.

- **Multiconfig Param:** The `Multiconfig Param Tolerance` specifies the tolerance for a parameter that is defined in the `Multiconfig Lookup Table`.

The above parameter can be easily added with the `Default Tolerances` in the `Tolerancing Workbench`. The display with the tolerance parameter information appears by pressing the button `Default Tolerances`. See figure 13.1. The type of parameter, its amplitude, and its distribution can be filled in.

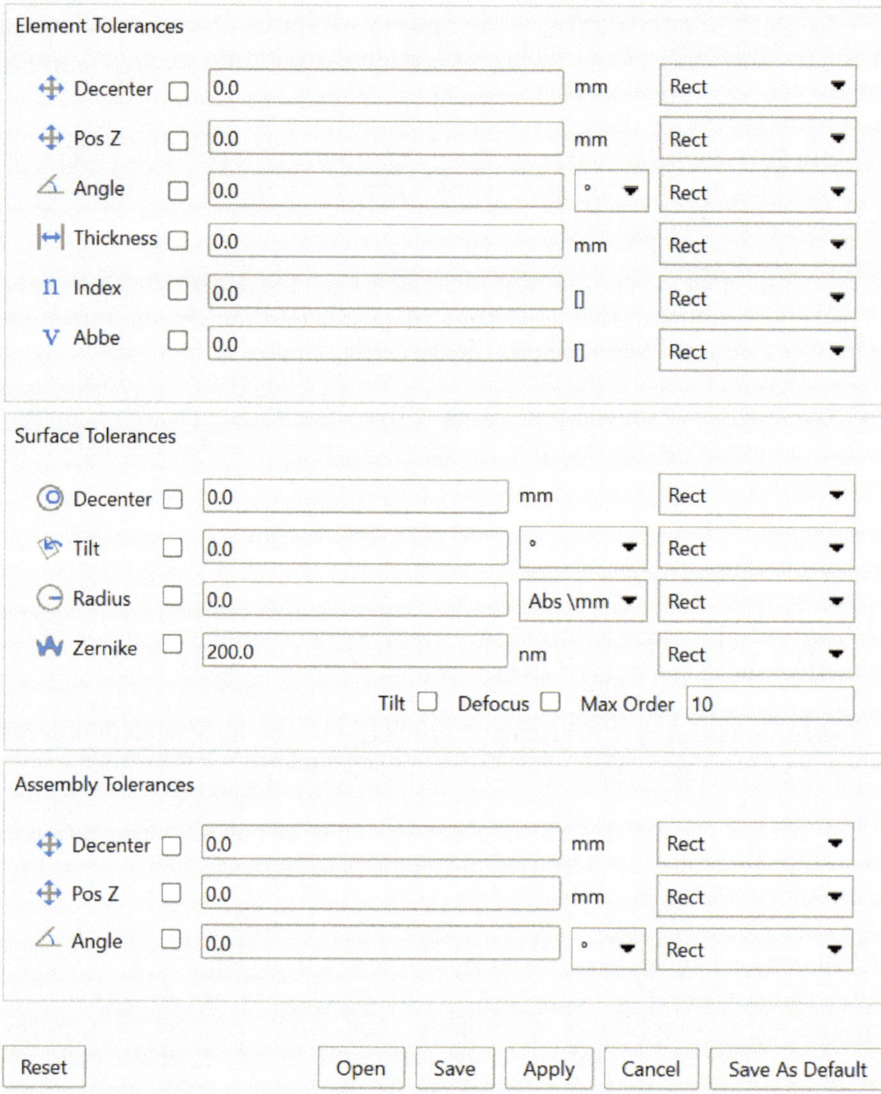

FIGURE 13.1
The display with the tolerance parameter information appears by pressing the button `Default Tolerances`. The type of parameter, its amplitude, and its distribution can be filled in.

13.3.3 Compensators

In `Quadoa optical CAD`, users have the capability to model various alignment strategies, ranging from straightforward focus adjustments to intricate alignment procedures, utilizing compensators.

The activation state of a `Variable` can be individually configured for each `Compensator`. To select the current `Compensator` to work on, users utilize the `Combo-Box` located above the `Optical Design Editor`, choosing the `Compensator` from the drop-down menu.

In cases where multiple `Compensators` are defined, they are applied sequentially from top to bottom in the list. This feature enables the modeling of complex alignment procedures, where adjustments are made one after another.

For any `Tolerance` with `Compensators`, the process operates as follows: Initially, the system, including the tolerances, is generated. Subsequently, the `Compensators` are applied one after another, optimizing the Variables assigned to each `Compensator` using the `Compensators Merit Function`. Finally, the system is evaluated using the assigned Merit Function to assess its performance.

13.3.4 Monte-Carlo Tolerance Analysis

The `Monte-Carlo Analysis` serves to provide a statistical estimation of the as-built performance of an optical system. Additionally, it can generate random exemplary systems, enabling further analysis of the optical properties within the range of possible outcomes. During the `Monte-Carlo Analysis`, a random value is assigned to each tolerance defined for the model, with the value and distribution determined by the `Tolerance Parameters`. Each system is then evaluated using the assigned `Merit Function` for assessing system performance. The user can specify the number of systems to be generated for statistical analysis. Moreover, it is possible to save all outcomes or only the best and worst results for further investigation.

13.3.5 Sensitivity Tolerance Analysis

The `Sensitivity Analysis` is utilized to pinpoint tolerances that exert a significant impact on the overall system performance, or to generally estimate the effect of each individual tolerance on the system. It generates a perturbed system for each tolerance and assesses its performance using the assigned `Merit Function` for evaluation. Subsequently, it produces a report displaying a sorted list of the impact of each individual tolerance. The `Random Distribution` for each tolerance is disregarded in this analysis, meaning that the report consistently illustrates the effect on performance under the assumption that the tolerance is off by $\pm$ the value specified by the value parameter.

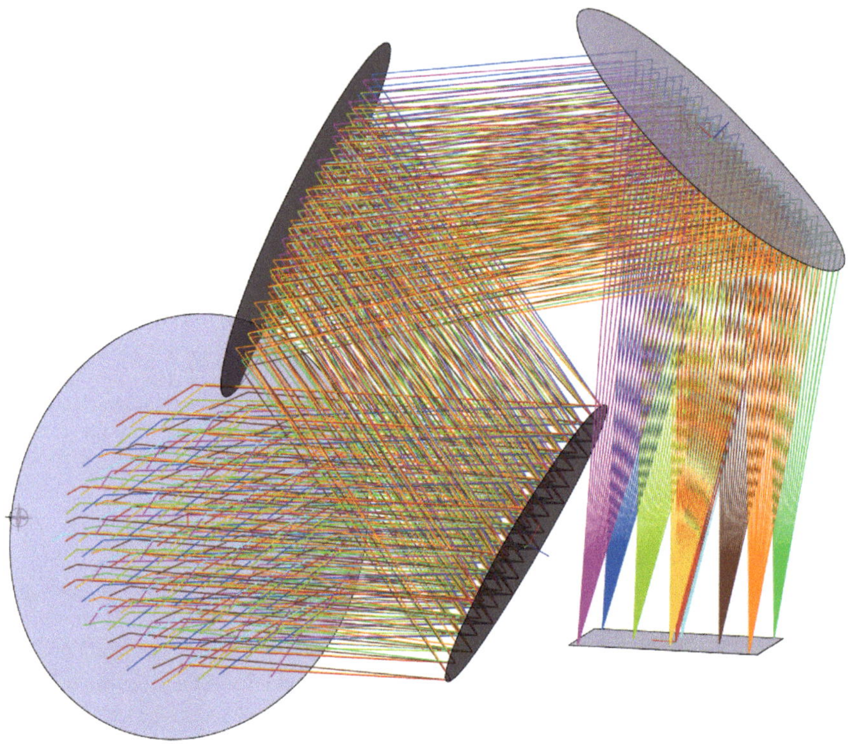

FIGURE 13.2
Ray-tracing of the optical system `10xe3`. The system consists of four mirrors. The first component in the system is a freeform mirror, which serves to fold the light. This mirror, which we'll call the folding mirror, bends the light by 45° degrees. Following this, there are three additional freeform mirrors, each with their sagitta described using the `XYZ Polynomial Form`. The sensor is positioned behind the second mirror.

13.4 Example of Tolerances

In this section, we are going to analyze the tolerances of an optical design with `Tolerancing Workbench` within `Quadoa optical CAD`. The design is mostly diffracted limited so during this tolerance analysis we are not going to consider any optimization in the design.

The file of the optical system for this section is presented in the supplementary material of this book. The name of the file is `10xe3`. We are going to refer this design to by the name of the file, `10xe3`.

The optical system `10xe3` is a freeform mirror base camera. The ray-tracing of the optical system `10xe3` is displayed in figure 13.2. The first mirror of the

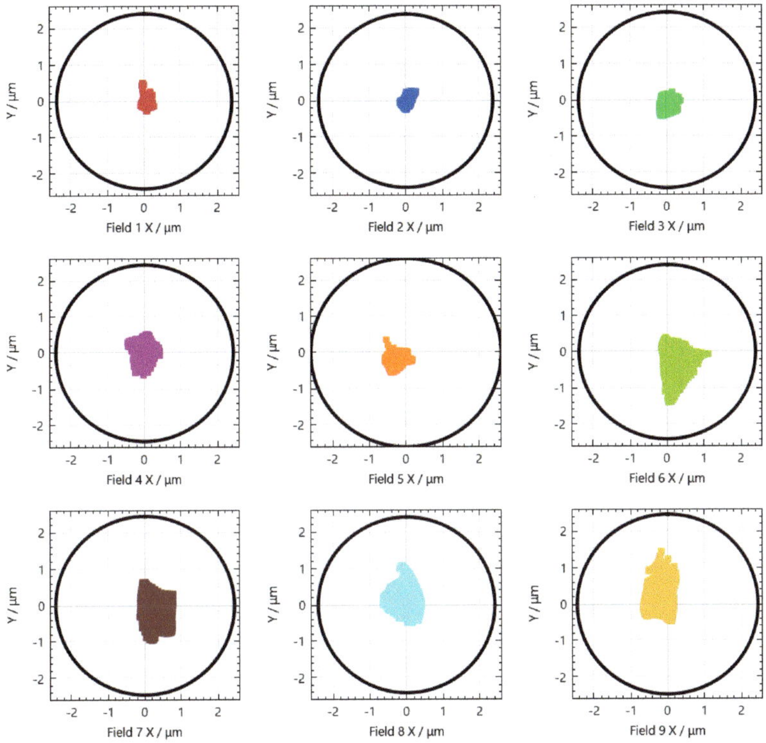

FIGURE 13.3
Spot Diagram of 10xe3.

system is a freeform mirror, which its function is to fold the light. Here, we are going to call the folding mirror, it bends the light 45°. Then, the three following three freeform mirrors. All the mirrors saggita is described with `XYZ Polynomial Form`. Finally, the sensor is located behind the second mirror.

The system has a field of view of 8.6° × 7.5°. The source has a wavelength of 550nm. It is recommended to the user to check the position of the mirrors and the center in the file `10xe3`, since we are going to focus on the tolerance analysis.

The system is diffracted-limited, this can be confirmed by observing the `Spot Diagram Plot` of the system is presented in figure 13.3. Also, the `FFT MTF Plot` of `10xe3` shows a diffracted-limited behavior, figure 13.4.

Now we know that our system is diffracted-limited, so it has a good performance overall all, but how sensitive its too manufacturing errors or misalignment's between the mirrors? Let's go to the `System Setup Table` and in the `Tolerancing` option type the numbers that appear in figure 13.5. For this example, all the distributions of the tolerance values will be `Rect` and the range will be 0.002, with the expectation of `Zernike` which will be 12nm.

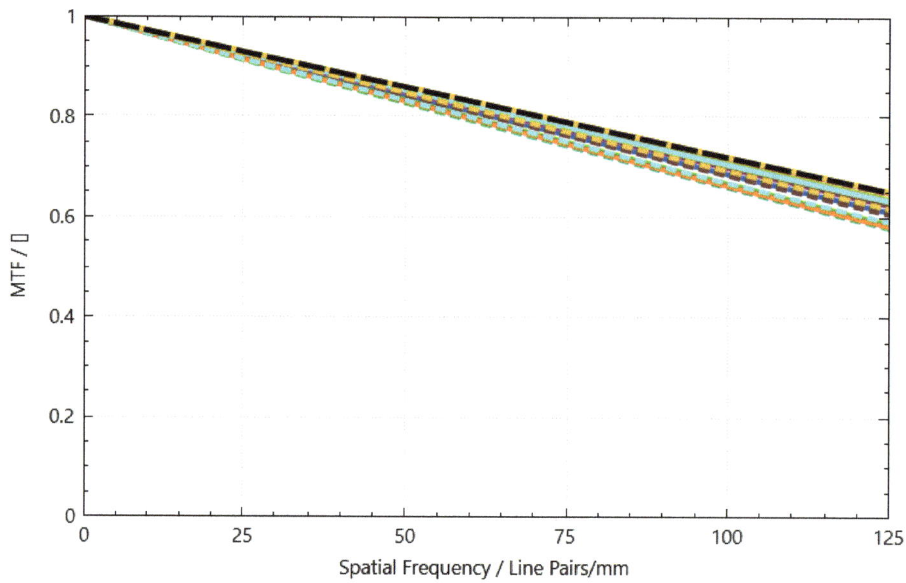

FIGURE 13.4
`FFT MTF Plot` of `10xe3`.

FIGURE 13.5
Tolerance specifications of `10xe3`.

For the tolerance related to distances, `Position`, `Surface Decenter`, `Radius` 0.002 is in millimeters. For the tolerance related to angles `Angel`, `Surface Tilt` is in degrees. See figure 13.5.

Since we are working with mirrors we do not consider `Thickness`, `Index` n, and `Abbe's Number` V parameters.

The `Sensitivity Analysis Result` is shown in figure 13.6. It shows the top offender parameters and how they change the merit function of the system,

Quadoa 24.05

Sensitivity Analysis Result

Tolerance Type	Object ID	Pertubation +	Target Val	Target Change	Pertubation -	Target Val	Target Change
Surface Decenter Y	Image	0.002	2.0377e-05	2e-05	-0.002	2.0377e-05	2e-05
Surface Tilt Y	Mirror 1	0.002	1.8121e-05	1.7744e-05	-0.002	1.812e-05	1.7743e-05
Surface Tilt X	Folding Mirror	0.002	1.7482e-05	1.7105e-05	-0.002	1.7481e-05	1.7104e-05
Surface Decenter X	Mirror 3	0.002	1.597e-05	1.5593e-05	-0.002	1.5964e-05	1.5587e-05
Surface Tilt Y	Mirror 2	0.002	1.1836e-05	1.1459e-05	-0.002	1.1838e-05	1.1461e-05
Surface Decenter Y	Mirror 1	0.002	1.1363e-05	1.0986e-05	-0.002	1.1367e-05	1.099e-05
Surface Tilt X	Mirror 1	0.002	1.092e-05	1.0543e-05	-0.002	1.0921e-05	1.0544e-05
Surface Tilt Y	Folding Mirror	0.002	9.1973e-06	8.8203e-06	-0.002	9.1964e-06	8.8194e-06
Zernike Z8	Mirror 1	1.2e-05	8.8516e-06	8.4746e-06	-1.2e-05	8.8495e-06	8.4726e-06
Surface Decenter X	Mirror 2	0.002	8.7647e-06	8.3878e-06	-0.002	8.774e-06	8.397e-06
Zernike Z7	Mirror 1	1.2e-05	6.6115e-06	6.2346e-06	-1.2e-05	6.6096e-06	6.2326e-06
Surface Tilt X	Mirror 2	0.002	5.7337e-06	5.3568e-06	-0.002	5.7338e-06	5.3569e-06
Surface Tilt Y	Object	0.002	4.7558e-06	4.3788e-06	-0.002	4.756e-06	4.379e-06
Surface Tilt X	Object	0.002	4.6532e-06	4.2763e-06	-0.002	4.6534e-06	4.2765e-06
Zernike Z7	Folding Mirror	1.2e-05	4.138e-06	3.7611e-06	-1.2e-05	4.1466e-06	3.7697e-06
Surface Tilt Y	Mirror 3	0.002	3.9072e-06	3.5302e-06	-0.002	3.9072e-06	3.5303e-06
Zernike Z10	Mirror 1	1.2e-05	3.3062e-06	2.9293e-06	-1.2e-05	3.3291e-06	2.9522e-06
Surface Decenter X	Mirror 1	0.002	3.2589e-06	2.882e-06	-0.002	3.25e-06	2.8731e-06
Zernike Z8	Mirror 2	1.2e-05	2.6425e-06	2.2656e-06	-1.2e-05	2.6154e-06	2.2385e-06
Surface Decenter Y	Mirror 3	0.002	2.0952e-06	1.7182e-06	-0.002	2.0926e-06	1.7157e-06
Zernike Z6	Mirror 1	1.2e-05	2.0437e-06	1.6667e-06	-1.2e-05	2.0082e-06	1.6312e-06
Zernike Z9	Mirror 1	1.2e-05	1.9932e-06	1.6163e-06	-1.2e-05	2.0099e-06	1.6329e-06
Zernike Z8	Folding Mirror	1.2e-05	1.9254e-06	1.5485e-06	-1.2e-05	1.9171e-06	1.5402e-06
Zernike Z7	Mirror 2	1.2e-05	1.5468e-06	1.1698e-06	-1.2e-05	1.5383e-06	1.1614e-06
Surface Tilt X	Mirror 3	0.002	1.3063e-06	9.2936e-07	-0.002	1.3055e-06	9.2857e-07
Zernike Z5	Mirror 1	1.2e-05	1.0541e-06	6.7712e-07	-1.2e-05	1.0261e-06	6.492e-07
Zernike Z10	Folding Mirror	1.2e-05	1.0124e-06	6.3547e-07	-1.2e-05	1.0174e-06	6.4043e-07
Zernike Z3	Mirror 1	1.2e-05	9.8901e-07	6.1207e-07	-1.2e-05	9.9774e-07	6.208e-07
Zernike Z8	Mirror 3	1.2e-05	9.4336e-07	5.6642e-07	-1.2e-05	9.6019e-07	5.8325e-07
Zernike Z6	Folding Mirror	1.2e-05	7.6227e-07	3.8533e-07	-1.2e-05	7.6412e-07	3.8718e-07
Zernike Z9	Folding Mirror	1.2e-05	7.4906e-07	3.7212e-07	-1.2e-05	7.6327e-07	3.8633e-07
Surface Decenter X	Image	0.002	7.2984e-07	3.529e-07	-0.002	7.275e-07	3.5056e-07
Zernike Z7	Mirror 3	1.2e-05	6.4282e-07	2.6588e-07	-1.2e-05	6.5397e-07	2.7703e-07
Zernike Z5	Folding Mirror	1.2e-05	6.3517e-07	2.5823e-07	-1.2e-05	6.2359e-07	2.4665e-07
Zernike Z3	Folding Mirror	1.2e-05	6.0475e-07	2.2781e-07	-1.2e-05	5.9872e-07	2.2178e-07
Zernike Z10	Mirror 2	1.2e-05	5.6814e-07	1.912e-07	-1.2e-05	5.6106e-07	1.8411e-07
Surface Decenter Y	Mirror 2	0.002	5.6745e-07	1.9051e-07	-0.002	5.6624e-07	1.893e-07
Zernike Z9	Mirror 2	1.2e-05	5.5874e-07	1.818e-07	-1.2e-05	5.4979e-07	1.7285e-07
Zernike Z6	Mirror 2	1.2e-05	5.3985e-07	1.6291e-07	-1.2e-05	5.5487e-07	1.7793e-07
Zernike Z5	Mirror 2	1.2e-05	5.4066e-07	1.6372e-07	-1.2e-05	5.5337e-07	1.7642e-07
Zernike Z10	Mirror 3	1.2e-05	5.3211e-07	1.5517e-07	-1.2e-05	5.3871e-07	1.6176e-07
Zernike Z3	Mirror 2	1.2e-05	5.1362e-07	1.3668e-07	-1.2e-05	5.0981e-07	1.3287e-07
Zernike Z6	Mirror 3	1.2e-05	5.0999e-07	1.3305e-07	-1.2e-05	5.0391e-07	1.2697e-07
Zernike Z5	Mirror 3	1.2e-05	4.9176e-07	1.1482e-07	-1.2e-05	4.888e-07	1.1186e-07
Zernike Z9	Mirror 3	1.2e-05	4.8488e-07	1.0794e-07	-1.2e-05	4.9017e-07	1.1323e-07
Surface Decenter Y	Folding Mirror	0.002	4.7608e-07	9.914e-08	-0.002	4.7455e-07	9.7605e-08

FIGURE 13.6
The `Sensitivity Analysis Result` of `10xe3`. The first column shows the tolerance type, and the second one is the object `ID`. The third column is the `Max` positive perturbation. The fourth column is the value of the merit function considering the `Max` perturbation. The fifth column is the change in the merit function. The next three columns are similar but for the `Max` negative perturbation.

actually in figure only shows some of the top offender parameters. The list is in order from the most offensive to the least. The first column shows the tolerance type, and the second one is the object `ID`. The third column is the `Max` positive perturbation. The fourth column is the value of the merit function considering the `Max` perturbation. The fifth column is the change in the merit function. The next three columns are similar but for the `Max` negative perturbation. We can see that the top nine are related to a `Decenter` or a `Tilt`.

The information presented in figure 13.6 is very important for the manufacturers so they can identify the most sensitive parts of the optical system. In this example, we put some standard ranges for tolerances, but in practice, the ranges depend a lot on the design itself and the capabilities of the manufacturer.

The next step is to apply this modification at the same time to our system and see how much it is affected. The distribution of the values is done by `Monte-Carlo Analysis`. So, on the `Tolerancing` tab, we press on `Monte-Carlo Analysis`. For this example, we do not use `Compensators`, we set `Number of Samples` to 10. Also, we allow the system to `Save Best and Worst` design. We press `OK`, `Quadoa optical CAD` will compute 10 different designs changing the parameters added to in the `Tolerancing` tab of `System Setup`, and it will not at any `Compensators` to improve the merit function after the change.

The `Monte-Carlo Simulation Result` is displayed in figure 13.7. The figure shows the value of the merit function before the `Monte-Carlo Analysis`. This value is reported as `Nominal Target Value`. It also shows the mean and the standard deviation of the merit function of the 10 designs. It is also reported the best and worst merit functions with their respective changes.

On the `Monte-Carlo Simulation Result`, it also displays a list of all the designs that were part of the simulation. In this case 10 designs. They are shown in a table format highlighting the respective merit function and changes in the merit function. The table also highlights the best and worst configurations.

Usually the most important is to see how worst is the worst so let us open the file, which is already saved since we allow the `Save Best and Worst` option.

The first important thing to see from this file is that all the `Surfaces` in the `Optical design` editor had suffered some rotation concerning the original design in `10xe3`. See figure 13.8.

Let's expand the `Optical Design Editor` to see the properties of each `Surface`, figure 13.9. The `Surfaces` were modified by adding a `Zernike Form` following the `Monte-Carlo Analysis`. The `Zernike Form` is used to model errors related to manufacturing. The Zernike terms implemented are according to the ones we defined in the in `Tolerancing` tab of `System Setup`. Well, now it is time to analyze how much the change of the merit function affects the performance of the optical system. Our first comparison

Quadoa 24.05

Monte Carlo Simulation Result

Nominal Target Value:	3.76941e-07
Mean Result:	3.59972e-05
Standard Deviation:	4.62047e-05
Best Monte Carlo Run:	4.87783e-06
Best Change:	4.50089e-06
Worst Monte Carlo Run:	0.00017208
Worst Change:	0.000171703

Simulation Run	Target Val	Target Change
1	1.68374e-05	1.64605e-05
2 (best)	4.87783e-06	4.50089e-06
3	6.23811e-06	5.86117e-06
4	2.85677e-05	2.81908e-05
5	2.07751e-05	2.03982e-05
6 (worst)	0.00017208	0.000171703
7	2.86025e-05	2.82256e-05
8	2.52439e-05	2.48669e-05
9	3.26037e-05	3.22268e-05
10	2.41463e-05	2.37693e-05

FIGURE 13.7
`Monte-Carlo Simulation Result` of `10xe3`.

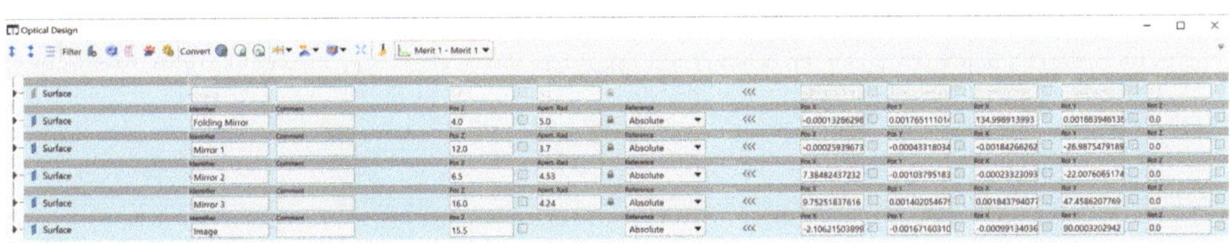

FIGURE 13.8
The `Surfaces` in the `Optical Design Editor` had suffered some rotation concerning the original design in `10xe3` after the `Monte-Carlo Analysis`. The rotations model errors concerning the alignment.

will be between figure 13.3 and figure 13.10. Figure 13.3 is the `Spot Diagram Plot` of the original design, while in figure 13.10 is the `Spot Diagram Plot` of the worst design after the tolerances. From figure 13.10, we can see that the system is still close to being diffracted-limited.

FIGURE 13.9
The `Surfaces` in the `Optical Design Editor` had suffered the addition of a `Zernike Form` after the `Monte-Carlo Analysis`. The `Zernike Form` models errors concerning the manufacturing.

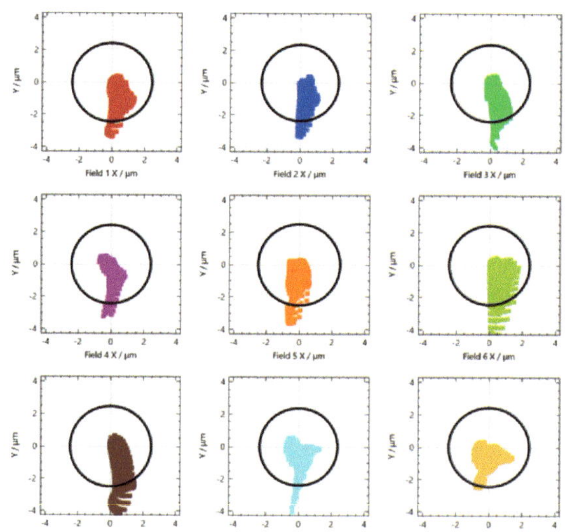

FIGURE 13.10
`Spot Diagram Plot` of the worst design after `Monte-Carlo Analysis`.

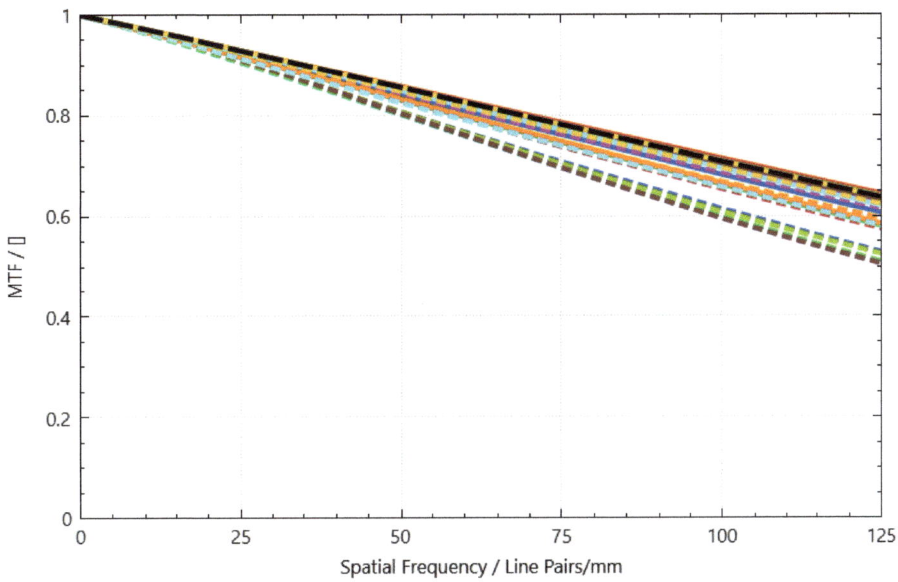

FIGURE 13.11
`FFT MTF Plot` of the worst design after the `Monte-Carlo Analysis`. The system is still close to being diffracted-limited for most of the fields. However, the performance has decreased for the tangential direction due that spots are smaller in the horizontal direction than in the vertical direction.

Observing the shape of the spots in figure 13.10, we can see that the system suffers from additional coma and astigmatism due the misalignments and deformations.

The next comparison will be between figure 13.4 and figure 13.11. Figure 13.4 is the `FFT MTF Plot` of the original design, while in figure 13.11 is the `FFT MTF Plot` of the worst design after the `Monte-Carlo Analysis`. From figure 13.11, we can see that the system is still close to being diffracted-limited for most of the fields. However, the performance has decreased for the tangential direction. This can be predicted with the `Spot Diagram Plot` of figure 13.10. Notice that in the horizontal direction, the spots are smaller than the vertical direction.

Overall, we can consider that the worst design after the `Monte-Carlo Analysis` is not as bad as we thought. I invite the reader to analyze now the best design. It is important to remember that this step in the design process of a lens usually works with a multidisciplinary team that involves experts in manufacturing, optics, mechanics, logistics, and mass production. So the discussions of the possible tolerance values in the `Sensitivity Analysis` and `Monte-Carlo Analysis` are not trivial and they change from one design to another.

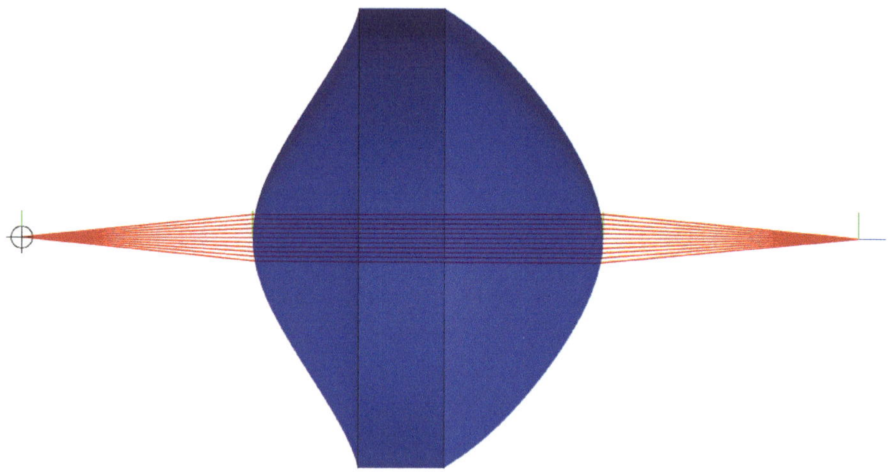

FIGURE 13.12
Ray-tracing of aspherical singlet lens.

13.5 Final Considerations

Once the tolerance analysis has been done. The technical drawings must be addressed. `Quadoa optical CAD` with `Technical Drawings ISO 10110 Dialog` facilitates the creation of technical drawings of lenses compliant with `ISO 10110`, which can then be saved as `PDF` files.

To select the `Element` for which to generate the drawing, simply click on the corresponding icon in the menu bar. You can edit the table specifying the lens properties and tolerances by clicking on the cells. The entire drawing content will be saved along with the model. To save the drawing, use the icon. To reset the tolerances for the technical drawing, click on the respective icon. The PDF file can be exported.

In figure 13.12 we present an example of a singlet. Its corresponding `ISO 10110` drawing is presented in figure 13.13. The `ISO 10110` drawing was generated with `Technical Drawings ISO 10110 Dialog` in `Quadoa optical CAD`. Figure 13.13 shows the specifications of the lens presented in figure 13.12. The left side is for the first surface and the right side is for the second surface. The specifications at the top of the `ISO 10110` drawing are the radius, conical constants, and the aspherical coefficients.

In the bottom of the `ISO 10110` drawing is presented the radius once more, the material and the aperture. The tolerance parameters by default are empty but can be modified by the user.

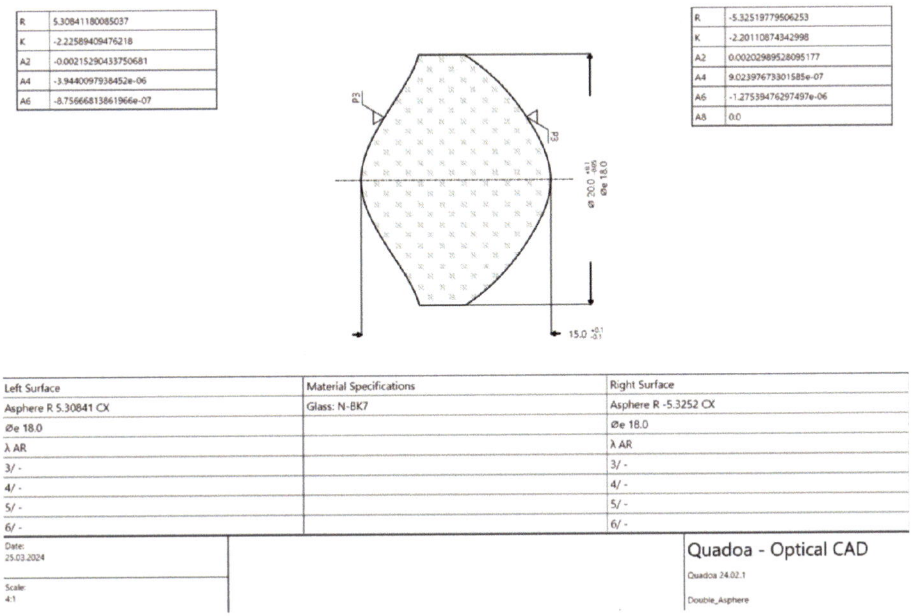

FIGURE 13.13
`ISO 10110` drawing of the lens of the design presented in figure 13.12. The left side is for the first surface and the right side is for the second surface. The specifications at the top are the radius, conical constants, and the aspherical coefficients. In the bottom of the drawing are presented the radius once more, the material, and the aperture.

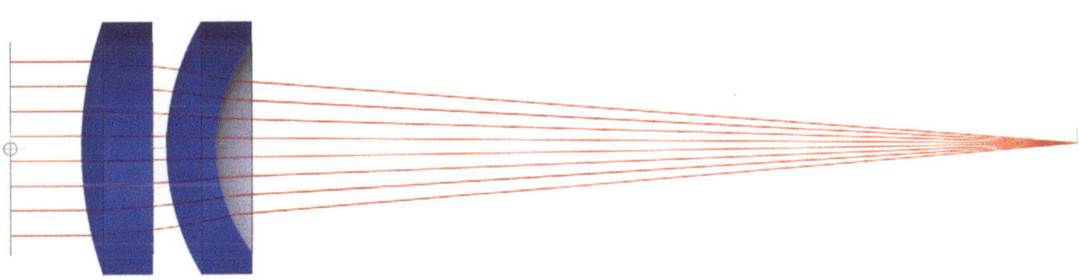

FIGURE 13.14
Ray-tracing of a doublet.

The next example is a doublet presented in figure 13.14. It's corresponding `ISO 10110` drawing is presented in figure 13.15. From the `ISO 10110` drawing we can confirm that is the second lens of the doublet. This lens is made of two spherical surfaces and `N-F2` glass material.

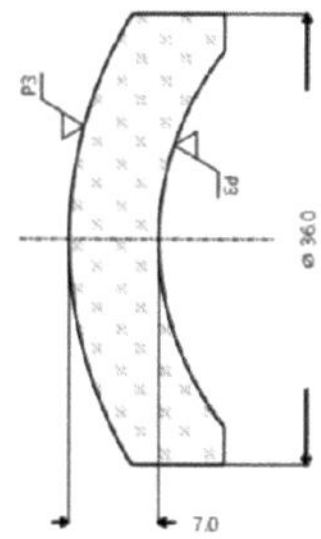

Left Surface	Material Specifications	Right Surface
R 35.4346 CX	Glass: N-F2	R 24.6138 CC
øe 36.0		øe 30.0
λ AR		λ AR
3/ -		3/ -
4/ -		4/ -
5/ -		5/ -
6/ -		6/ -

Date: 25.03.2024		Quadoa - Optical CAD
Scale: 2:1		Quadoa 24.02.1
		Lens_Drawing

FIGURE 13.15
`ISO 10110` drawing of the second lens of the design presented in figure 13.14.

13.6 Summary

In this chapter, we have seen the main components of the tolerance analysts and the tools provided by `Quadoa optical CAD` to perform it. Finally, we also look for the `ISO 10110` format for drawing lenses and how to do it in `Quadoa optical CAD`.

Bibliography

[1] R. Kingslake and R. B. Johnson, *Lens Design Fundamentals.* Academic Press, 2009.

[2] I. Livshits and V. Vasilyev, "Q and a tutorial on optical design," *Advanced Optical Technologies*, vol. 2, no. 1, pp. 31–39, 2013.

[3] D. Malacara, *Handbook of Optical Engineering.* CRC Press, 2001.

[4] D. Malacara-Hernández and Z. Malacara-Hernández, *Handbook of Optical Design.* CRC Press, 2016.

[5] H. Sun, *Lens Design: A Practical Guide.* CRC Press, 2016.

[6] R. K. Luneburg and M. Herzberger, *Mathematical Theory of Optics.* University of California Press, 1964.

[7] R. G. González-Acuña, H. A. Chaparro-Romo, and J. C. Gutiérrez-Vega, *Analytical Lens Design.* IOP Publishing, 2020.

[8] M. Born and E. Wolf, *Principles of Optics: Electromagnetic Theory of Propagation, Interference and Diffraction of Light.* Elsevier, 2013.

[9] H. Gross, W. Singer, M. Totzeck, F. Blechinger, and B. Achtner, *Handbook of Optical Systems*, vol. 1. Wiley Online Library, 2005.

[10] D. Dilworth, *Lens Design.* IOP Publishing, 2018.

[11] H. Nakajima, *Optical Design Using Excel: Practical Calculations for Laser Optical Systems.* John Wiley & Sons, 2015.

[12] D. C. O'Shea and J. L. Bentley, *Designing Optics Using CODE V.* SPIE Press, 2018.

[13] D. C. O'Shea and J. L. Bentley, *Designing Optics Using Zemax OpticStudio.* SPIE Press, 2024.

[14] D. Stephenson, "Using microsoft excel as a pre-processor for code v optimization of air spaces when building camera lenses," in *Optifab 2013*, vol. 8884, pp. 328–333, SPIE, 2013.

[15] *Optic Studio User Manual 19.4.* Zemax LCC, 2019.

[16] C. Wynne and P. Wormell, "Lens design by computer," *Applied Optics*, vol. 2, no. 12, pp. 1233–1238, 1963.

[17] M. Laikin, *Lens Design.* CRC Press, 2018.

[18] *Quadoa Optical CAD Software Manual v.* Quadoa Optical Systems, 2022.

[19] T. L. Heath *et al.*, *The Thirteen Books of Euclid's Elements.* Courier Corporation, 1956.

[20] I. Mueller, *Philosophy of Mathematics and Deductive Structure of Euclid's" Elements".* Dover Publications, 1981.

[21] G. De Young, "The arabic textual traditions of euclid's elements," *Historia Mathematica*, vol. 11, no. 2, pp. 147–160, 1984.

[22] M. Herzberger, *Modern Geometrical Optics.* Interscience, 1958.

[23] J. Braat and P. Török, *Imaging Optics.* Cambridge University Press, 2019.

[24] R. G. González-Acuña and H. A. Chaparro-Romo, *Stigmatic Optics.* IOP Publishing, 2020.

[25] H. A. Buchdahl, *An Introduction to Hamiltonian Optics.* Courier Corporation, 1993.

[26] V. Lakshminarayanan, A. Ghatak, and K. Thyagarajan, *Lagrangian Optics.* Springer Science & Business Media, 2013.

[27] R. S. Westfall, *Never at Rest: A Biography of Isaac Newton.* Cambridge University Press, 1983.

[28] R. S. Westfall, *The Life of Isaac Newton.* Cambridge University Press, 1994.

[29] M. S. Mahoney, *The Mathematical Career of Pierre de Fermat, 1601-1665.* Princeton University Press, 1994.

[30] H. Gross, F. Blechinger, and B. Achtner, *Handbook of Optical Systems, Survey of Optical Instruments*, vol. 4. Wiley-VCH, 2008.

[31] H. Gross, H. Zügge, M. Peschka, and F. Blechinger, *Handbook of Optical Systems, Aberration Theory and Correction of Optical Systems*, vol. 3. Wiley-VCH, 2007.

[32] J. W. Goodman, *Introduction to Fourier Optics.* Roberts and Company publishers, 2005.

[33] L. Mandel and E. Wolf, *Optical Coherence and Quantum Optics.* Cambridge University Press, 1995.

[34] R. G. González-Acuña and S. Thibault, "General diffraction integral for converging freeform vectorial fields," *Physica Scripta*, vol. 98, no. 11, p. 115503, 2023.

[35] R. G. González-Acuña and S. Thibault, "Exact vectorial model for non-paraxial focusing of freeform wavefronts," *Optics Express*, vol. 30, no. 13, pp. 23656–23663, 2022.

[36] R. G. González-Acuña and S. Thibault, "Focusing of an aberrated radially polarized beam by microscope objectives," *Optical Engineering*, vol. 63, no. 3, pp. 035108–035108, 2024.

[37] A. Vijayakumar and S. Bhattacharya, *Design and Fabrication of Diffractive Optical Elements with MATLAB.* SPIE Press, 2017.

[38] R. A. Chipman, W.-S. T. Lam, and G. Young, *Polarized Light and Optical Systems.* CRC Press, 2018.

[39] D. H. Goldstein, *Polarized Light.* CRC Press, 2017.

[40] D. C. Dilworth, *Lens Design: Automatic and Quasi-Autonomous Computational Methods and Techniques.* IOP Publishing, 2020.

[41] A. E. Conrady, *Applied Optics and Optical Design, Part One.* Courier Corporation, 2013.

[42] M. C. Rocha, *Aplanatic Lens Correctors for Imaging Optics.* PhD thesis, National University OF Ireland Galway, 2022.

[43] O. N. Stavroudis, *The Mathematics of Geometrical and Physical Optics: the K-Function and its Ramifications.* John Wiley & Sons, 2006.

[44] O. Stavroudis, *The Optics of Rays, Wavefronts, and Caustics*, vol. 38. Elsevier, 2012.

[45] R. G. González-Acuña and H. A. Chaparro-Romo, "General formula for bi-aspheric singlet lens design free of spherical aberration," *Applied Optics*, vol. 57, no. 31, pp. 9341–9345, 2018.

[46] R. Kingslake and R. B. Johnson, *Lens Design Fundamentals.* Academic press, 2009.

[47] K. Niu and C. Tian, "Zernike polynomials and their applications," *Journal of Optics*, vol. 24, no. 12, p. 123001, 2022.

[48] C. McAlinden, M. McCartney, and J. Moore, "Mathematics of zernike polynomials: a review," *Clinical & Experimental Ophthalmology*, vol. 39, no. 8, pp. 820–827, 2011.

[49] V. Lakshminarayanan and A. Fleck, "Zernike polynomials: a guide," *Journal of Modern Optics*, vol. 58, no. 7, pp. 545–561, 2011.

[50] C. S. Williams and O. A. Becklund, *Introduction to the Optical Transfer Function*, vol. 112. SPIE Press, 2002.

[51] A. Yabe, *Optimization in Lens Design.* SPIE Press, 2018.

[52] K. Höschel and V. Lakshminarayanan, "Genetic algorithms for lens design: a review," *Journal of Optics*, vol. 48, no. 1, pp. 134–144, 2019.

[53] M. Isshiki, H. Ono, K. Hiraga, J. Ishikawa, and S. Nakadate, "Lens design: global optimization with escape function," *Optical Review*, vol. 2, pp. 463–470, 1995.

[54] I. Ono, S. Kobayashi, and K. Yoshida, "Global and multi-objective optimization for lens design by real-coded genetic algorithms," in *International Optical Design Conference*, pp. LMC–2, Optica Publishing Group, 1998.

[55] Z. Tang, M. Sonntag, and H. Gross, "Ant colony optimization in lens design," *Applied Optics*, vol. 58, no. 23, pp. 6357–6364, 2019.

[56] C. Velzel, *A Course in Lens Design*, vol. 183. Springer, 2014.

[57] J. Sasián, *Introduction to Lens Design.* Cambridge University Press, 2019.

[58] J. M. Geary, *Introduction to Lens Design: with Practical ZEMAX Examples.* Willmann-Bell Richmond, 2002.

[59] J. H. McLeod, "The axicon: a new type of optical element," *Journal of the Optical Society of America*, vol. 44, no. 8, pp. 592–597, 1954.

[60] R. G. González-Acuña and J. C. Guitiérrez-Vega, "Generalization of the axicon shape: the gaxicon," *Journal of the Optical Society of America*, vol. 35, no. 11, pp. 1915–1918, 2018.

[61] Z. Jaroszewicz, A. Burvall, and A. T. Friberg, "Axicon-the most important optical element," *Optics & Photonics News*, vol. 16, no. 4, pp. 34–39, 2005.

[62] G. Forbes, "Manufacturability estimates for optical aspheres," *Optics Express*, vol. 19, no. 10, pp. 9923–9942, 2011.

[63] G. W. Forbes, "Shape specification for axially symmetric optical surfaces," *Optics Express*, vol. 15, no. 8, pp. 5218–5226, 2007.

[64] K. P. Thompson, F. Fournier, J. P. Rolland, and G. Forbes, "The forbes polynomial: a more predictable surface for fabricators," in *Optical Fabrication and Testing*, p. OTuA6, Optica Publishing Group, 2010.

[65] B. C. Platt and R. Shack, "History and principles of shack-hartmann wavefront sensing," 2001.

[66] P. Hariharan, *Basics of Interferometry.* Elsevier, 2010.

[67] D. Malacara, "Twyman–green interferometer," *Optical shop testing*, pp. 46–96, 2007.

[68] A. Bauer and J. P. Rolland, "Roadmap for the unobscured three-mirror freeform design space," *Optics Express*, vol. 29, no. 17, pp. 26736–26744, 2021.

[69] A. Bauer, E. M. Schiesser, and J. P. Rolland, "Starting geometry creation and design method for freeform optics," *Nature Communications*, vol. 9, no. 1, pp. 1–11, 2018.

[70] J. C. Papa, J. M. Howard, and J. P. Rolland, "Three-mirror freeform imagers," in *Optical Design and Engineering VII*, vol. 10690, p. 106901D, International Society for Optics and Photonics, 2018.

[71] F. Duerr and H. Thienpont, "Freeform imaging systems: Fermat's principle unlocks "first time right" design," *Light: Science & Applications*, vol. 10, no. 1, pp. 1–12, 2021.

[72] J. Chaves, *Introduction to Nonimaging Optics.* CRC press, 2016.

[73] J. C. Miñano, R. Mohedano, and P. Benítez, "Nonimaging optics," *The Optics Encyclopedia*, pp. 1–34, 2007.

[74] R. G. González-Acuña and H. A. Chaparro-Romo, *Optical Path Theory: Fundamentals to Freeform Adaptive Optics.* IOP Publishing, 2022.

[75] R. G. González-Acuna, H. A. Chaparro-Romo, and J. C. Gutiérrez-Vega, "General formula to design freeform singlet free of spherical aberration and astigmatism," *Applied Optics*, vol. 58, no. 4, pp. 1010–1015, 2019.

[76] S. George H, *Optical Design of Microscopes*, vol. 1. SPIE Press, 2010.

[77] A. Baranne and F. Launay, "Cassegrain: un célèbre inconnu de l'astronomie instrumentale," *Journal of Optics*, vol. 28, no. 4, p. 158, 1997.

[78] J. Nasmyth, *James Nasmyth, Engineer: an Autobiography.* John Murray, 1885.

[79] J. P. Gardner, J. C. Mather, M. Clampin, R. Doyon, M. A. Greenhouse, H. B. Hammel, J. B. Hutchings, P. Jakobsen, S. J. Lilly, K. S. Long, *et al.*, "The james webb space telescope," *Space Science Reviews*, vol. 123, pp. 485–606, 2006.

[80] J. Kalirai, "Scientific discovery with the james webb space telescope," *Contemporary Physics*, vol. 59, no. 3, pp. 251–290, 2018.

[81] P. A. Sabelhaus and J. E. Decker, "An overview of the james webb space telescope (jwst) project," *Optical, Infrared, and Millimeter Space Telescopes*, vol. 5487, pp. 550–563, 2004.

[82] J. P. Gardner, J. C. Mather, M. Clampin, R. Doyon, K. A. Flanagan, M. Franx, M. A. Greenhouse, H. B. Hammel, J. B. Hutchings, P. Jakobsen, *et al.*, "The james webb space telescope," in *Astrophysics in the Next Decade: The James Webb Space Telescope and Concurrent Facilities*, pp. 1–29, Springer, 2009.

[83] V. Blahnik and O. Schindelbeck, "Smartphone imaging technology and its applications," *Advanced Optical Technologies*, vol. 10, no. 3, pp. 145–232, 2021.

[84] D. M. Lyons, "Image spectrometry with a diffractive optic," in *Imaging Spectrometry*, vol. 2480, pp. 123–131, SPIE, 1995.

[85] J. W. Olesik and G. M. Hieftje, "Optical imaging spectrometers," *Analytical Chemistry*, vol. 57, no. 11, pp. 2049–2055, 1985.

[86] M. J. Golay, "Multi-slit spectrometry," *Journal of the Optical Society of America*, vol. 39, no. 6, pp. 437–444, 1949.

[87] W. W. Szymanski, A. Nagy, and A. Czitrovszky, "Optical particle spectrometry—problems and prospects," *Journal of Quantitative Spectroscopy and Radiative Transfer*, vol. 110, no. 11, pp. 918–929, 2009.

[88] B. Kress, "Optical design of next-gen augmented reality: Ahead of chairing the ar/vr/mr industry days at spie photonics west, microsoft's bernard kress considers the optimal way of designing optics for augmented and mixed reality headsets," *Electro Optics*, no. 290, pp. 12–17, 2019.

[89] B. C. Kress and C. Peroz, "Optical architectures for displays and sensing in augmented, virtual, and mixed reality (ar, vr, mr)," in *Proceedings of SPIE*, vol. 11310, p. 1131001, 2020.

[90] B. C. Kress and W. J. Cummings, "Optical architecture of hololens mixed reality headset," in *Digital Optical Technologies 2017*, vol. 10335, pp. 124–133, SPIE, 2017.

[91] B. Kress, "Digital optics as key enabling technologies to achieve small form factor ar display systems," in *Digital Optical Technologies 2023*, vol. 12624, pp. 108–119, SPIE, 2023.

[92] D. Malacara, *Optical Shop Testing*, vol. 59. John Wiley & Sons, 2007.

Index